Lecture Notes in Computer Science 14935

Founding Editors

Gerhard Goos
Juris Hartmanis

Editorial Board Members

Elisa Bertino, *Purdue University, West Lafayette, IN, USA*
Wen Gao, *Peking University, Beijing, China*
Bernhard Steffen, *TU Dortmund University, Dortmund, Germany*
Moti Yung, *Columbia University, New York, NY, USA*

The series Lecture Notes in Computer Science (LNCS), including its subseries Lecture Notes in Artificial Intelligence (LNAI) and Lecture Notes in Bioinformatics (LNBI), has established itself as a medium for the publication of new developments in computer science and information technology research, teaching, and education.

LNCS enjoys close cooperation with the computer science R & D community, the series counts many renowned academics among its volume editors and paper authors, and collaborates with prestigious societies. Its mission is to serve this international community by providing an invaluable service, mainly focused on the publication of conference and workshop proceedings and postproceedings. LNCS commenced publication in 1973.

Harold Mouchère · Anna Zhu
Editors

Document Analysis and Recognition – ICDAR 2024 Workshops

Athens, Greece, August 30–31, 2024
Proceedings, Part I

Editors
Harold Mouchère
Nantes Université
Nantes, France

Anna Zhu
Wuhan University of Technology
Wuhan, China

ISSN 0302-9743   ISSN 1611-3349 (electronic)
Lecture Notes in Computer Science
ISBN 978-3-031-70644-8   ISBN 978-3-031-70645-5 (eBook)
https://doi.org/10.1007/978-3-031-70645-5

© The Editor(s) (if applicable) and The Author(s), under exclusive license to Springer Nature Switzerland AG 2024

This work is subject to copyright. All rights are solely and exclusively licensed by the Publisher, whether the whole or part of the material is concerned, specifically the rights of translation, reprinting, reuse of illustrations, recitation, broadcasting, reproduction on microfilms or in any other physical way, and transmission or information storage and retrieval, electronic adaptation, computer software, or by similar or dissimilar methodology now known or hereafter developed.
The use of general descriptive names, registered names, trademarks, service marks, etc. in this publication does not imply, even in the absence of a specific statement, that such names are exempt from the relevant protective laws and regulations and therefore free for general use.
The publisher, the authors and the editors are safe to assume that the advice and information in this book are believed to be true and accurate at the date of publication. Neither the publisher nor the authors or the editors give a warranty, expressed or implied, with respect to the material contained herein or for any errors or omissions that may have been made. The publisher remains neutral with regard to jurisdictional claims in published maps and institutional affiliations.

This Springer imprint is published by the registered company Springer Nature Switzerland AG
The registered company address is: Gewerbestrasse 11, 6330 Cham, Switzerland

If disposing of this product, please recycle the paper.

Foreword

We are honoured to welcome you to the proceedings of ICDAR 2024, the 18th IAPR International Conference on Document Analysis and Recognition, which took place in Athens, the beautiful and historic capital of Greece. ICDAR 2024 marked the start of the annual basis for the ICDAR series.

ICDAR 2024 was the 18th edition of a longstanding conference series that has come of age, sponsored by the International Association for Pattern Recognition (IAPR). It is the premier international event for scientists and practitioners in document analysis and recognition. This field continues to play an important role in document understanding and recognition.

The IAPR TC10/11 technical committees endorse the conference. The very first ICDAR was held in St. Malo, France in 1991, followed by Tsukuba, Japan (1993), Montreal, Canada (1995), Ulm, Germany (1997), Bangalore, India (1999), Seattle, USA (2001), Edinburgh, UK (2003), Seoul, South Korea (2005), Curitiba, Brazil (2007), Barcelona, Spain (2009), Beijing, China (2011), Washington, DC, USA (2013), Nancy, France (2015), Kyoto, Japan (2017), Sydney, Australia (2019), Lausanne, Switzerland (2021) and San Jose, USA (2023).

Keeping with its tradition from past years, ICDAR 2024 featured a three-day main conference, including several competitions to challenge the field and a pre-conference slate of workshops, tutorials, and a doctoral consortium. The conference was held in Athens, Greece on September 2–4, 2024, and the pre-conference tracks on August 30 till September 1, 2024.

The highlights of the conference included keynote talks by the recipient of the IAPR/ICDAR Outstanding Achievements Award, and distinguished speakers: Jürgen Schmidhuber, Director of the AI Initiative at KAUST, Swiss AI Lab IDSIA, and the University of Lugano, Switzerland; Maria Kamilaki, Acting Director-General of e-Administration, Library & Publications of the Hellenic Parliament, Greece; and Cheng-Lin Liu, State Key Laboratory of Multimodal Artificial Intelligence Systems, Institute of Automation of the Chinese Academy of Sciences, China.

A total of 263 papers were submitted to the main conference (plus 35 papers to the ICDAR-IJDAR journal track), with 52 papers accepted for oral presentation (plus 17 IJDAR track papers) and 92 for poster presentation. We would like to express our deepest gratitude to our Program Committee Chairs, featuring three distinguished researchers from academia, Elisa Barney Smith, Liangrui Peng, and Marcus Liwicki, who did a phenomenal job in overseeing a comprehensive reviewing process and who worked tirelessly to put together a very thoughtful and interesting technical program for the main conference. We are also very grateful to the members of the Program Committee for their high-quality peer reviews. We extend our gratitude to our competition chairs, George Retsinas and Xiang Bai, for overseeing the competitions.

The pre-conference featured 6 excellent workshops, 4 value-filled tutorials, and the doctoral consortium. We would like to thank Harold Mouchère and Anna Zhu, the workshop chairs, Vincent Christlein and Alicia Fornés, the tutorial chairs, and KC Santosh and Andreas Fischer, the doctoral consortium chairs, for their efforts in putting together a wonderful pre-conference program. We would like to thank and acknowledge the hard work put in by our Publication Chairs, Giorgos Sfikas and Christophoros Nikou, who worked diligently to compile the camera-ready versions of all the papers and organize the conference proceedings with Springer. Many thanks are also due to our sponsorship, awards, industry, and publicity chairs for their support of the conference.

Finally, we would like to thank our many financial sponsors for their support and the conference attendees and authors, for helping make this conference a success. We sincerely hope that all attendees had an enjoyable conference, a wonderful stay in Athens, and fruitful academic exchanges with their colleagues.

August 2024
Basilis Gatos
Vassilis Katsouros
Foteini Simistira Liwicki

Preface

Welcome to the proceedings of the 18th International Conference on Document Analysis and Recognition (ICDAR 2024). ICDAR is the premier international event for scientists and practitioners involved in document analysis and recognition. This iteration is the first iteration in an even year, marking the beginning of the series becoming an annual event.

This year, we received 263 conference paper submissions. In order to create a high-quality scientific program for the conference, we recruited 159 regular and 32 senior program committee (PC) members. Each paper received at least 2 single-blind reviews, with most papers receiving 3 or more reviews. In addition, senior PC members who oversaw the review phase for typically 8–10 submissions took care of consolidating reviews and suggested paper decisions in their meta-reviews. Based on the information provided in both the reviews and the prepared meta-reviews we PC Chairs then selected 144 submissions (54.8%) for inclusion in the scientific program of ICDAR 2024. From the accepted papers, 52 were selected for oral presentation, and 92 for poster presentation.

In addition to the papers submitted directly to ICDAR 2024, we continued the tradition of teaming up with the International Journal of Document Analysis and Recognition (IJDAR) and organized a special journal issue. The journal-track submissions underwent the same rigorous review process as regular IJDAR submissions. The ICDAR PC Chairs served as Guest Editors and oversaw the review process. From the 35 manuscripts submitted to the journal track, 17 were accepted and were published in a Special Issue of IJDAR entitled "Advanced Topics of Document Analysis and Recognition." In addition, all papers accepted in the journal track were included as oral presentations in the conference program with a slightly extended presentation time (25 instead of 20 minutes including Q&A).

With the transition of ICDAR to an annual conference, the alternate year workshops DAS and ICFHR also made some changes. DAS decided to now be colocated with ICDAR. Handwritten text recognition is now a prominent topic in the DAR community, so the organizers and attendees of ICFHR 2023 decided to become part of the main ICDAR conference. Two sessions in ICDAR 2024 were on Frontiers of Handwriting Recognition and a third was devoted to Chinese text recognition. The popular Workshop on Historical Document Imaging and Processing (HIP) decided to remain an alternate year workshop, but in odd years. The main ICDAR conference had 2 sessions devoted to Historical Document Analysis. Scene text recognition, music, tables and charts continue to be topics of interest. Visual Question Answering, Document Understanding and NLP are rising topics gaining enough papers to devote sessions to them. As over the years many new machine learning techniques have been developed by researchers in the DAR community, a track also was devoted to papers that focused more on the methods than on any particular application. This year, nine scientific competitions were held in conjunction with ICDAR. A session was devoted to presenting the results.

As ICDAR 2024 was held with in-person attendance, all papers were presented by their authors during the conference. Exceptions were only made for authors who could not attend the conference for unavoidable reasons. Such oral presentations were then provided by synchronous video presentations. Posters of authors that could not attend were presented by recorded teaser videos, in addition to the physical posters.

Three keynote talks were given by Jürgen Schmidhuber, Director of the AI Initiative at KAUST, Swiss AI Lab IDSIA, and University of Lugano, Switzerland; Maria Kamilaki, Acting Director-General of e-Administration, Library & Publications of the Hellenic Parliament, Greece; and Cheng-Lin Liu, State Key Laboratory of Multimodal Artificial Intelligence Systems, Institute of Automation of the Chinese Academy of Sciences, China and the recipient of the IAPR/ICDAR Outstanding Achievements Award. We thank them for the valuable insights and inspiration that their talks provided for participants.

Finally, we would like to thank everyone who contributed to the preparation of the scientific program of ICDAR 2024, namely the authors of the scientific papers submitted to the journal track and directly to the conference, reviewers for journal-track papers, and both our regular and senior PC members. We also thank the Springer staff and the ICDAR 2024 Publication Chairs, who oversaw the creation of these proceedings.

August 2024

Elisa Barney Smith
Marcus Liwicki
Liangrui Peng

Foreword from ICDAR 2024 Workshop Chairs

We extend a warm welcome to the proceedings of the ICDAR 2024 Workshops, which were organized as part of the 18th International Conference on Document Analysis and Recognition (ICDAR) held in Athens, Greece from August 30 to September 4, 2024. The workshops were conducted prior to the commencement of the main conference, on the 30th and 31st of August, 2024. All of the workshops were conducted in person, as was the main conference.

The ICDAR conference comprised five workshops, which addressed a range of document image analysis and recognition topics, as well as related subjects such as natural language processing, computational paleography, and digital humanities. In total, 46 papers were submitted for consideration, and 30 were accepted, representing a global acceptance rate of 65%.

This volume compiles the edited papers from the five workshops. We extend our sincerest gratitude to the ICDAR general chairs for entrusting us with the responsibility of organizing the workshops and to the publication chairs for their invaluable assistance in publishing this volume. We also express our profound appreciation to the workshop organizers for their invaluable contributions to this pivotal event in our field. Finally, we acknowledge and thank all the workshop presenters and authors.

September 2024

Harold Mouchère
Anne Zhu

Organization

Organizing Committee

General Chairs

Basilis Gatos — NCSR "Demokritos", Greece
Vassilis Katsouros — Athena Research Center, Greece
Foteini Simistira Liwicki — Luleå University of Technology, Sweden

Program Committee Chairs

Elisa Barney Smith — Luleå University of Technology, Sweden
Marcus Liwicki — Luleå University of Technology, Sweden
Liangrui Peng — Tsinghua University, China

Workshop Chairs

Harold Mouchère — Nantes Université, France
Anna Zhu — Wuhan University of Technology, China

Competition Chairs

George Retsinas — National Technical University of Athens, Greece
Xiang Bai — Huazhong Univ. of Sci. & Technology, China

Tutorial Chairs

Vincent Christlein — University of Erlangen-Nuremberg, Germany
Alicia Fornés — Universitat Autònoma de Barcelona, Spain

Publication Chairs

Giorgos Sfikas — University of West Attica, Greece
Christophoros Nikou — University of Ioannina, Greece

Doctoral Consortium Chairs

Andreas Fischer	Univ. of App. Sci. & Arts Western Switzerland, Switzerland
K. C. Santosh	University of South Dakota, USA

Awards Chairs

Michael Blumenstein	University of Technology Sydney, Australia
Ioannis Pratikakis	Democritus University of Thrace, Greece

Posters/Demo Chairs

Umapada Pal	Indian Statistical Institute, India
Momina Moetesum	National University of Sciences & Technology, Pakistan
Kenny Davila	DePaul University, USA

Sponsorship Chairs

Markus Weber	Wacom, USA
Xu-Cheng Yin	University of Science and Technology Beijing, China

Industry Chairs

Dimosthenis Karatzas	Universitat Autònoma de Barcelona, Spain
Srirangaraj Setlur	University at Buffalo, USA
Errui Ding	Baidu Inc., China

Local Organization Chairs

Anastasios Kesidis	University of West Attica, Greece
Kosmas Kritsis	Athena Research Center, Greece
Elena Galifianaki	NCSR "Demokritos", Greece
Pelagia Drosaki	NCSR "Demokritos", Greece

Publicity Chairs

Panagiotis Kaddas	NCSR "Demokritos", Greece
Vassilis Papavassiliou	Athena Research Center, Greece
Elena Galifianaki	NCSR "Demokritos", Greece

Program Committee

Senior Program Committee Members

Apostolos Antonacopoulos	University of Salford, UK
Anurag Bhardwaj	eBay Research Labs, USA
Michael Blumenstein	University of Technology, Sydney, Australia
Jean-Christophe Burie	L3I - Université de La Rochelle, France
Bertrand Coüasnon	Irisa/Insa, France
Mickaël Coustaty	Laboratoire L3i - La Rochelle Université, France
David Doermann	University at Buffalo, USA
Véronique Eglin	LIRIS-INSA de Lyon, France
Gernot Fink	TU Dortmund, Germany
Andreas Fischer	University of Fribourg, Switzerland
Alicia Fornés	Computer Vision Center, UAB, Spain
Liangcai Gao	Peking University, China
Nicholas Howe	Smith College, USA
C. V. Jawahar	CVIT, IIIT, Hyderabad, India
Lianwen Jin	South China University of Technology, China
Dimosthenis Karatzas	CVC, Universitat Autónoma de Barcelona, Spain
Koichi Kise	Osaka Metropolitan University, Japan
Bart Lamiroy	Université de Reims Champagne-Ardenne, France
Cheng-Lin Liu	CASIA, China
Lu Liu	Lazada, Singapore
Josep Llados	CVC, Universitat Autónoma de Barcelona, Spain
Daniel Lopresti	Lehigh University, USA
R. Manmatha	University of Massachusetts, Amherst, USA
Angelo Marcelli	Università di Salerno, Italy
Simone Marinai	University of Florence, Italy
Jean-Marc Ogier	University of La Rochelle, France
Wataru Ohyama	Tokyo Denki University, Japan
Marçal Rusiñol	AllRead Machine Learning Technologies, Spain
Robert Sablatnig	TU Wien, Austria
Faisal Shafait	National University of Sciences and Technology, Pakistan
Seiichi Uchida	Kyushu University, Japan
Jerod Weinman	Grinnell College, USA
Richard Zanibbi	Rochester Institute of Technology, USA
Yu Zhou	Nankai University, Japan

Program Committee Members

Irfan Ahmad
Alireza Alaei
Musab Al-Ghadi
Eric Anquetil
Vlad Atanasiu
Muhammad Naseer Bajwa
Byron Bezerra
Ujjwal Bhattacharya
Jean-Luc Bloechle
Alceu Britto
Rina Buoy
Jorge Calvo-Zaragoza
Cristina Carmona-Duarte
Sukalpa Chanda
Clément Chatelain
Bidyut B. Chaudhuri
Joseph Chazalon
Shanxiong Chen
Jin Chen
Youssouf Chherawala
Vincent Christlein
Christian Clausner
Mark Clement
Florence Cloppet
Kenny Davila
Claudio De Stefano
Abhisek Dey
Sounak Dey
Antoine Doucet
Fadoua Drira
Mounîm A. El Yacoubi
Jonathan Fabrizio
Francesco Fontanella
Yasuhisa Fujii
Akio Fujiyoshi
Rajib Ghosh
Romain Giot
Lluis Gomez
Petra Gomez-Krämer
Daichi Haraguchi
Sheng He
Nina S. T. Hirata
Qiang Huo

Donato Impedovo
Brian Kenji Iwana
Maham Jahangir
Aashi Jain
Mohammed Javed
Jobin K. V.
Ehsanollah Kabir
Karim Kalti
Lei Kang
Slim Kanoun
Christopher Kermorvant
Yousri Kessentini
Florian Kleber
Pramod Kompalli
Aurélie Lemaitre
Hongjun Li
Zhouhui Lian
Lingyu Liang
Minghui Liao
Laurence Likforman
Rafael Lins
Chang Liu
Yuliang Liu
Muhammad Muzzamil Luqman
Nam Tuan Ly
Sriganesh Madhvanath
Nishatul Majid
Carlos David Martinez Hinarejos
Maroua Mehri
Carlos Mello
Ronaldo Messina
Evangelos Milios
Zuheng Ming
Tomo Miyazaki
Momina Moetesum
Hussein Mohammed
Ajoy Mondal
Harold Mouchère
Shobharani N.
Nibal Nayef
Clemens Neudecker
Hung Tuan Nguyen
Shinichiro Omachi

Umapada Pal
Shivakumara Palaiahnakote
Thierry Paquet
Mohammad Tanvir Parvez
Antonio Parziale
Marco Peer
Dezhi Peng
Vincent Poulain D'Andecy
Ioannis Pratikakis
Irina Rabaev
Jean-Yves Ramel
Oriol Ramos-Terrades
Frédéric Rayar
Kaspar Riesen
Christophe Rigaud
Verónica Romero
Henry A. Rowley
Joan Andreu Sanchez
Ravi Kiran Sarvadevabhatla
Martin Schall
Amina Serir
Anuj Sharma
Ying Sheng
Nicolas Sidère
Steven Simske
Sukhdeep Singh
Daniel Stoekl Ben Ezra
Tonghua Su
Xiangdong Su
Suresh Sundaram
Salvatore Tabbone
Sandeep Tata
Christopher Tensmeyer
Kengo Terasawa
Iuliia Tkachenko
Ruben Tolosana
Alejandro Toselli
Xiao Tu
Oliver Tüselmann
Huy Quang Ung
Szilard Vajda
Ernest Valveny
Ekta Vats
Ruben Vera-Rodriguez
Enrique Vidal

Lars Vögtlin
Yanwei Wang
Qiu-Feng Wang
Da-Han Wang
Yang Xue
Chun Yang
Mingkun Yang
Berrin Yanikoglu
Fei Yin
Qi Zeng
Heng Zhang
Yanming Zhang
Guangwei Zhang
Yuchen Zheng
Anna Zhu
Majid Ziaratban
Chandranath Adak
Oluwatosin Adewumi
Akshay Agarwal
Peeta Basa Pati
Khadiravana Belagavi
Asma Bensalah
Mohammad Idrees Bhat
Mélodie Boillet
Victoria Bourgeais
Iheb Brini
Francisco J. Castellanos
Francesco Castro
Apurba Chakraborty
Xu Chen
Denis Coquenet
Simon Corbillé
Aravinda Cv
Tiziana D'Alessandro
Avijit Dasgupta
Julien Delaunay
Vincenzo Dentamaro
Shubhang Desai
Alessandra Scotto di Freca
Moises Diaz
Ray Ding
Kalvin Dobler
Biyi Fang
Yuhang Fu
Gilad Fuchs

Cristiano Garcia
Vincenzo Gattulli
Loann Giovannangeli
Nathalie Girard
Tongkun Guan
Ahmed Hamdi
Raphaela Heil
Andre Hochuli
Kai Hu
Ludvig Hult
Syed Mohammad Baqir Husain
Nushrat Hussain
Aman Jaiswal
Mahdi Jampour
Nanfeng Jiang
Wang Jiawei
Michael Jungo
Wafa Khlif
Florian Kordon
Omar Krichen
Ahana Kundu
Songze Li
Zhixin Liu
Dongliang Luo
Puneet Mathur
Lin Meng
Elmokhtar Mohamed Moussa
Omar Moured
Emanuele Nardone
Emanuel Orler

Glen Pouliquen
Zhidong Qiao
Xingming Qu
Sachin Raja
Bulla Rajesh
Yann Ricquebourg
Antonio Ríos-Vila
Hugo Romat
Anna Scius-Bertrand
Gianfranco Semeraro
Mathias Seuret
Yilin Shi
Yongxin Shi
Mohamed Ali Souibgui
Yann Soullard
Maksym Taranukhin
Solène Tarride
Stacey Taylor
Vishvesh Trivedi
David Villanova-Aparisi
Manuel Villarreal Ruiz
Jiawei Wang
Xuewen Wang
Minghui Xia
Yejing Xie
Fuxiang Yang
Zhenhua Yang
Yan Zheng
Peijun Zou

Contents – Part I

ADAPDA

Domain Adaptation for Handwriting Trajectory Reconstruction from IMU Sensors 3
Florent Imbert, Romain Tavenard, Yann Soullard, and Eric Anquetil

TrOCR Meets Language Models: An End-to-End Post-correction Approach ... 12
Yung-Hsin Chen and Phillip B. Ströbel

LayeredDoc: Domain Adaptive Document Restoration with a Layer Separation Approach 27
Maria Pilligua, Nil Biescas, Javier Vazquez-Corral, Josep Lladós, Ernest Valveny, and Sanket Biswas

Normalized vs Diplomatic Annotation: A Case Study of Automatic Information Extraction from Handwritten Uruguayan Birth Certificates 40
Natalia Bottaioli, Solène Tarride, Jérémy Anger, Seginus Mowlavi, Marina Gardella, Antoine Tadros, Gabriele Facciolo, Rafael Grompone von Gioi, Christopher Kermorvant, Jean-Michel Morel, and Javier Preciozzi

ARPC

Diminutives in Political Discourse – The Case of Serbian and Slovenian 59
Milena Oparnica

Loghi: An End-to-End Framework for Making Historical Documents Machine-Readable 73
Rutger van Koert, Stefan Klut, Tim Koornstra, Martijn Maas, and Luke Peters

Open Parliamentary Data as a Tool for Linguistic Research: Exploring the 'Greek Language Question' in the *Journal of Parliamentary Debates* 89
Maria Kamilaki

Digitization of Written Parliamentary Questions from the Historical
Archive (1974–1977) of the Hellenic Parliament 103
   Fotios Fitsilis, Basilis Gatos, Konstantinos Palaiologos,
   Panagiotis Kaddas, Charalambis Kyrkos, Maria-Eleni Georgoulea,
   Yiannis Armenakis, Christina Tasouli, George Mikros,
   Olivier Rozenberg, and Eleni Kiousi

MANPU

Retrieving and Analyzing Translations of American Newspaper Comics
with Visual Evidence 125
   Jacob Murel and David A. Smith

Investigating Neural Networks and Transformer Models for Enhanced
Comic Decoding 138
   Eleanna Kouletou, Vassilis Papavassiliou, and Vassilis Katsouros

Comics Datasets Framework: Mix of Comics Datasets for Detection
Benchmarking 154
   Emanuele Vivoli, Irene Campaioli, Mariateresa Nardoni,
   Niccolò Biondi, Marco Bertini, and Dimosthenis Karatzas

A Comprehensive Gold Standard and Benchmark for Comics Text
Detection and Recognition 168
   Gürkan Soykan, Deniz Yuret, and Tevfik Metin Sezgin

Toward Accessible Comics for Blind and Low Vision Readers 198
   Christophe Rigaud, Jean-Christophe Burie, and Samuel Petit

Quantitative Evaluation Based on CLIP for Methods Inhibiting Imitation
of Painting Styles 216
   Motoi Iwata, Keito Okamoto, and Koichi Kise

Spatially Augmented Speech Bubble to Character Association via Comic
Multi-task Learning 231
   Gürkan Soykan, Deniz Yuret, and Tevfik Metin Sezgin

ComicBERT: A Transformer Model and Pre-training Strategy
for Contextual Understanding in Comics 257
   Gürkan Soykan, Deniz Yuret, and Tevfik Metin Sezgin

Author Index 283

Contents – Part II

IWCP

An Interpretable Deep Learning Approach for Morphological Script Type Analysis 3
Malamatenia Vlachou-Efstathiou, Ioannis Siglidis, Dominique Stutzmann, and Mathieu Aubry

Detecting and Deciphering Damaged Medieval Armenian Inscriptions Using YOLO and Vision Transformers 22
Chahan Vidal-Gorène and Aliénor Decours-Perez

Optimizing HTR and Reading Order Strategies for Chinese Imperial Editions with Few-Shot Learning 37
Marie Bizais-Lillig, Chahan Vidal-Gorène, and Boris Dupin

Mind the Gap: Analyzing Lacunae with Transformer-Based Transcription 57
Jaydeep Borkar and David A. Smith

NeuroPapyri: A Deep Attention Embedding Network for Handwritten Papyri Retrieval 71
Giuseppe De Gregorio, Simon Perrin, Rodrigo C. G. Pena, Isabelle Marthot-Santaniello, and Harold Mouchère

MONSTERMASH: Multidirectional, Overlapping, Nested, Spiral Text Extraction for Recognition Models of Arabic-Script Handwriting 87
Danlu Chen, Jacob Murel, Taimoor Shahid, Xiang Zhang, Jonathan Parkes Allen, Taylor Berg-Kirkpatrick, and David A. Smith

A New Framework for Error Analysis in Computational Paleographic Dating of Greek Papyri 102
Giuseppe De Gregorio, Lavinia Ferretti, Rodrigo C. G. Pena, Isabelle Marthot-Santaniello, Maria Konstantinidou, and John Pavlopoulos

Automated Dating of Medieval Manuscripts with a New Dataset 119
Boraq Madi, Nour Atamni, Vasily Tsitrinovich, Daria Vasyutinsky-Shapira, Jihad El-Sana, and Irina Rabaev

Image-to-Image Translation Approach for Page Layout Analysis
and Artificial Generation of Historical Manuscripts 140
  Chahan Vidal-Gorène and Jean-Baptiste Camps

VINALDO

A Multimodal Framework For Structuring Legal Documents 163
  Thibaud Real and Pauline Chavallard

Reformulating Key-Information Extraction as Next Sentence Prediction
for Hierarchical Data 175
  Ashish Kubade, Prathyusha Akundi, and Bilal Arif Syed Mohd

HPSegNet: A Method for Handwritten and Printed Text Separation
in Document Images 184
  Yu Chao, Changsong Liu, Liangrui Peng, and Yanwei Wang

Ablation Study of a Multimodal Gat Network on Perfect Synthetic
and Real-world Data to Investigate the Influence of Language Models
in Invoice Recognition 199
  Lukas-Walter Thiée

Author Index 213

ADAPDA

ADAPDA 2024 Preface

Document Analysis (DA) technologies are becoming increasingly pervasive in our daily lives due to the digitization of documents (both in the cultural and industrial domains) and the widespread use of paper tablets, pads, and smartphones to take notes and sign documents. In this respect, high-performing DA algorithms are needed that are able to deal with digitized documents from different writers, in different languages (including ancient languages, modern slang terms, or writer-preferred abbreviations and symbols), and with different visual characteristics (due to the paper support and the writing tool), often very peculiar to the application domain. In this respect, domain adaptation and automatic personalization strategies are worth investigating to boost the performance of DA techniques in the scenarios mentioned above, which are of great cultural, practical, and economic interest.

The Automatically Domain-Adapted and Personalized Document Analysis (ADAPDA) workshop, which was in its second edition, aims at gathering expertise and novel ideas for personalized DA tasks such as Handwritten Text Recognition, Handwritten Text Generation, Writer Identification, Writer and Signature Verification, and Handwriting Analysis. In particular, it welcomes contributions on training and adaptation strategies of writer, language, and visual-specific models, new benchmarks, data collection strategies to explore the mentioned tasks in a personalized setting, and related works on personalized DA. To this end, authors were encouraged to submit either short or long papers presenting ongoing projects, datasets, final or preliminary results, as well as innovative methodologies and tools. From five submissions, we selected four long papers from authors in three different countries, working both in academia and industry and coming from both the online and offline handwriting communities. Each paper received two single-blind reviews, timely provided by members of the Program Committee, who we take the chance to thank for their dedication.

The workshop consisted of a positional talk in which the organizers presented their views on the topics of the workshop, an oral presentation of each accepted paper, and a discussion session based on these talks and the issues that they raised. The participants were also asked for topics for the next workshop on the same subject, with the aim of building a fruitful conversation that we hope will continue in future ADAPDA editions, with which we plan to have an annual focus on the problems of adaptations and system evolutivity, alongside the large-scale systems that are going to become the trend.

August 2024

Silvia Cascianelli
Christopher Kermorvant
Eric Anquetil
Rita Cucchiara

Domain Adaptation for Handwriting Trajectory Reconstruction from IMU Sensors

Florent Imbert[1(✉)], Romain Tavenard[2], Yann Soullard[2], and Eric Anquetil[1]

[1] IRISA, Universite De Rennes, INSA Rennes, Rennes, France
florent.imbert@irisa.fr
[2] IRISA, Universite Rennes 2, Rennes, France

Abstract. Digital pens are commonly used to write on digital devices, providing the handwriting trace and enhancing human-computer interation. This study focuses on a digital pen equipped with kinematic sensors, allowing users to write on any surface while simultaneously preserving a digital trajectory of handwriting. This technology holds significant potential as a valuable educational tool, particularly in classrooms where it can facilitate the process of learning to write. A major issue is based on the difference in captured signals between adults and children. For similar handwriting trace, we have large differences in sensor signals due to differences in speed and confidence in the handwriting gesture of children. To address this, we investigate a domain adaptation approach to build a unified intermediate feature representation aimed at facilitating the trajectory reconstruction. We demonstrate the interest of domain adaptation methods in leveraging existing knowledge for application in different contexts. Specifically, we compare our domain adaptation approach with two other methods: training the model from scratch and fine-tuning the model.

Keywords: Domain Adaptation · Online Handwriting · Trajectory Reconstruction · Digital Pen · Inertial Measurement Units · Deep Neural Network

1 Introduction

This work focuses on reconstructing digital handwriting trajectories using the Digipen stylus [3]. The Digipen is equipped with 2 accelerometers, 1 gyroscope and a force sensor, and it can be used to write on any surface. While Inertial Measurement Units (IMU) are commonly used in tracking systems due to their cost-effectiveness, their signals are often imprecise due to high noise levels, presenting challenges. This issue is particularly prominent in the context of handwriting trajectory reconstruction, where precision is crucial, especially for e-learning purposes that require accurate feedback.

Although the Digipen stylus can be used on any surface and by different users, state-of-the art works [9,10] have experimented handwriting reconstruction approaches from data written by adults on tablets. Using the Digipen in another experimental context, e.g. on data written by children, or on another surface, e.g. on paper, leads to very different input signals. On the one hand, children's handwriting are of variable speed depending of the assertiveness in the handwriting gesture. On the other hand, handwriting on paper produces noisier signals due to friction than when the user writes on a tablet. This leads us to consider a domain adaptation method for dealing with the different domains of data.

In this context, we present a domain adaptation approach designed to enhance the adaptability of our model to deal with the different data sources. Initially trained on data acquired from tablets by adults, our model aims to effectively handle a additional types of data: handwriting acquired from tablets by children. In fact, while collecting labelled adult data on a tablet is not a problem, it's more complicated to collect data from children (contact a school). This approach is expected to broaden the applicability of handwriting trajectory reconstruction, making it more versatile and robust across different user groups. To our knowledge, no domain adaptation method addresses handwriting trajectory reconstruction from different sources, e.g. from adults vs children.

2 Related Works

Handwriting Trajectory Reconstruction. Traditional approaches [5,7], which do not rely on deep learning, often struggle with sensor noise and error accumulation. These methods typically involve a series of preprocessing steps, such as applying low-pass filters to remove high-frequency noise and using coordinate transformation matrices to adjust for sensor orientation. However, the cumulative effect of noise and small errors during these preprocessing steps can lead to significant inaccuracies in the reconstructed trajectories over time.

Deep learning methods offer a more advanced approach to handling IMU data by learning and adapting to noise patterns, potentially providing more accurate trajectory reconstructions. For instance, [6] applied multi-task learning for joint classification and trajectory reconstruction. [4] explored magnetic signals, while [10] addressed the Stabilo Digipen, using linear interpolation to align pen and tablet signals as preprocessing in training. This alignment is crucial for matching sequences of variable sizes and having a point-to-point matching for training the network. A Convolutional Neural Network is used for online handwriting trajectory reconstruction.

Recently, [9] improved on [10] by proposing a complete pipeline for online handwriting trajectory reconstruction (Fig. 1). Specifically, they used a specific preprocessing based on the Dynamic Time Warping (DTW) to align the ground truth with the input signal, which has the advantage of preserving the dynamics of handwriting, unlike the linear interpolation used in [10] and use a TCN model to extract local context information.

Domain Adaptation Methods. Domain adaptation (DA) adjusts a model trained on one domain (source) to perform well on a different but related domain

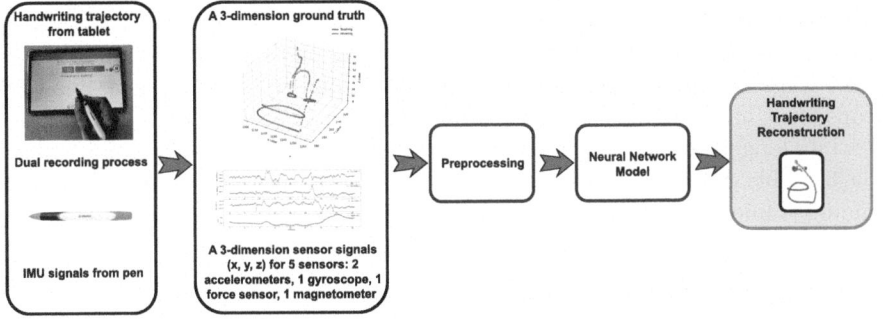

Fig. 1. Pipeline of handwriting reconstruction.

(target), mainly when labeled data are rare or unavailable in the target domain. The goal is to minimize domain shift, or differences in data distribution, between the source and target domains. DA can be:

- Supervised: Labeled data from the target domain are available.
- Semi-Supervised: Both labeled and unlabeled data from the target domain are available.
- Unsupervised: Only unlabeled data from the target domain are available.

Domain adaptation approaches can be broadly categorized into divergence-based, adversarial-based, and reconstruction-based methods. Divergence-based DA minimizes statistical differences between source and target distributions using measures like Maximum Mean Discrepancy (MMD) or Kullback-Leibler (KL) divergence [8]. These methods focus on reducing the distribution gap to improve model performance on the target domain. Adversarial-based DA, inspired by Generative Adversarial Networks (GAN), employs a discriminator and feature extractor to create domain-indistinguishable features through adversarial training [1]. This approach leverages the power of adversarial learning to align features from different domains, making them indistinguishable to a domain classifier. Reconstruction-based DA, uses autoencoders to maintain reconstruction ability across domains, thereby creating robust, domain-agnostic feature representations [2]. By ensuring that features can be accurately reconstructed regardless of the domain, these methods enhance the model's robustness and adaptability. Together, these approaches help models generalize better to target domains by aligning data distributions, creating domain-invariant features, or ensuring robust feature representations, thereby improving performance and reliability in diverse settings.

3 DANN-Based Method for Handwriting Reconstruction

To get a shared feature space for different data sources, we explore a an adversarial-based DA method, named Domain-Adversarial Training of Neural Networks (DANN).

Motivation. The variability of handwriting sources complicates the task of handwriting reconstruction. Children handwriting poses a unique challenge due to the ongoing development of graphomotor skills, resulting in dynamic and inconsistent handwriting patterns as children learn to write (Fig. 2). This translates into longer signal sequences than adults and a wider range of possible values (Fig. 3). This difference makes it difficult the handwriting reconstruction using a model trained on tablet-acquired data.

DANN as a Solution. Domain-Adversarial Training of Neural Networks [1] are specialized neural network architectures designed to address the challenge of domain shift, where a model trained on one domain (source) is expected to perform well on a different but related domain (target). These networks operate by learning features that are domain-invariant, meaning they are useful and generalizable across both domains. This is typically achieved through a shared feature extractor on which additional components are built: a domain classifier and a task-specific classifier. The domain classifier is trained to determine the domain of the input data, whereas the task-specific classifier focuses on predicting the label of the dedicated task. In our context, the task-specific classifier is trained to reconstruct the handwriting trajectory.

Training involves a twist (Fig. 4): the domain classifier's gradients are reversed during backpropagation, which encourages the feature extractor to generate features that are indistinguishable between domains, thus fooling the domain classifier. This technique, known as adversarial training, helps the network to minimize the representation gap between the source and target domains, leading to better performance on the target domain without requiring extensive labeled data from it.

Application. In this work, the reconstruction part of the DANN is the TCN-based network from [9] which is currently the state-of-the art model for handwriting reconstruction from the Digipen sensors (Fig. 5). We named it baseline model (BM) in the following. Then we have slice the baseline model as follows: the 4 blocks of the non-causal TCN is the feature extractor (in green in Fig. 4). Each TCN block is composed of 2 convolutions with dilation 1 and 2 respectively and a kernel size of 3. The next two dense layers refer to the label predictor (in blue in Fig. 4) which in our context corresponds to the trajectory reconstruction. The domain classifier (in pink in the Fig. 4), is made up of a max polling layer followed by two dense layers.

In our case study, we aim to transition from adult handwriting on a tablet (source) to children handwriting on a tablet (target). During training, we provide mixed matches to two model branches: (1) the feature extractor and the label predictor (green + blue in Fig. 4) pre-trained on adult data, and (2) a feature extractor and domain classifier (green + pink in Fig. 4).

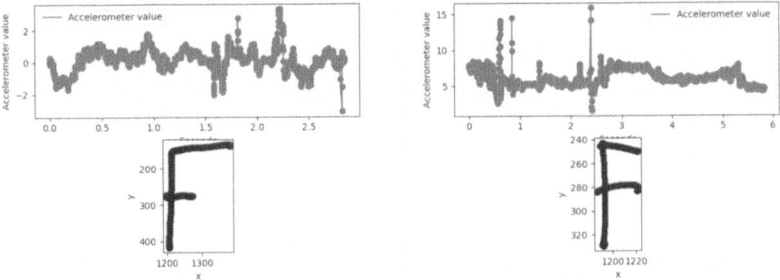

Fig. 2. Visualization of the x component of the Digipen's rear accelerometer over time in seconds, from left to the right: a F from adult on tablet, a F from children on tablet.We notice that the same pattern is written but not with the same intensity due to the different level of automation of handwriting (adult vs child)

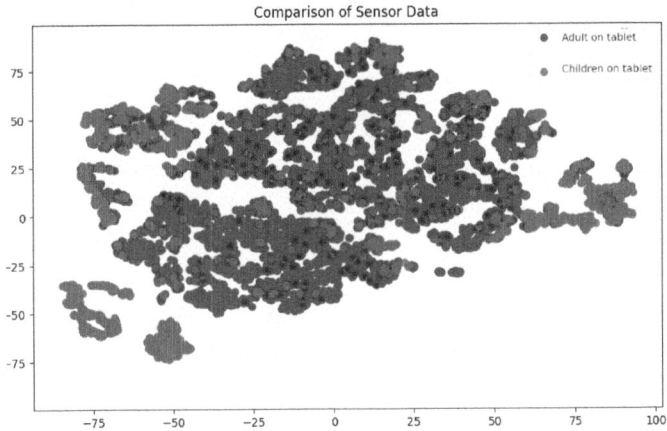

Fig. 3. Data visualization with the Multidimensional Scaling (MDS) method, we can see that children's data on tablets (in red) takes on a wider range of values due to their handwriting which is still being learned.

4 Experimental Results

We experiment our approach on two datasets, one for adult tablet data (9629 samples), another for children tablet data (3910 samples). Each datasets contains characters, words, word groups, equations, and shapes. Following [9], we compute the Fréchet distance to evaluate the quality of reconstruction on children's characters. We trained a DANN with children and adult data, and compared it qualitatively (Fig. 6) and quantitatively (Table 1) to the following methods: baseline model trained on adult data and fine-tuned on children data and the baseline model trained from scratch on children data.

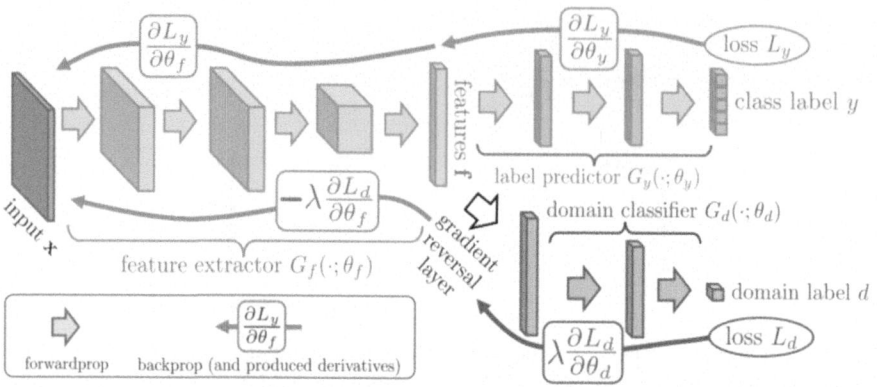

Fig. 4. DANN from [1].

Table 1. Comparison of reconstruction methods between training from scratch, fine tuning and domain adaptation method on children test data with the Fréchet distance.

Model	BM			DANN
Pretraining Data			Adult	Adult
Training Data	Adult	Children	Children	Children
Fréchet distance	0.470	**0.345**	0.348	0.349

Table 1 shows that methods integrating children data perform the best overall. Specifically, all the methods trained on children's data improve the Baseline Model trained only on adult data. The advantage of DANN is that it keeps a common representation of adult and children's features, and Table 2 shows that it performs well in both areas, making it a 2-in-1 solution. Unlike from scratch and fine tuning models (Table 1), which lose out on performance in the domain where they are not learned (Table 2). Regarding the qualitative analysis (Fig. 6), we observe that the trajectory reconstruction using the DANN is quite satisfactory and it seems closest to the ground truth than using the other approaches on those examples. The unique representation shared by the two data sources seems help the model the model in the trajectory reconstruction, especially on the hovering part. Additionally, this work is still in progress and the DANN has potential for further improvement.

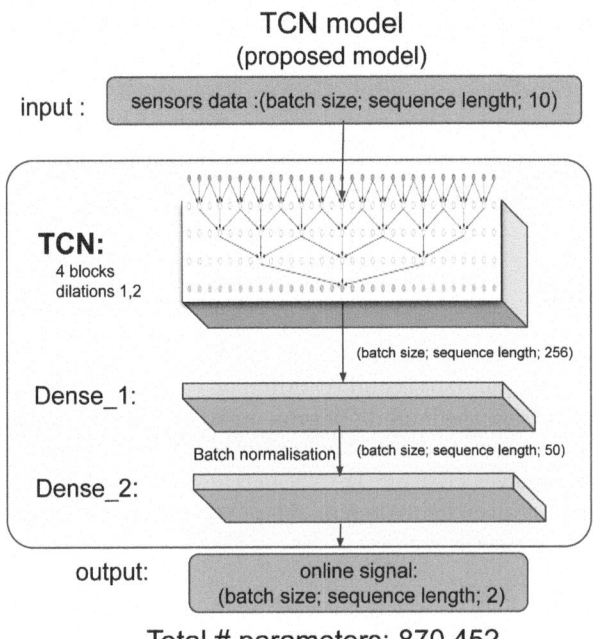

Fig. 5. TCN model from [9]. Also named baseline model (BM).

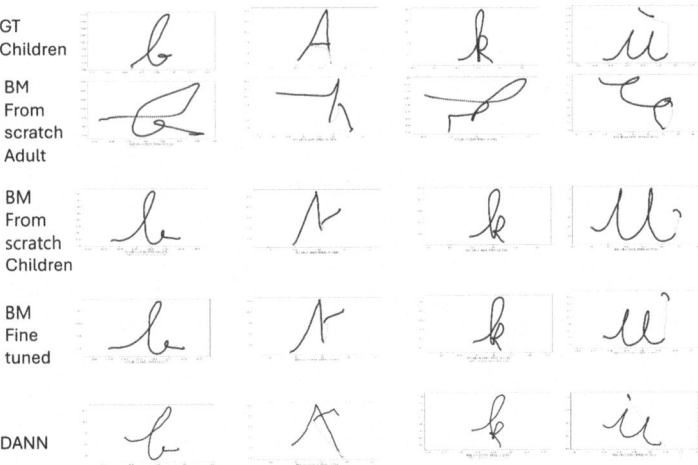

Fig. 6. Comparison of reconstruction methods on children test data, on the first line the ground truth, on the second the Baseline Model trained on adult data, on the third the Baseline Model trained on child data, then the Baseline Model trained on adult data and fine-tuned on child data, on the last line the DANN pretrain on adult data.

Table 2. Comparison of reconstruction methods between training from scratch, fine tuning and domain adaptation method on adult test data with the Fréchet distance.

Model	BM		DANN	
Pretraining Data			Adult	Adult
Training Data	Adult	Children	Children	Children
Fréchet distance	**0.332**	0.378	0.386	0.364

5 Conclusion

This paper shows the benefits of retaining knowledge from one domain (adults on tablet) and moving on to a second (children on tablet). Our areas for improvement include optimizing the lambda parameter to adaptively manage the weight of each branch over time, which could enhance DANN performance. Another area for improvement involves creating DANN batches and studying factors such as padding. We will also investigate domain adaptation from data acquired on tablet to data written on paper.

Acknowledgments. This project is financed by the KIHT French-German bilateral ANR-21-FAI2-0007-01 project and these four partners, IRISA, KIT, Learn & Go and Stabilo. This work was performed using HPC resources from GENCI-IDRIS (Grant 2021-AD011013148).

Disclosure of Interests. The authors have no competing interests to declare that are relevant to the content of this article.

References

1. Ganin, Y., et al.: Domain-adversarial training of neural networks. In: Csurka, G. (ed.) Domain Adaptation in Computer Vision Applications. ACVPR, pp. 189–209. Springer, Cham (2017). https://doi.org/10.1007/978-3-319-58347-1_10
2. Ghifary, M., Kleijn, W.B., Zhang, M., Balduzzi, D., Li, W.: Deep reconstruction-classification networks for unsupervised domain adaptation. In: Leibe, B., Matas, J., Sebe, N., Welling, M. (eds.) ECCV 2016. LNCS, vol. 9908, pp. 597–613. Springer, Cham (2016). https://doi.org/10.1007/978-3-319-46493-0_36
3. Harbaum, T., et al.: KIHT: kaligo-based intelligent handwriting teacher. In: DATE 2024. Valencia, Spain (Mar 2024)
4. Liu, Y., Huang, K., Song, X., et al.: Maghacker: eavesdropping on stylus pen writing via magnetic sensing from commodity mobile devices. In: Proceedings of the 18th International Conference on Mobile Systems, Applications, and Services. MobiSys '20, ACM (2020). https://doi.org/10.1145/3386901.3389030
5. Miyagawa, T., Yonezawa, Y., Itoh, K., Hashimoto, M.: Handwritten pattern reproduction using pen acceleration and angular velocity. IEICE Trans. (2000)
6. Ott, F., Rügamer, D., Heublein, L., et al.: Joint classification and trajectory regression of online handwriting using a multi-task learning approach. In: Proceedings of the IEEE/CVF Winter Conference on Applications of Computer Vision, pp. 266–276 (2022)

7. Pan, T.Y., Kuo, C.H., Hu, M.C.: A noise reduction method for IMU and its application on handwriting trajectory reconstruction. In: 2016 IEEE International Conference on Multimedia and Expo Workshops (ICMEW), pp. 1–6 (2016). https://doi.org/10.1109/ICMEW.2016.7574685
8. Shen, J., Qu, Y., Zhang, W., Yu, Y.: Wasserstein distance guided representation learning for domain adaptation. In: Proceedings of the AAAI Conference on Artificial Intelligence, vol. 32 (2018)
9. Swaileh, W., Imbert, F., Soullard, Y., Tavenard, R., Anquetil, É.: Online handwriting trajectory reconstruction from kinematic sensors using temporal convolutional network. Int. J. Doc. Analysis Recogn. (IJDAR), 1–14 (2023)
10. Wehbi, M., Luge, D., Hamann, T., et al.: Surface-free multi-stroke trajectory reconstruction and word recognition using an IMU-enhanced digital pen. Sensors. **22**(14), 5347 (2022)

TrOCR Meets Language Models: An End-to-End Post-correction Approach

Yung-Hsin Chen and Phillip B. Ströbel(✉)

Department of Computational Linguistics, University of Zurich, Zürich, Switzerland
{yung-hsin.chen,phillip.stroebel}@uzh.ch

Abstract. This study aims to enhance handwritten text recognition (HTR) performance and domain adaptability by combining an optical character recognition (OCR) model with a language model (LM) that serves as a corrector. This integration addresses three principal challenges: over-correction, which compromises text authenticity; poor domain adaptation; and the scarcity of annotated images. We explore the synergy between TrOCR, a state-of-the-art OCR model, and CharBERT, a BERT-based LM. A novel aspect of our research involves introducing common errors made by the recogniser into the LM, enabling it to consider these errors during correction, thereby improving overall performance. Our findings reveal that the hybrid TrOCR-CharBERT model effectively balances visual and linguistic information, preserving the authenticity of the original texts. Furthermore, the model is able to adapt to historical data even when the recogniser is trained solely on contemporary data, mitigating the need for a large number of annotated historical handwritten images.

Keywords: Handwritten Text Recognition · Domain Adaptation · Annotated Data Scarcity Mitigation

1 Introduction

OCR has become a key tool for digitising handwritten documents [33]. While OCR tasks for modern printed materials are typically straightforward, digitising historical texts or handwritten documents introduces complex challenges. Inadequate OCR can significantly affect downstream tasks such as text classification, named entity recognition [7], and information retrieval [15], leading to poor data utility.

One of the main challenges of OCR is the poor quality of images, which can include issues such as heterogeneous character heights, ink smears, and ink bleed-through [17]. Other challenges include the dynamics of languages and the lack of resources. Consequently, post-OCR correction is crucial to overcoming these limitations and enhancing the accuracy of digitised data [29]. Several post-OCR correction methods have been applied, showing significant improvements. Most of these methods function sequentially rather than in an end-to-end manner. However, allowing backpropagation to influence both the recogniser and the

LM for correction concurrently can yield better results than training them separately [11]. This integrated approach enables the LM to leverage insights from its own processing and feedback from the recogniser, facilitating more accurate corrections.

This study seeks to assess the effectiveness of integrating a recogniser with an LM to enhance performance and to enable the composite model to adapt to historical data even when trained on modern data. We have selected TrOCR [19], a state-of-the-art (SOTA) OCR model known for its advanced text recognition capabilities, and CharBERT [25], a variant of the BERT [5] model with additional character-level processing. In addition, we propose a novel approach to integrate the common errors made by TrOCR into CharBERT. By leveraging these technologies, our approach substantially improves the accuracy and reliability of text digitisation processes. The model enhances accuracy and reliability and adapts to different time periods of English, even if the recogniser is trained solely on contemporary English. This approach significantly reduces the need for annotated OCR images.

To test domain adaptability, we train a CharBERT variant (referred to as CharBERT$_{\text{HISTORICAL}}$) on a historical dataset rather than the contemporary dataset originally used to pre-train CharBERT. We aim to validate the adaptability of different composite models with LMs trained on different domains of datasets. To test the effect of integrating common OCR errors into CharBERT, we introduce common errors made by TrOCR into the training process of CharBERT to enable it to learn to correct them. This variant of CharBERT is referred to as CharBERT$_{\mathcal{P}_{ij}}$.

The variants of CharBERT are trained with a substantially smaller amount of data than the original pre-trained CharBERT due to computational resources and time limitations. To ensure a fair comparison of the composite model with different variants of CharBERT, we trained a CharBERT using the same setup as the pre-trained CharBERT but with a smaller amount of data to serve as the baseline for other variants of CharBERT. This CharBERT will be referred to as CharBERT$_{\text{SMALL}}$.

2 Related Work

The OCR process includes several stages, such as preprocessing, segmentation, feature extraction, and recognition [12]. While traditional models typically handle these stages separately, modern OCR models operate on an end-to-end basis [2,9,27]. These end-to-end models streamline the text recognition process by integrating all these stages into a single, continuous workflow. This approach leverages advanced machine learning techniques, particularly deep learning.

Modern text recognition approaches employ convolutional neural networks (CNNs) [36] and long short-term memory networks [3] to enhance accuracy. Transformer models [35], originally developed for natural language processing tasks, have also been successfully adapted for OCR applications. Models such as TrOCR [19], which combines a Vision Transformer [6] with pre-trained LMs like

RoBERTa [24], have demonstrated remarkable effectiveness in extracting text from images. Additionally, the incorporation of attention mechanisms in OCR models has enabled the network to focus on specific parts of an image sequentially, mimicking the human reading process. Examples include the Attention-based Scene Text Recognizer [32] and STAR-Net [23], both of which have substantially enhanced OCR accuracy.

Post-OCR correction refers to the methods and processes used to correct errors in text after it has been converted from images (such as scanned documents or photos) to editable and searchable text data. This stage is crucial because OCR technology, despite its advancements, often makes mistakes due to various challenges such as poor image quality, complex layouts, unusual fonts, or difficult handwriting.

Post-OCR correction is an effective tool but comes with its own set of challenges, such as distorted outputs from OCR processes or domain-specific terminology within datasets. To address this, [13] proposed a RoBERTa model employing a self-supervised pre-training approach to predict masked sections of medical texts. The authors discuss and tackle the inherent challenges posed by the varying accuracy of OCR technology, especially in recognising texts that contain medical terminologies and are often scanned from physical documents where text may be skewed or obscured.

Other research proposed solving OCR error correction as a machine translation problem [1,26]. [26] employed two deep learning models: a word-based sequence-to-sequence model and a character-based model. Results show that character-based models, which allow for the correction of individual characters, handle words not seen during training more effectively. While word-based models struggle with unseen words, character-based models perform well across different datasets [4,11].

3 Data

3.1 Data for Composite Model Training

The data used in this study includes the George Washington (GW) handwritten dataset [8] and the Joseph Hooker (JH) handwritten dataset [30]. These datasets serve as benchmarks for developing and evaluating handwriting recognition systems. We selected the GW and JH datasets because they represent different centuries of English and various topics, making them ideal for our experiments. Detailed information about these datasets is presented in Table 1.

With its extensive collection of 19th-century botanical writings and correspondence, the JH dataset provides a unique challenge for HTR technologies due to scientific terminology and personal handwriting styles. Similarly, the GW dataset, consisting of an array of 18th-century materials, including letters, diaries, and official documents, poses unique challenges due to the use of archaic words and phrases.

Table 1. GW and JH dataset statistics

Metric	GW Dataset	JH Dataset
Text lines	656	6,916
Training data (lines)	329	5,532
Validation data (lines)	168	691
Test data (lines)	163	693
Tokens	4,850	38,831
Types	1,456	8,308
Unique characters	68	84
Average # of characters per line	40.23	28.45
Average # of tokens per line	7.39	5.62
Percentage of non-characters[1]	21%	22.4%

[1] Characters other than A-Z, a-z, 0-9

Image Processing. In this study, images are initially resized to 384 × 384 pixels to comply with the input requirements of the pre-trained TrOCR model. Following resizing, the images undergo normalisation. The normalisation specifies the mean for each of the three colour channels (Red, Green, Blue), all set to 0.5. Together, resizing and normalising data ensure that no single pixel range overly influences the network due to its scale.

3.2 Data for CharBERT Variants

The training process for $CharBERT_{SMALL}$ and $CharBERT_{\mathcal{P}_{ij}}$ involved randomly sampled sentences from Wikipedia (1.13 GB) with 167M words. The only difference is that the errors introduced in the inputs of $CharBERT_{SMALL}$ are random, while the errors introduced in $CharBERT_{\mathcal{P}_{ij}}$ follow the probability of OCR errors occurring in the OCR model outputs, which will be discussed shortly in Sect. 5. On the other hand, $CharBERT_{HISTORICAL}$ is trained on 637MB of literature from the 16th to 19th centuries [10,14,18,20–22,28].

$CharBERT_{SMALL}$ serves as a baseline for comparisons with two other implementations of CharBERT: $CharBERT_{\mathcal{P}_{ij}}$ – which incorporates common OCR errors into the model – and $CharBERT_{HISTORICAL}$ – which is trained on historical English (from the 16th and 19th centuries) rather than contemporary English.

4 Recogniser and Language Model

TrOCR [19] is a Transformer-based model that focuses on the text recognition part of the OCR task, converting images to text. It has been selected as our primary OCR model for text recognition due to its SOTA performance and its capability to adapt to new handwriting styles with little data (see, e. g., [34]). As an end-to-end model, TrOCR simplifies the processing pipeline by eliminating the need for separate image processing and feature extraction steps. This allows it to be easily fine-tuned in conjunction with LMs. TrOCR will serve as the baseline against the composite model of TrOCR combined with CharBERT in this study.

CharBERT [25] is an enhancement of BERT designed to address issues in byte-pair encoding (BPE) [31] used by pre-trained LMs like BERT and RoBERTa. CharBERT takes text as input during inference and outputs two representative embeddings: a token embedding and a character embedding.

CharBERT introduces two techniques during the pre-training stage: 1) employing a dual-channel architectural approach for both subword and character information and 2) utilising noisy language modelling (NLM) for unsupervised representation learning. The first technique processes and fuses subword and character-level information, ensuring a more robust representation in case of typos. The second technique involves introducing character-level noise into words, and training the model to correct these errors.

CharBERT will function as the corrector in our post-OCR correction system. It meets our requirements by featuring character-level processing and a BERT-like architecture. Additionally, given its pre-training tasks, CharBERT is particularly effective at correction tasks, making it highly suitable for our post-OCR correction needs. Moreover, we can easily integrate common OCR errors into the CharBERT pre-training task, NLM.

5 Composite Model Architecture

The composite model is designed to integrate TrOCR and CharBERT. During the inference phase, the decoder output is recycled back as input. Before this recycled input is fed back into the decoder, it undergoes correction and refinement by CharBERT. Consequently, CharBERT is positioned between the decoder input and the decoder stacks as illustrated in Fig. 1. However, integrating these systems requires several adaptations to the models due to the following reasons: 1) The TrOCR decoder accepts token IDs as input, whereas CharBERT outputs embeddings; 2) CharBERT requires textual inputs, but the TrOCR decoder input is a tensor; 3) The embedding representations of TrOCR do not align with those of CharBERT; 4) The input to the TrOCR decoder is a single tensor, while CharBERT produces dual-channel outputs (token and character channel outputs). The following sections will discuss these adaptations in detail.

Adapted TrOCR. The TrOCR decoder is specifically designed to accept token IDs as input, which are then mapped to embeddings. However, CharBERT outputs embeddings rather than token IDs, leading to a compatibility issue. To resolve this, we reposition the embedding layer from the TrOCR decoder to precede CharBERT. This adjustment ensures that token IDs are initially converted to the TrOCR embedding, which are then input into CharBERT for correction.

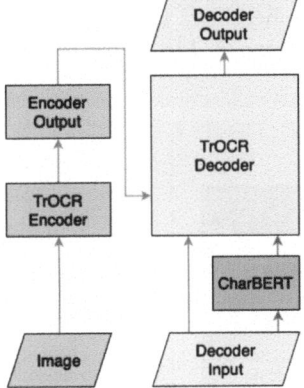

Fig. 1. Workflow in the composite model combining TrOCR and CharBERT.

Consequently, the adapted TrOCR can accept embeddings directly, bypassing the need for token IDs. This adaptation is shown in Fig. 2

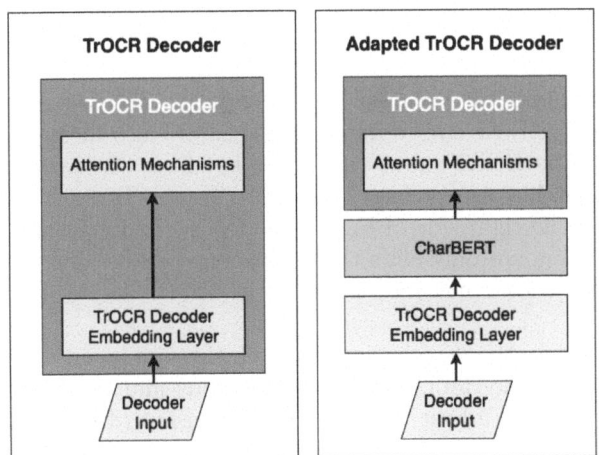

Fig. 2. Comparison of the TrOCR Decoder architecture.

Adapted CharBERT. According to the modifications described in the adapted TrOCR, the TrOCR decoder input is now an embedding, which should be processed by CharBERT for correction. However, the original CharBERT only accepts text as input. Therefore, we have redesigned this revised model so that CharBERT no longer converts text into IDs and then into embeddings; instead, it receives pre-processed embeddings directly. This adaptation allows both token and character embeddings to be processed through their respective channels in CharBERT as shown in Fig. 3.

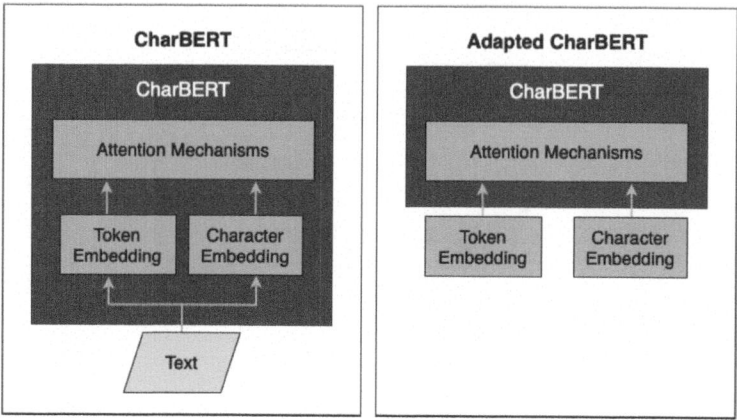

Fig. 3. Comparison of the CharBERT architecture

Tensor Transformation Module. Not only do the models represent the same text differently, but their embedding dimensions are also incompatible, further complicated by CharBERT's dual-channel embeddings. To overcome this issue, we designed an architecture referred to as the Tensor Transformation Module, as illustrated on the right side of Fig. 4. The CNN and feedforward neural network (FFNN) layers not only align the dimensions between the tensors but also learn to map the contextual information from TrOCR embeddings to those of CharBERT.

In the first stage, the decoder input passes through a series of CNN layers, interspersed with LeakyReLU activation functions and batch normalisations, specifically designed to adjust the sequence dimension (dim=1). Subsequently, the output from the first stage is processed through FFNN layers in the second stage to modify the embedding dimension (dim=2).

The Tensor Transformation Module is specifically designed to convert the TrOCR decoder input into CharBERT token and character channel inputs. This module is also critical in the Tensor Combine Module.

Tensor Combine Module. CharBERT produces two separate tensors – token and character representations – while the TrOCR decoder requires a single tensor input. To address this, we design the Tensor Combine Module to merge the two output tensors from CharBERT into a single tensor. Additionally, a residual connection from the original TrOCR decoder embedding is added. This residual connection helps to reuse features from the original TrOCR decoder embedding and prevents gradient vanishing. The architecture of the Tensor Combine Module is shown on the left side of Fig. 4.

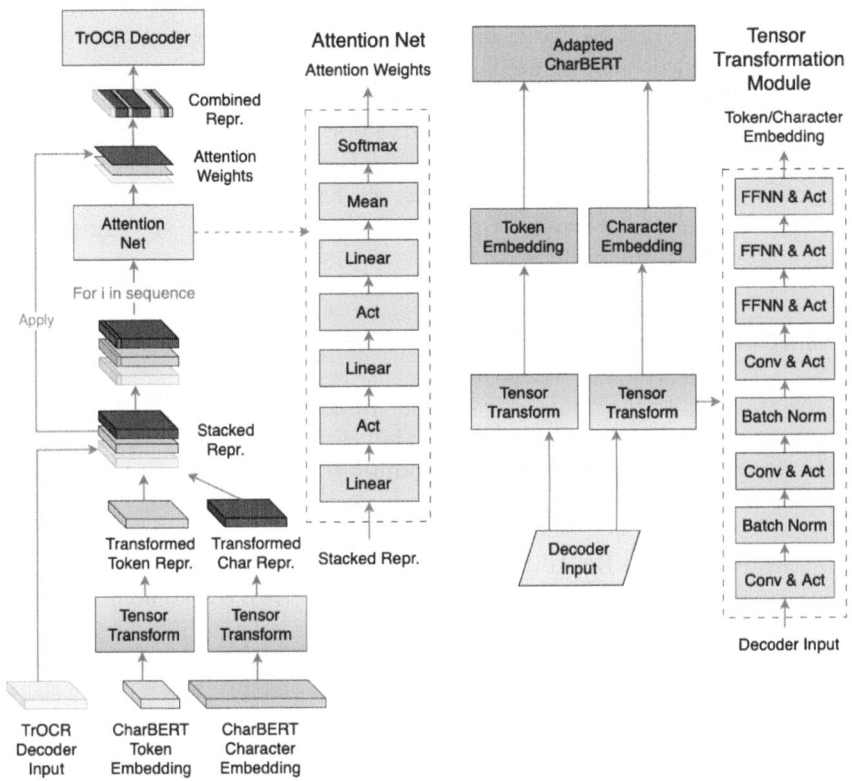

Fig. 4. Architecture of the Tensor Combine Module (left) and the Tensor Transformation Module (right).

Firstly, the two outputs from CharBERT undergo a transformation via the Tensor Transformation Module to match the original input size of the TrOCR decoder. Then, the Tensor Combine Module dynamically allocates attention weights to each word across the three input tensors. It incorporates linear layers as the attention network. This strategy is particularly effective for non-spatial input types such as text embedding.

Common Error Incorporation. In our methodology, we enhance the training process by specifically targeting commonly misrecognised characters, such as ".″ and ",″ or "O" and "o," to reduce the likelihood of these errors in future recognitions. To achieve this, we begin by determining the transition probability $\mathcal{P}_{ij}$, where i represents the correct character that has been erroneously recognised as character j. We obtain $\mathcal{P}_{ij}$ by calculating the frequency of each character misrecognised by TrOCR on the GW and JH datasets. Then, according to this probability, we incorporate errors into the text during CharBERT NLM training.

Training. CharBERT$_{\mathcal{P}_{ij}}$ To train CharBERT$_{\mathcal{P}_{ij}}$, we adhere to the training methods outlined in the original CharBERT paper, but with a smaller amount of data. In the original CharBERT training, the model is pre-trained by randomly adding, deleting, or swapping characters within the input text to simulate typical errors, thereby training CharBERT to correct them. For our specific application focusing on OCR corrections, we have modified this approach by replacing the random swapping of characters with common misrecognised OCR errors according to the probability, $\mathcal{P}_{ij}$.

6 Experiments and Analysis

For the training of the composite model, we use Adam [16] as the optimiser and cross-entropy for loss computation. The learning rate is set to 1e−5 with a weight decay parameter of 1e−5. The composite model is trained with all the TrOCR parameters frozen. Each experiment utilises one A100 GPU with 80GB RAM.

In this study, CharBERT is initially trained on different datasets to learn fundamental language patterns. Subsequently, CharBERT is combined with the recogniser and trained on handwritten datasets. This approach enables the CharBERT to consider both its own knowledge and TrOCR's predictions when generating the output text, adjusting its predictions by considering TrOCR's decisions.

6.1 Baseline Model

We use the pre-trained handwritten large TrOCR as a baseline and evaluate on the GW and the JH datasets. Additionally, we use a fine-tuned version of TrOCR on each of these datasets for further comparison. The results in terms of word error rates (WER) and character error rates (CER) of this analysis are shown in Table 2.

Upon examining the GW dataset TrOCR outputs without fine-tuning, we see that TrOCR tends to over-correct the text. As illustrated below, TrOCR autocorrects "Expamples" (an original misspelling by George Washington) to "Examples," the correct form of the word. Additionally, it completes the truncated word "ar" as "arm," without presuming the word was inadvertently cut short. A few images (fewer than 10) in the JH dataset contain printed rather than handwritten letters. TrOCR recognised "THE Camp." instead of the correct label "THE CAMP." Although TrOCR's handwritten model can recognise printed letters, it struggles with correct capitalisation.

Table 2. Baseline and composite model results in word and character error rates (WER, CER).

Dataset	Model	TrOCR Fine-Tuned	WER	CER
GW	TrOCR	False	37.76	15.40
GW	TrOCR	True	14.44	4.78
GW	TrOCR-CharBERT	False	12.84	5.88
JH	TrOCR	False	91.31	58.57
JH	TrOCR	True	36.97	20.28
JH	TrOCR-CharBERT	False	35.50	21.33

```
Test on GW Dataset Without Fine-Tuning
label:  est occasion for Expamples, will be morally im
output: cut occasion for Examples, will be morally in-

Test on GW Dataset After Fine-Tuning
label:  that were expected in; and to wait the ar
output: That were expected in; and to wait the arm

Test on JH Dataset Without Fine-Tuning
label:  THE CAMP.
output: THE Camp.
```

The tendency to over-correct is particularly noticeable at the end of sentences where the last word is truncated. TrOCR often attempts to complete these cut-off words, or it may substitute them with a different word that, while seemingly appropriate, is irrelevant to the original token image. In some cases, TrOCR even transforms the incomplete word into a non-existent word. Notably, this tendency to over-correct persists even after fine-tuning.

6.2 Composite Model (TrOCR-CharBERT)

The composite model significantly outperforms the baseline model and achieves more precise post-corrections. Unlike TrOCR alone, which may over-correct or erroneously complete words, this hybrid approach maintains the authenticity of the original images. For example, TrOCR-CharBERT correctly recognises "ar" and "Expamples" without over-correcting them as TrOCR does. By integrating CharBERT, the model leverages both visual information and linguistic knowledge, enabling it to make more informed decisions about when to amend the text and when to preserve the original input. This is also effective in the JH dataset, where the combined model correctly recognizes "THE CAMP.," accurately handling printed letters without the capitalization errors seen with TrOCR alone. The result of this analysis is shown in Table 2.

TrOCR-CharBERT substantially reduces the number of over-corrections, as shown in the Fig. 5. This figure illustrates the percentage of outcomes for unfinished word scenarios within the GW dataset, comparing the fine-tuned TrOCR and the TrOCR-CharBERT. In the case of the GW testing dataset, which includes 30 labels[1] ending with unfinished words, the fine-tuned TrOCR model correctly transcribes only 5 unfinished words, whereas the TrOCR-CharBERT model correctly transcribes 10. The categories "Complete word," "Other word," and "Not a word" indicate whether the model attempted to complete the unfinished words, substituted them with a different word it deemed fit, or transformed them into non-words, respectively. The pie charts reveal that TrOCR-CharBERT significantly reduces the instances of attempting to complete words erroneously, demonstrating its ability to preserve text authenticity more accurately.

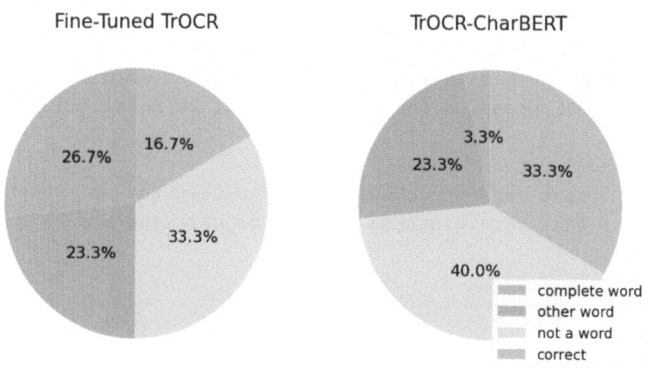

Fig. 5. Comparison of over-correction for TrOCR and TrOCR-CharBERT

6.3 Validating Model Domain Adaptability

To assess the model's domain adaptability, we compare the performance of both TrOCR-CharBERT$_{SMALL}$ and TrOCR-CharBERT$_{HISTORICAL}$ to determine the extent of the composite model's ability to adapt to the domain-specific characteristics of different datasets. The result of the analysis is shown in Table 3.

TrOCR is trained on contemporary English and is frozen during the experiment. Despite this, there is still a performance boost when TrOCR-CharBERT$_{HISTORICAL}$ is applied to the GW dataset, which is not contemporary English. This indicates that the recogniser can be trained to recognise general English character glyphs, and the LM can adapt to different domains of image data by training on that specific domain corpora. This can greatly reduce the need for annotated OCR images.

---

[1] The number of labels ending with unfinished words is counted by the author.

Table 3. Model Domain Adaptability Results

Training Dataset	LM Training Data	WER	CER
GW	Contemporary English	13.88	6.51
GW	15th – 18th Century English	13.18	6.05

6.4 TrOCR-CharBERT$_{\mathcal{P}_{ij}}$ Analysis

This analysis evaluates the positive effect of integrating common errors identified in TrOCR outputs into the training of CharBERT, referred to as CharBERT$_{\mathcal{P}_{ij}}$. We show the result of this analysis in Table 4 and illustrate the 4 most commonly misrecognised characters in Fig. 6.

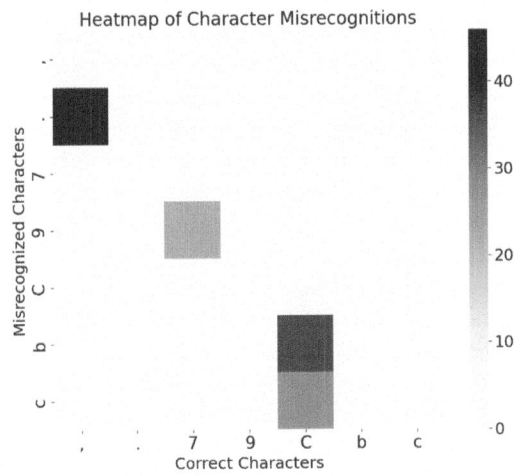

Fig. 6. Top 4 most commonly misrecognised characters.

The results suggest that incorporating knowledge about common OCR mistakes into the model helps refine its predictions. This refinement is more pronounced in the GW dataset, indicating that the nature of errors in this dataset may be more systematically addressable. While the performance improvement is less marked for the JH dataset, WER still decreases. Interestingly, the CER slightly increases, indicating that while some errors are corrected, new ones may be introduced due to the complexity and variability in the JH dataset.

Integrating common OCR mistakes into the training process enhances model performance, particularly for more homogeneous datasets like GW. Using CharBERT trained with more data can expect even better performance than the TrOCR-CharBERT model above.

Table 4. TrOCR-CharBERT$_{\mathcal{P}_{ij}}$ Results

Training Dataset	Model	$\mathcal{P}_{ij}$	WER	CER
GW	TrOCR-CharBERT$_{\text{SMALL}}$	False	13.88	6.51
GW	TrOCR-CharBERT$_{\mathcal{P}_{ij}}$	True	12.94	6.03
JH	TrOCR-CharBERT$_{\text{SMALL}}$	False	34.42	21.60
JH	TrOCR-CharBERT$_{\mathcal{P}_{ij}}$	True	33.95	21.86

7 Conclusion

Combining the recogniser with the LM allows the LM to access image information, correct words more accurately, and prevent over-correction. This helps preserve the authenticity of the texts in the images. In addition, the composite model adapts to different data domains while only the LM is trained on that specific text domain. This reduces the need for annotated historical text images. Furthermore, the composite structure allows us to integrate common OCR errors into the LM training process, improving error rates by making it more aware of frequent recognition mistakes. Thus, integrating a recogniser and an LM remains a valid and promising approach, offering benefits worth further exploration. Future research should support multilingual scripts and reduce the model's computational requirements to improve efficiency and applicability.

Disclosure of Interests. The authors declare that there are no conflicts of interest regarding the publication of this paper. This research received no specific grant from funding agencies in the public, commercial, or not-for-profit sectors. The views and opinions expressed in this paper are those of the authors and do not necessarily reflect the official policy or position of any affiliated agency of the authors.

References

1. Amrhein, C., Clematide, S.: Supervised OCR error detection and correction using statistical and neural machine translation methods. J. Lang. Technol. Comput. Linguist. (JLCL) **33**(1), 49–76 (2018)
2. Belay, B., Habtegebrial, T., Meshesha, M., Liwicki, M., Belay, G., Stricker, D.: Amharic OCR: an end-to-end learning. Appl. Sci. **10**(3), 1117 (2020)
3. Breuel, T.M., Ul-Hasan, A., Al-Azawi, M.A., Shafait, F.: High-performance OCR for printed English and Fraktur using LSTM networks. In: 2013 12th International Conference on Document Analysis and Recognition, pp. 683–687. IEEE (2013)
4. Chen, Y.H., Zhou, Y.: Enhancing OCR Performance Through Post-OCR Models: Adopting Glyph Embedding for Improved Correction. arXiv preprint arXiv:2308.15262 (2023)
5. Devlin, J., Chang, M-W., Lee, K., Toutanova, K.: BERT: pre-training of deep bidirectional transformers for language understanding. In: Burstein, J., Doran, C., Solorio, T., (eds) Proceedings of the 2019 Conference of the North American Chapter of the Association for Computational Linguistics: Human Language Technologies, Volume 1 (Long and Short Papers), pp. 4171–4186 (2019)

6. Dosovitskiy, A., et al.: An image is worth 16x16 words: transformers for image recognition at scale. In: International Conference on Learning Representations (2021)
7. Ehrmann, M., Hamdi, A., Pontes, E.L., Romanello, M., Doucet, A.: Named entity recognition and classification in historical documents: a survey. ACM Comput. Surv. **56**(2), 1–47 (2023)
8. Fischer, A., Keller, A., Frinken, V., Bunke, H.: Lexicon-free handwritten word spotting using character HMMs. Pattern Recogn. Lett. **33**(7), 934–942 (2012)
9. Huang, J.: A multiplexed network for end-to-end, multilingual OCR. In: Proceedings of the IEEE/CVF Conference on Computer Vision and Pattern Recognition, pp. 4547–4557 (2021)
10. Huber, M., Nissel, M., Puga, K.: Old bailey corpus 2.0. hdl:11858/00-246C-0000-0023-8CFB-2, Licensed under a Creative Commons Attribution-NonCommercial-ShareAlike 4.0 International License (2016)
11. Kang, L., Riba, P., Villegas, M., Fornés, A., Rusiñol, M.: Candidate fusion: integrating language modelling into a sequence-to-sequence handwritten word recognition architecture. Pattern Recogn. **112**, 107790 (2021)
12. Karthick, K., Ravindrakumar, K.B., Francis, R., Ilankannan, S.: Steps involved in text recognition and recent research in OCR; a study. Int. J. Recent Technol. Eng. **8**(1), 2277–3878 (2019)
13. Karthikeyan, S., de Herrera, A.G.S., Doctor, F., Mirza, A.: An OCR post-correction approach using deep learning for processing medical reports. IEEE Trans. Circuits Syst. Video Technol. **32**(5), 2574–2581 (2021)
14. Karthikeyan, S., de Herrera, A.G.S., Doctor, F., Mirza, A.: An OCR post-correction approach using deep learning for processing medical reports. IEEE Trans. Circuits Syst. Video Technol. **32**(5), 2574–2581 (2021)
15. Kettunen, K., Keskustalo, H., Kumpulainen, S., Pääkkönen, T., Rautiainen, J.: OCR quality affects perceived usefulness of historical newspaper clippings - a user study. In: Proceedings of the 18th Italian Research Conference on Digital Libraries (IRCDL 2022), CEUR-WS.org, Padova, Italy, CEUR Workshop Proceedings (2022)
16. Kinga, D., Adam, J.B.: A method for stochastic optimization. CoRR, abs/1412.6980 (2014)
17. Kurar Barakat, B., Cohen, R., Droby, A., Rabaev, I., El-Sana, J.: Learning-free text line segmentation for historical handwritten documents. Appl. Sci. **10**(22), 8276 (2020)
18. Kytö, M. and Culpeper, J.: A corpus of english dialogues 1560-1760 (CED), Literary and Linguistic Data Service (2006)
19. Li, M., et al.: TrOCR: transformer-based optical character recognition with pre-trained models. In Proc. AAAI Conf. Artif. Intell. **37**, 13094–13102 (2023)
20. Literary and Linguistic Data Service. The lampeter corpus of Early Modern English tracts. Literary and Linguistic Data Service
21. Literary and Linguistic Data Service. Pamphlets of the American revolution : [selections]/edited by bernard bailyn. Literary and Linguistic Data Service (1994)
22. Literary and Linguistic Data Service. The english language of the north-west in the late modern english period: a corpus of late 18c prose. Literary and Linguistic Data Service (2003)
23. Liu, W., Chen, C., Wong, K.Y.K., Su, Z., Han, J.: Star-net: a spatial attention residue network for scene text recognition. In: BMVC, vol. 2, p. 7 (2016)
24. Liu, Y., et al.: Roberta: a robustly optimized BERT pretraining approach. arXiv preprint arXiv:1907.11692 (2019)

25. Ma, W., Cui, Y., Si, C., Liu, T., Wang, S., Hu, G.: CharBERT: character-aware pre-trained language model. In: Scott, D., Bel, N., Zong, C., (eds) In: Proceedings of the 28th International Conference on Computational Linguistics, pp. 39–50, Barcelona, Spain , December 2020. International Committee on Computational Linguistics
26. Mokhtar, K., Bukhari, S.S., Dengel, A.: OCR error correction: State-of-the-art vs an NMT-based approach. In 2018 13th IAPR International Workshop on Document Analysis Systems (DAS), pp. 429–434. IEEE (2018)
27. Neudecker, C.,et al.: OCR-D: an end-to-end open source OCR framework for historical printed documents. In: Proceedings of the 3rd International Conference on Digital Access to Textual Cultural Heritage, pp. 53–58 (2019)
28. Nevalainen, T.: Parsed corpus of early english correspondence (PCEEC). Literary and Linguistic Data Service (2006)
29. Nguyen, T.T.H., Jatowt, A., Coustaty, M., Doucet, A.: Survey of post-OCR processing approaches. ACM Comput. Surv. (CSUR) **54**(6), 1–37 (2021)
30. Schaefer, J., Litvine, A.: Joseph hooker HTR model (June 2023)
31. Sennrich, R., Haddow, B., Birch, A.: Neural machine translation of rare words with subword units. In Erk, K., Smith, N.A. (eds), Proceedings of the 54th Annual Meeting of the Association for Computational Linguistics, Vol. 1, pp. 1715–1725, Berlin, Germany. Association for Computational Linguistics (2016)
32. Shi, B., Yang, M., Wang, X., Lyu, P., Yao, C., Bai, X.: ASTER: an attentional scene text recognizer with flexible rectification. IEEE Trans. Pattern Anal. Mach. Intell. **41**(9), 2035–2048 (2018)
33. Singh, A., Bacchuwar, K., Bhasin, A.: A survey of OCR applications. Int. J. Mach. Learn. Comput. **2**(3), 314 (2012)
34. Ströbel, P.B., Hodel, T., Boente, W., Volk, M.: The adaptability of a transformer-based OCR model for historical documents. In: Document Analysis and Recognition - ICDAR 2023 Workshops: San José, CA, USA, 24-26 August 2023, Proceedings, Part I, pp. 34–48 (2023)
35. Vaswani, A.,et al.: Attention is all you need. In: Advances in Neural Information Processing Systems, vol. 30 (2017)
36. Zhang, H., Liu, D., Xiong, Z.: CNN-based text image super-resolution tailored for OCR. In 2017 IEEE Visual Communications and Image Processing (VCIP), pp. 1–4. IEEE (2017)

LayeredDoc: Domain Adaptive Document Restoration with a Layer Separation Approach

Maria Pilligua[1,2]($\boxtimes$), Nil Biescas[1,2]($\boxtimes$), Javier Vazquez-Corral[1,2], Josep Lladós[1,2], Ernest Valveny[1,2], and Sanket Biswas[1,2]($\boxtimes$)

[1] Computer Vision Center, Universitat Autònoma de Barcelona, Catalonia, Spain
{mpilligua,jvazquez,josep,ernest,sbiswas}@cvc.uab.cat,
nbiescas@autonoma.cat
[2] Computer Science Department, Universitat Autònoma de Barcelona, Catalonia, Spain

Abstract. The rapid evolution of intelligent document processing systems demands robust solutions that adapt to diverse domains without extensive retraining. Traditional methods often falter with variable document types, leading to poor performance. To overcome these limitations, this paper introduces a text-graphic layer separation approach that enhances domain adaptability in document image restoration (DIR) systems. We propose *LayeredDoc*, which utilizes two layers of information: the first targets coarse-grained graphic components, while the second refines machine-printed textual content. This hierarchical DIR framework dynamically adjusts to the characteristics of the input document, facilitating effective domain adaptation. We evaluated our approach both qualitatively and quantitatively using a new real-world dataset, *LayeredDocDB*, developed for this study. Initially trained on a synthetically generated dataset, our model demonstrates strong generalization capabilities for the DIR task, offering a promising solution for handling variability in real-world data. Our code is accessible on this GitHub(https://github.com/mpilligua/LayeredDoc).

Keywords: Document Image Restoration · Layer Separation · Domain Adaptation · Text-Graphic Separation

1 Introduction

Documents, whether they be maps, architectural plans, historical manuscripts, identity documents or administrative paperwork, can be understood as compositions of semantically rich layers. For instance, maps contain layers of geographical information such as mountains, rivers, and roads. Architectural drawings comprise layers that detail structural components, electrical setups, and plumbing systems. Similarly, historical manuscripts include background degradation and foreground components (text and graphic symbols) which are essential for content interpretation as time evolves. The concept of layers also applies to forensic

---

M. Pilligua and N. Biescas—Authors have equally contributed to the Work.

documents (e.g. passports, ID cards) because of the multiple objects they have as security control (e.g. textures, ultraviolet marks, holograms etc.). Administrative documents are also structured in layers, comprising primary textual content with additional elements like stamps and annotations superimposed. These layers, like transparent overlays, collectively shape the document's meaning. Their individual recognition characterizes the document within its domain: a particular seal contextualizes the text of the page where it appears. Our hypothesis is that the separation of certain layers of information from the base text is crucial in the adaptation of solving downstream document understanding tasks, mainly Optical Character Recognition (OCR) [17,27], Handwritten Text Recognition (HTR) [14,15] and Document Layout Analysis (DLA) [4,5] over multiple domains.

In the task of document image restoration (DIR), a significant challenge arises from the complex interplay between textual and graphic components within the page, which often include noisy artifacts (e.g. ink stains, smears), postage marks (due to stamps, seals), degradation effects (e.g. bleedthrough, show-through), background variations (e.g. lighting, appearance, shadows), geometric distortion or warps, blurs and watermarks. However, most state-of-the-art (SOTA) approaches have been designed for a specific domain (mostly historical documents) to perform a specific restoration task (mostly document image binarization). This focus on narrow used cases has left a considerable gap in adaptability, especially when dealing with contemporary documents that exhibit a diverse array of artifacts and layout complexities. Consequently, systems that excel in historical document restoration may falter when applied to modern documents, which often feature different types of paper quality, printing techniques, and digital noise. *Our approach aims to bridge this gap by introducing a flexible, domain-adaptive framework that leverages advanced denoising and feature extraction techniques to robustly handle both historical and modern documents across various document processing tasks.* This adaptability is crucial for developing more comprehensive document analysis systems that are capable of performing consistently well across a broad spectrum of document conditions and eras.

In this work, **Layer Separation** emerges as a powerful tool to address these challenges by decomposing a document image into distinct layers, each representing different information types such as text, background, and graphics. This segregation not only facilitates enhanced noise reduction and clarity, but could also potentially improve OCR systems by isolating text from disruptive background/foreground elements. In the computer vision literature, layer-wise image decomposition has been employed in tasks such as foreground segmentation [31], reflection removal [18], watermark removal [22] and image deraining [23]. Inspired by how humans can interact with documents [21] suffering multiple degradations by inferring decompositions between the textual and non-textual components (which include graphical objects, watermarks, and stained artifacts), we propose a simplified two-layered separation of a document image which satisfies the following criteria: (i) the reconstructed layers, which when recombined should yield the original input document. (ii) The reconstructed layers should be independent of each other (uncorrelated), as they contain their intrinsic properties. This eventually serves as an effective self-supervision strategy to learn robust, generic representations by disentangling the textual and non-textual properties into two separate layers.

The overall contributions of this work can be divided into 3 folds: (1) A novel text-graphic Layer Separation approach for DIR has been introduced that utilizes a two-layer information processing system. (2) A hierarchical self-supervised DIR framework that allows for effective domain generalization without the need for retraining, making it highly efficient for use in environments where document types and conditions vary frequently. (3) We developed a manually curated real-world evaluation dataset specifically designed for testing the efficacy of unified multi-task DIR frameworks.

2 Related Work

2.1 Document Image Restoration

Many approaches have been proposed to address the enhancement of documents (both handwritten and machine-printed) which suffer several kinds of artefacts/defects such as bleed-through, show-through, faint characters, contrast variations and so on. The work from [6,16] maps images from the degraded domain to the enhanced one using end-to-end CNN-based autoencoders. Other techniques [12,28,29] used conditional-Generative Adversarial Network (c-GAN) based approaches to design a generator which produces the enhanced version of the document while the discriminator assesses the quality of binarization. Lately, an end-to-end ViT autoencoder was proposed in [26] to capture high-level global features using self-attention for binarizing degraded documents. Other prominent DIR tasks in the literature mainly include document deblurring [11], document dewarping [8,9], document deshadowing [1,20], document relighting or illumination correction [10,33] where restoration models were mainly trained for a specific document enhancement. Efforts to unify all the aforementioned DIR tasks and propose a generalist framework have been lately undertaken by DocRes [34]. In this work, inspired by [34] we propose a more generalist task of Layer Separation for domain adaptive DIR which shows potential to create a unified and flexible DIR framework.

2.2 Domain Adaptation for Document Analysis Systems

Since the data used for pre-training is essentially different from the target domain, domain-adaptive strategies are needed to be considered. In recent years, cross-domain generalization has been studied in DLA task. Li et al.. [19] proposed a benchmark suite that evaluates domain transfer from documents in English language to Chinese and vice-versa. Recently, Banerjee et al. [2,3] proposed a contrastive denoising training objective where a model trained on large collection from scientific domain [35] to other low-resource domains like magazines [7].

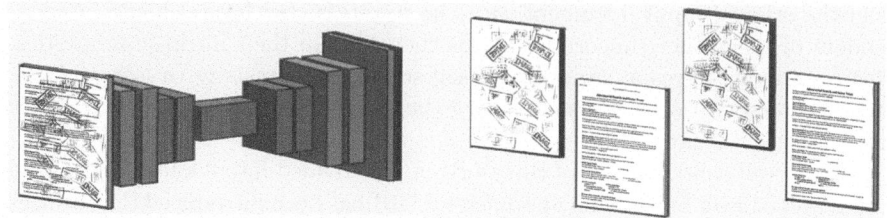

Fig. 1. General pipeline of **LayeredDoc**. We consider the standard architecture of image restoration models and propose to output two image layers instead of the standard single image in those methods. The first layer aims to output the text parts of the document, while the second layer aims to output the overlaid objects.

In the handwriting space, Kang et al. [15] devised a writer adaptation approach which automatically adjusts a generic handwritten word recognizer, fully trained with synthetic fonts, towards a new incoming writer. Later, they extended the work towards generating synthetic handwritten text lines [13]. Synthetic data generation strategies could indeed overcome challenges related to: (i) document source domain: from modern printed to degraded historical samples; (ii) different handwriting styles: from single to multiple writer collections; and (iii) language. Inspired by [13,15], we generate synthetic data which has been used to train our proposed DIR system. This provides a practical and generic approach to deal with unseen and real-world document collections without requiring any retraining or fine-tuning.

3 Methodology

In this section, we will explore the proposed adaptation of restoration models for the task of document layer separation, as well as the learning objectives used in the architecture shown in Fig. 1.

Our main goal is to adapt current natural image restoration models, mainly based on the U-Net [24] architecture, by using a simple yet effective modification in the last upsampling operation, enabling these models to also perform layer separation in documents.

3.1 Natural Image Restoration

In natural images restoration, given a degraded image $\mathbf{I} \in \mathbb{R}^{H \times W \times 3}$, current state-of-the-art restoration models such as Restormer [32] are composed of an encoder $E(x)$ and a decoder $D(x)$. The degraded image $\mathbf{I}$ is passed through the encoder $E(x)$, obtaining latent features $\mathbf{F}_0 \in \mathbb{R}^{\frac{H}{8} \times \frac{W}{8} \times C}$, where $\frac{H}{8} \times \frac{W}{8}$ denotes the spatial dimensions and C is the number of channels. Next, these shallow features $\mathbf{F}_0$ pass through the decoder $D(x)$, which includes several upsampling layers. Finally, a convolution layer is applied to the refined features to generate the output image $\mathbf{R} \in \mathbb{R}^{H \times W \times 3}$.

3.2 Layer Separation in Documents

In document image processing, the task of layer separation involves isolating distinct components of a document, such as text and overlaid objects, into separate layers. Mathematically, consider an input image $\mathbf{I} \in \mathbb{R}^{H \times W \times 3}$, where H and W denote the height and width of the image, and 3 represents the RGB color channels. The goal is to decompose this image into two separate layers: layer 0 ($\mathbf{L_0}$), which may represents the text elements, and layer 1 ($\mathbf{L_1}$), which may represent the overlaid objects.

To do so, we propose to modify natural image restoration architectures to perform text/graphics separation. Our proposed modification happens in the last upsampling layer of these models, as we propose to modify current restoration models to output a $\mathbf{R'} \in \mathbb{R}^{H \times W \times 6}$ tensor. The first 3 channels of this tensor compose an image that contains the first layer information ($\mathbf{L_0}$), while the last three channels correspond to the second layer $\mathbf{L_1}$. See Fig. 1 for a visual explanation.

3.3 Learning Objectives

As explained above, we propose that the models output a tensor $\mathbf{O} \in \mathbb{R}^{H \times W \times 6}$, where the first three channels correspond to the reconstructed layer 0 ($\mathbf{L_0}$) and the last three channels correspond to the reconstructed layer 1 ($\mathbf{L_1}$). Due to this, we can consider intermediate losses in each of the two layers.

We compute a L_1 loss for each of the layers $\mathbf{L_i}$. Mathematically,

$$\mathcal{L}_{layer_i}(\mathbf{I_i}, \mathbf{L}_i) = \|\mathbf{I_i} - \mathbf{L}_i\|_1, \tag{1}$$

where I_i represents the ground truth image for the layer i and L_i represents the predicted layer i.

Then, our final loss is just the addition of the losses for both layers:

$$\mathcal{L}_{final} = \mathcal{L}_{layer_0} + \mathcal{L}_{layer_1}. \tag{2}$$

Related Insights: Our dual loss approach facilitates the model's ability to learn distinct yet complementary tasks. By optimizing for both layer 0 and layer 1 reconstructions, the model gains a richer understanding of the underlying structure within the document images. Thanks to this, and without increasing the model's complexity, our model is able not only to enhance the performance of each individual task but also to leverage the inherent relationship between the two tasks to improve the overall model robustness.

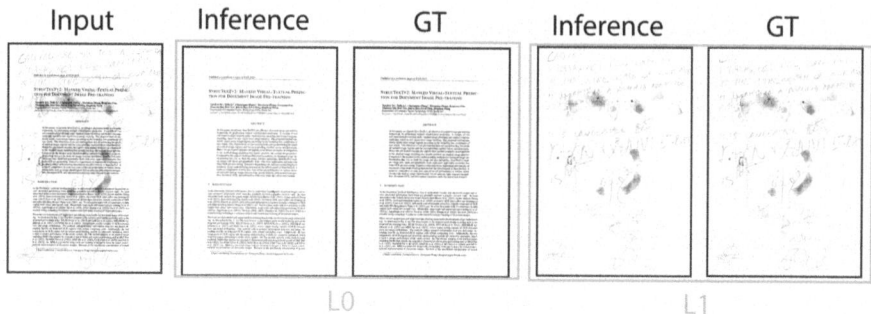

Fig. 2. Visual illustration of the layer separation done by our 6-channel LayeredDoc model vs the ground truth

4 Experiments

4.1 Datasets

Training Dataset. To train our model, we have derived a synthetic dataset. This dataset has the proper ground truth layers to perform supervised learning. We derived ground truth for layer 0 and layer 1, together with the input noisy image. This dataset aims at resembling real documents containing text with overlayed artifacts like seals and annotations. We selected a subset of Publaynet [35] with pages that do not contain figures. This set consists of documents with text information only, and therefore these documents build the corresponding Layer 0. Over these images, we have added stamps, signatures, barcodes, QR codes, and passport photos with random rotation, position, and alpha value - which changes the transparency of the image. We have also added random shadows and color changes to simulate that documents were not scanned but captured with a mobile phone.

LayeredDocDB. To test our model we crafted a dataset made from real documents. To create this dataset we first randomly added different objects to blank paper sheets. The objects consisted of stamps, signatures and background watermark styled words inspired from [25]. These sheets were printed and scanned to include a real-world copy-machine noise to them. Next, we digitally removed the background and, therefore, generated several Layer 1 (L_1) instances. To generate the final synthetic dataset, we overlayed the object images (L_1) to a set of documents coming from a variety of domains, not only those that could be found in Publaynet - scientific papers- to show that our model can adapt to unseen domains. In Fig. 3 we present some examples of our crafted dataset.

4.2 Evaluation Metrics

We evaluated our model using PSNR for color, PSNR for illumination, and SSIM.

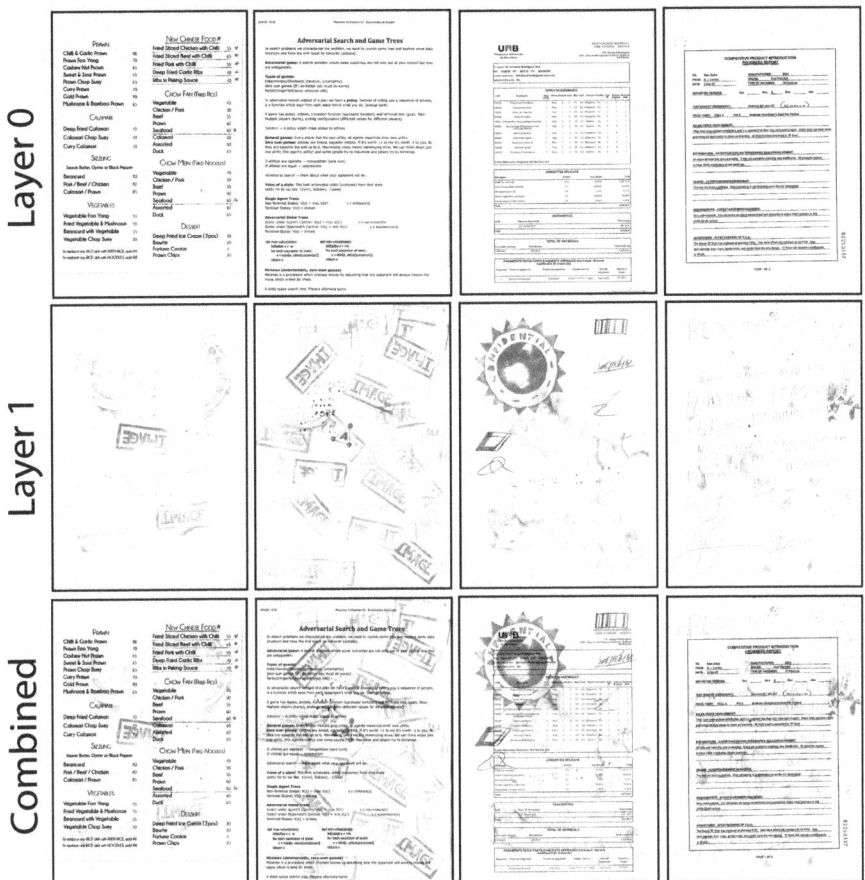

Fig. 3. Some examples of our manually crafted **LayeredDocDB dataset**, illustrating layer 0, layer 1 and the merged noisy image with both layers.

PSNR-Color: Peak Signal-to-Noise Ratio (PSNR) measures the accuracy of color reproduction in the image. Higher PSNR values indicate better color fidelity and overall image quality.

PSNR-Illumination: This metric assesses the preservation of intensity information in the image. Higher values mean better quality.

SSIM (Structural Similarity Index): SSIM [30] evaluates perceived image quality by comparing structural information, luminance, and contrast between the original and processed images. It provides a value between -1 and 1, where 1 indicates perfect similarity, reflecting high-quality image preservation.

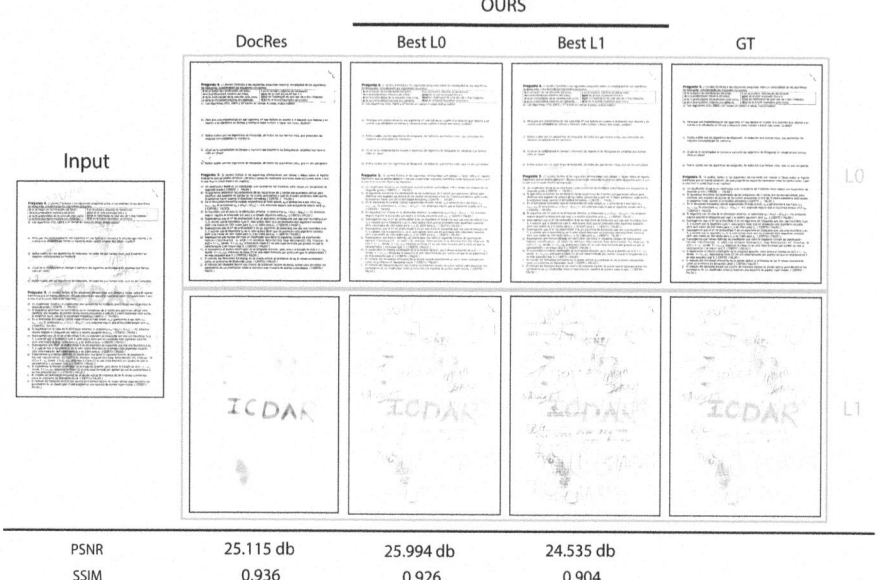

Fig. 4. Comparison between our LayeredDoc and the DocRes [34] approach. Our propose framework preserves the color in layer 1 as opposed to DocRes which in comparition puts the objects in gray scale.

4.3 Qualitative Analysis

Figure 2 shows an example of our model considering Restormer [32] as the baseline restoration architecture. Given a "noisy" document image (most-left column), we show the ground truth data and our results for both layers. As we can see, the model is able to correctly decompose the image, presenting the text part of the image on the layer 0 and the overlaid objects (stamps, background text) on the layer 1. This positive separation of text from other objects shall also help in further downstream tasks requiring an OCR, as this is easily confuse by this kind of noise.

The most similar alternative able to perform this type of layer separation is DocRes [34]. This model also leverages the Restormer image restoration model [32] by considering a textual prompt. In particular, the DocRes model performs document deshadowing, that can be understand as a similar problem to ours. Figure 4 shows a visual comparison between the result of DocRes and our results, together with the PSRN and SSIM values obtained for the images in the example. We can see that even though DocRes decently reconstructs the Layer 0 of the document, it fails to obtain a good performance on retrieving Layer 1. This is not the case for the two versions of our model: best $\mathbb{L}_0$ and best $\mathbb{L}_1$ in validation.

Finally, Fig. 5 compares the output of our proposed modification versus the output of the standard Restormer model that considers just a 3-channel output.

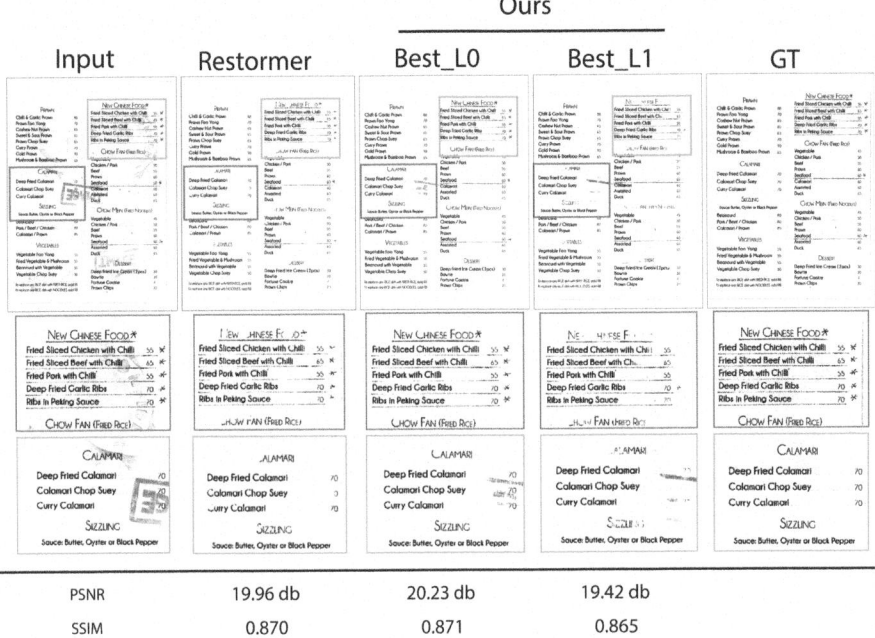

Fig. 5. Comparison between the proposed LayeredDoc model and the standard Restormer [32] approach.

In this figure, we can see how our layer separation model has reconstructed better the document. This is specially noticeable in the titles of the different sections of the menu. While our 6-channel model learns to accurately reconstruct the text, the standard 3-channel model seems to have difficulties and removes parts of the words. As in the previous case, the figure displays PSNR and SSIM results for the images.

4.4 Quantitative Analysis

We have also performed a quantitative evaluation comparing the proposed method with DocRes [34] on our synthetically created dataset. The complete summary of results has been depicted in Table 1. We can see that our model considering the checkpoint with the best L_0 value in validation outperforms both our model with best L_1 and the DocRes model. The results indicate we have a significant **+2 dB gain** over DocRes in PSNR metrics for both color and illumination. We also have a gain in the SSIM metric showing that LayeredDoc shows a lot of potential over prompt-learning based DocRes approach.

Table 1. Quantitative Results on our real-world dataset LayeredDocDB. Final scores computed by averaging over all the images. Results style: **best**, second best.

Method	PSNR(color) ↑	PSNR(ilum) ↑	SSIM ↑
DocRes [34]	21.2469	22.8686	0.9145
Ours (Best_L0)	**23.4026**	**25.0724**	**0.9273**
Ours (Best_L1)	21.8596	23.3913	0.9034

5 Conclusion and Future Work

In conclusion, this paper has successfully demonstrated the efficacy of a novel layer separation approach in enhancing the adaptability of document image restoration (DIR) systems across diverse domains. By implementing a dual-layer information processing system, LayeredDoc effectively handles different types of document degradations and complexities without the need for any retraining. The introduction of the LayeredDocDB, a new real-world dataset, further validates our method, showing significant promise in practical applications with its strong generalization capabilities.

Looking ahead, the future scope of our work could explore the integration of more sophisticated self-supervised frameworks to refine layer separation techniques, potentially increasing accuracy and reducing computational demands. Additionally, expanding the LayeredDocDB to include more varied document types and languages could enhance the robustness and applicability of the DIR system, making it even more effective in global document processing scenarios. Lastly, this work holds a lot of promise to generate more robust and generalist models that could transfer to multiple document understanding tasks in real-world scenario.

Acknowledgment. This work was partially supported by "The European Lighthouse on Safe and Secure AI - ELSA" funded by the European Union's Horizon Europe programme under grant agreement No 101070617; the Spanish projects PID2021-128178OB-I00 and PID2021-126808OB-I00 funded by MCIN/AEI/10.13039/50110 0011033, ERDF "A way of making Europe"; and the Catalan projects 2021-SGR-01499, and 2021-SGR-01559, funded by the Generalitat de Catalunya. S. Biswas is supported by the PhD Scholarship from AGAUR (2023 FI-3-00223), and N. Biescas and M. Pilligua are supported by the CVC Rosa Sensat Student Fellowship. The Computer Vision Center is part of the CERCA Program/Generalitat de Catalunya.

References

1. Bako, S., Darabi, S., Shechtman, E., Wang, J., Sunkavalli, K., Sen, P.: Removing shadows from images of documents. In: Lai, S.-H., Lepetit, V., Nishino, K., Sato, Y. (eds.) ACCV 2016. LNCS, vol. 10113, pp. 173–183. Springer, Cham (2017). https://doi.org/10.1007/978-3-319-54187-7_12
2. Banerjee, A., Biswas, S., Lladós, J., Pal, U.: Swindocsegmenter: an end-to-end unified domain adaptive transformer for document instance segmentation. In: International Conference on Document Analysis and Recognition, pp. 307–325. Springer (2023). https://doi.org/10.1007/978-3-031-41676-7_18
3. Banerjee, A., Biswas, S., Lladós, J., Pal, U.: Semidocseg: harnessing semi-supervised learning for document layout analysis. Int. J. Document Anal. Recogn. (IJDAR), pp. 1–18 (2024)
4. Biswas, S., Banerjee, A., Lladós, J., Pal, U.: Docsegtr: an instance-level end-to-end document image segmentation transformer. arXiv preprint arXiv:2201.11438 (2022)
5. Biswas, S., Riba, P., Lladós, J., Pal, U.: Beyond document object detection: instance-level segmentation of complex layouts. Int. J. Document Anal. Recogn. (IJDAR) **24**(3), 269–281 (2021)
6. Calvo-Zaragoza, J., Gallego, A.J.: A selectional auto-encoder approach for document image binarization. Pattern Recogn. **86**, 37–47 (2019)
7. Clausner, C., Antonacopoulos, A., Derrick, T., Pletschacher, S.: Icdar2019 competition on recognition of early indian printed documents–reid2019. In: 2019 International Conference on Document Analysis and Recognition (ICDAR), pp. 1527–1532. IEEE (2019)
8. Das, S., Ma, K., Shu, Z., Samaras, D.: Learning an isometric surface parameterization for texture unwrapping. In: European Conference on Computer Vision, pp. 580–597. Springer (2022). https://doi.org/10.1007/978-3-031-19836-6_33
9. Das, S., Ma, K., Shu, Z., Samaras, D., Shilkrot, R.: Dewarpnet: single-image document unwarping with stacked 3D and 2D regression networks. In: Proceedings of the IEEE/CVF International Conference on Computer Vision, pp. 131–140 (2019)
10. Das, S., Sial, H.A., Ma, K., Baldrich, R., Vanrell, M., Samaras, D.: Intrinsic decomposition of document images in-the-wild. arXiv preprint arXiv:2011.14447 (2020)
11. Hradiš, M., Kotera, J., Zemcık, P., Šroubek, F.: Convolutional neural networks for direct text deblurring. In: Proceedings of BMVC, vol. 10 (2015)
12. Jemni, S.K., Souibgui, M.A., Kessentini, Y., Fornés, A.: Enhance to read better: a multi-task adversarial network for handwritten document image enhancement. Pattern Recogn. **123**, 108370 (2022)
13. Kang, L., Riba, P., Rusinol, M., Fornes, A., Villegas, M.: Content and style aware generation of text-line images for handwriting recognition. IEEE Trans. Pattern Anal. Mach. Intell. **44**(12), 8846–8860 (2021)
14. Kang, L., Riba, P., Rusiñol, M., Fornés, A., Villegas, M.: Pay attention to what you read: non-recurrent handwritten text-line recognition. Pattern Recogn. **129**, 108766 (2022)
15. Kang, L., Rusinol, M., Fornés, A., Riba, P., Villegas, M.: Unsupervised writer adaptation for synthetic-to-real handwritten word recognition. In: Proceedings of the IEEE/CVF Winter Conference on Applications of Computer Vision, pp. 3502–3511 (2020)
16. Kang, S., Iwana, B.K., Uchida, S.: Complex image processing with less data-document image binarization by integrating multiple pre-trained u-net modules. Pattern Recogn. **109**, 107577 (2021)

17. Kim, G., e al.: Ocr-free document understanding transformer. In: European Conference on Computer Vision. pp. 498–517. Springer (2022). https://doi.org/10.1007/978-3-031-19815-1_29
18. Li, C., Yang, Y., He, K., Lin, S., Hopcroft, J.E.: Single image reflection removal through cascaded refinement. In: Proceedings of the IEEE/CVF Conference on Computer Vision and Pattern Recognition, pp. 3565–3574 (2020)
19. Li, K., et al.: Cross-domain document object detection: Benchmark suite and method. In: Proceedings of the IEEE/CVF Conference on Computer Vision and Pattern Recognition, pp. 12915–12924 (2020)
20. Li, Z., Chen, X., Pun, C.M., Cun, X.: High-resolution document shadow removal via a large-scale real-world dataset and a frequency-aware shadow erasing net. In: 2023 IEEE/CVF International Conference on Computer Vision (ICCV), pp. 12415–12424. IEEE (2023)
21. Liang, J., Doermann, D., Li, H.: Camera-based analysis of text and documents: a survey. IJDAR **7**, 84–104 (2005)
22. Liu, Y., Zhu, Z., Bai, X.: Wdnet: watermark-decomposition network for visible watermark removal. In: Proceedings of the IEEE/CVF Winter Conference on Applications of Computer Vision, pp. 3685–3693 (2021)
23. Ren, D., Zuo, W., Hu, Q., Zhu, P., Meng, D.: Progressive image deraining networks: a better and simpler baseline. In: Proceedings of the IEEE/CVF Conference on Computer Vision and Pattern Recognition, pp. 3937–3946 (2019)
24. Ronneberger, O., Fischer, P., Brox, T.: U-Net: convolutional networks for biomedical image segmentation. In: Navab, N., Hornegger, J., Wells, W.M., Frangi, A.F. (eds.) MICCAI 2015. LNCS, vol. 9351, pp. 234–241. Springer, Cham (2015). https://doi.org/10.1007/978-3-319-24574-4_28
25. Roy, P.P., Pal, U., Lladós, J.: Document seal detection using ght and character proximity graphs. Pattern Recogn. **44**(6), 1282–1295 (2011)
26. Souibgui, M.A., et al.: Docentr: an end-to-end document image enhancement transformer. arXiv preprint arXiv:2201.10252 (2022)
27. Souibgui, M.A., et al.: Text-diae: A self-supervised degradation invariant autoencoder for text recognition and document enhancement. In: Proceedings of the AAAI Conference on Artificial Intelligence, vol. 37, pp. 2330–2338 (2023)
28. Souibgui, M.A., Kessentini, Y.: De-gan: a conditional generative adversarial network for document enhancement. IEEE Transactions on Pattern Analysis and Machine Intelligence (2020)
29. Souibgui, M.A., Kessentini, Y., Fornés, A.: A conditional gan based approach for distorted camera captured documents recovery. In: Mediterranean Conference on Pattern Recognition and Artificial Intelligence, pp. 215–228. Springer (2020). https://doi.org/10.1007/978-3-030-71804-6_16
30. Wang, Z., Bovik, A.C., Sheikh, H.R., Simoncelli, E.P.: Image quality assessment: from error visibility to structural similarity. IEEE Trans. Image Proces. **13**(4), 600–612 (2004). http://dblp.uni-trier.de/db/journals/tip/tip13.html#WangBSS04
31. Yang, Y., Bilen, H., Zou, Q., Cheung, W.Y., Ji, X.: Learning foreground-background segmentation from improved layered GANs. In: Proceedings of the IEEE/CVF Winter Conference on Applications of Computer Vision, pp. 2524–2533 (2022)
32. Zamir, S.W., Arora, A., Khan, S., Hayat, M., Khan, F.S., Yang, M.H.: Restormer: efficient transformer for high-resolution image restoration. In: Proceedings of the IEEE/CVF Conference on Computer Vision and Pattern Recognition, pp. 5728–5739 (2022)

33. Zhang, J., Liang, L., Ding, K., Guo, F., Jin, L.: Appearance enhancement for camera-captured document images in the wild. IEEE Transactions on Artificial Intelligence (2023)
34. Zhang, J., Peng, D., Liu, C., Zhang, P., Jin, L.: Docres: a generalist model toward unifying document image restoration tasks. IEEE/CVF Conference on Computer Vision and Pattern Recognition (CVPR) (2024)
35. Zhong, X., Tang, J., Yepes, A.J.: Publaynet: largest dataset ever for document layout analysis. In: 2019 International Conference on Document Analysis and Recognition (ICDAR), pp. 1015–1022. IEEE (2019)

Normalized vs Diplomatic Annotation: A Case Study of Automatic Information Extraction from Handwritten Uruguayan Birth Certificates

Natalia Bottaioli[1,2,3](✉), Solène Tarride[4], Jérémy Anger[1],
Seginus Mowlavi[1], Marina Gardella[5], Antoine Tadros[1],
Gabriele Facciolo[1], Rafael Grompone von Gioi[1],
Christopher Kermorvant[4], Jean-Michel Morel[6], and Javier Preciozzi[2,3]

[1] Université Paris-Saclay, ENS Paris-Saclay, CNRS, Centre Borelli, Paris, France
natalia.bottaioli@gmail.com
[2] Facultad de Ingeniería, Universidad de la República, Montevideo, Uruguay
[3] Digital Sense, Montevideo, Uruguay
[4] TEKLIA, Paris, France
[5] IMPA, Rio de Janeiro, Brazil
[6] City University of Hong Kong, Hong Kong, China

Abstract. This study evaluates the recently proposed Document Attention Network (DAN) for extracting key-value information from Uruguayan birth certificates, handwritten in Spanish. We investigate two annotation strategies for automatically transcribing handwritten documents, fine-tuning DAN with minimal training data and annotation effort. Experiments were conducted on two datasets containing the same images (201 scans of birth certificates written by more than 15 different writers) but with different annotation methods. Our findings indicate that normalized annotation is more effective for fields that can be standardized, such as dates and places of birth, whereas diplomatic annotation performs much better for fields containing names and surnames, which can not be standardized.

Keywords: Automatic information extraction · Handwritten text recognition · Birth certificates transcription · Normalized and diplomatic annotation

1 Introduction

Civil Registration and Vital Statistics (CRVS) is defined by the United Nations as the "continuous, permanent, compulsory and universal recording of the occurrence and characteristics of vital events of the population in accordance with the law".[1] CRVS is strongly related to article 6 of the Universal Declaration of

---

[1] https://unstats.un.org/unsd/demographic-social/Standards-and-Methods/files/Handbooks/crvs/crvs-mgt-E.pdf.

Normalized vs Diplomatic Annotation 41

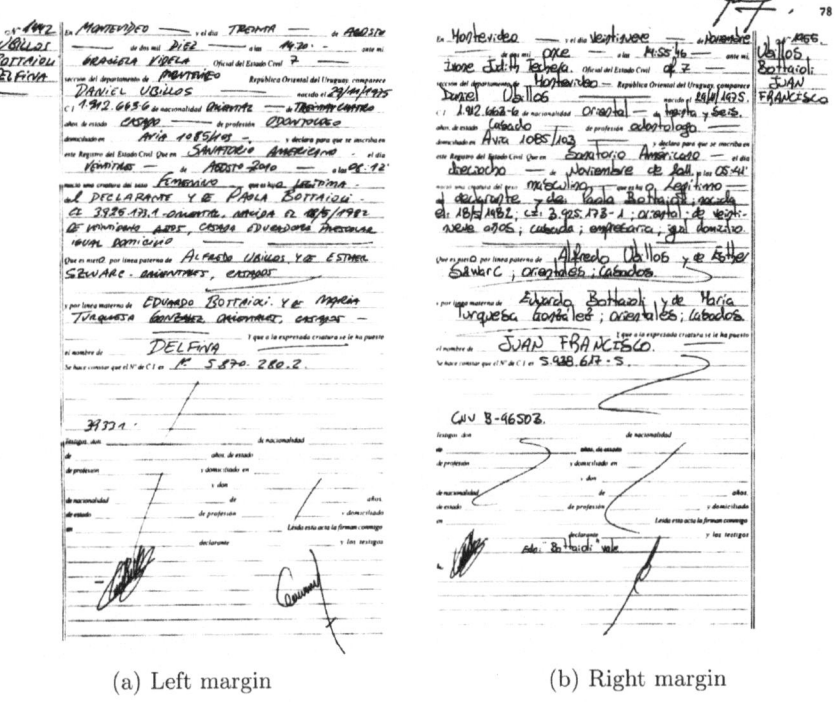

(a) Left margin (b) Right margin

Fig. 1. Two scanned Uruguayan birth certificates, one with the margin on the left and the other one on the right.

Human Rights, which states that everybody has the right to a legal identity.[2] Having a birth certificate gives to individuals legal evidence of their name, place and date of birth, and parents' names (among other data). In most countries, CRVS are key to ensure the right to have a legal identity, and this is clearly reflected by the fact that 67 of the 232 Sustainable Development Goals[3] are related to CRVS.[4] Yet, more than 1 billion people worldwide do not have a document that serves as a proof of legal identity.

Because most CRVS were created long before the advent of computers, the majority of their registries are paper-based handwritten documents. This huge amount of documents is only recently being scanned and integrated into digital systems. Belize, Jamaica and Perú are some of the countries in America that are nowadays in the process of scanning historical CRVS data.

Here, we will focus on the case of Uruguay, where CRVS responsibility is held by a National Civil Registry Office, named the *Dirección General del Reg-*

---

[2] https://www.undp.org/africa/blog/having-legal-identity-fundamental-human-rights.
[3] https://sdgs.un.org/es/goals.
[4] https://getinthepicture.org.

istro de Estado Civil.[5] Initially named *Registro de Estado Civil*, this office exists since 1879, when the government started keeping records of births, marriages, recognitions and deaths nationwide. In 2012, Uruguay's capital city, Montevideo, started registering birth certificates digitally, a process that was completed for the rest of the country only by 2022. Figure 1 shows two scanned birth certificates from Montevideo, one in which the margin is on the right and another one where the margin is on the left. It is worth mentioning that, due to the long period in which registries were generated manually, several different templates exist, and not all formats contain the same information. Even when the format is the same, the way of filling the data is not homogeneous in time.

A digitization process has been gradually taking place in some departments but not yet nationwide: books have been scanned and, therefore, digital images of vital events have been generated. In order to make scanned images searchable and linked to other relevant registries, this digitization process is generally followed by a manual transcription of, at least, part of the data. This is a very demanding and time-consuming task, needless to say prone to errors. It is in this context that automatically processing these registries becomes of great interest.

The process of automatically recognizing the text present in an image requires both the interpretation of handwritten text (known as handwritten text recognition) and the layout interpretation of the scanned document (known as document layout analysis). This process generally requires the adaptation of the existing methods to a specific context. In [11], authors propose a Document Attention Network (DAN) for handwritten document recognition, where the adaptation process does not require to manually segment the different fields present in an image. In fact, only the scanned images and the fields to be transcribed with their corresponding annotations are required.

The objective of this work is twofold. First, we want to understand whether DAN can be adapted to work on the specific problem of handwritten document recognition for Uruguayan birth certificates, using very few data for the adaptation process. This analysis is relevant since DAN was trained for a different context (letters) and a different language (French). Second, we want to analyze two different annotation strategies. The first one consists of using **normalized** annotations, directly obtained from the Civil Registry Office. The second one uses **diplomatic** transcriptions (i.e. those that transcribe each and every character, respecting upper and lower cases, accents, etc.). It is worth noting that in the first strategy, the data used to train the model was neither designed nor generated to train a deep learning model, while in the second one annotations were specifically generated for this task.

The rest of the article is organized as follows: Sect. 2 describes the state of the art regarding handwritten document processing. Section 3 reviews DAN, which is the model evaluated in this work. The Uruguayan birth certificates dataset is described in Sect. 4. Our two major experiments are presented in Sect. 5, while Sect. 6 discusses the results. Conclusions are drawn in Sect. 7, where future work is also described.

---

[5] https://www.gub.uy/ministerio-educacion-cultura/registro-civil.

2 Related Work

Handwritten document processing aims at producing some desired textual information from image inputs. It, therefore, sits at the intersection of computer vision (CV) and natural language processing (NLP). The difference in the modalities of these two fields has encouraged a sequential approach of the problem. The typical document processing pipeline involves one or several steps which perform transcription, i.e. mapping of the input image to a faithful textual output, followed by a natural language understanding step addressing a downstream task, e.g. key information extraction or named entity recognition. The large improvements brought by deep learning in CV [17] and NLP [19], while strengthening the different steps of the sequential document processing pipeline, have also recently enabled integrated approaches. In the latter, algorithms perform the desired downstream task directly from the image input in an end-to-end fashion.

The two aforementioned approaches, sequential and integrated, lie at both ends of a spectrum that we briefly describe here. The sequential approach decomposes the workflow into a pipeline of simple subtasks, most often text line detection, handwritten text recognition (HTR), and finally the desired natural language understanding task. Current text line detection methods [20,23] adapt the object detection literature, in particular CNNs [29], to the domain of document images. HTR methods [4,26] follow an encoder-decoder architecture, making use of decoders popularized in the speech-processing context such as hidden Markov models or recurrent neural networks with connectionist temporal classification [13]. Finally, the NLP step is often carried out by a pre-trained transformer encoder model [12] with simple decoder heads fine-tuned on the given task [3,27]. Variations on this workflow include fusing and jointly training the text line detection and HTR steps [5,37], or enriching the NLP step with the additional input modalities of layout and/or image besides text [15,24,35]. This added integration can bring specific benefits: for example, the former gains efficiency in the image embedding by using a single image encoder, and the latter augments the performance of the NLP step by allowing it to retain information from the input image or the intermediate text line detection output. More recently, the potential of transformers as decoders of visual embeddings [18] has allowed for fully end-to-end models for document processing [11,16].

An important part of the literature focuses on applying this body of research to settings arising from an industrial need [6,9,22,28,32,34]. This type of work contributes back to the theoretical aspects in two key respects. First, it addresses the need for evaluation protocols able to eliminate the influence of benchmark-specific optimizations [7], without expensive searches of the best training procedure for each method. Second, carrying out the entire document processing workflow enables the discovery of important insights beyond the issues of model design and training. For instance, a segmented approach benefits from the availability of powerful pre-trained algorithms [34] but runs the risk of errors accumulating across the steps [31]. Hence, the choice of each algorithm needs to be guided by its input domain, e.g. the NLP module needs to be robust to "recognition error noise" in the input [2,21]. Conversely, [25,34] highlight the influence of the training dataset creation philosophy on the performance of HTR models and, thus, of sequential pipelines. Indeed, not only do annotations need to contain full transcriptions (even if a downstream key information extraction task only concerns a subset of that information), but transcriptions also need to stay faithful to the text (this is known as "diplomatic annotation"). Diplomatic annotation becomes an issue when many database creators apply standardization to dates and names, expand abbreviations or introduce them, sometimes even without following consistent guidelines [8]. By contrast, integrated approaches can deal with partial annotation of training data [31]. In addition, it has been shown that their multi-modality yields an increase in language modeling capabilities [33], which could translate into better flexibility with respect to unfaithfully annotated datasets.

3 Model Description

For our experiments, we use DAN [1,11]. We chose it for its off-the-shelf ability to be fine-tuned on data with a minimal amount of annotation effort, as it does not require bounding box-like annotation of text in the full image. Being an end-to-end architecture, it also allows us to study the influence of different annotation strategies in the context of an integrated approach to document processing.

The architecture of DAN follows the encoder-decoder design. The encoder is the fully convolutional network described in [10], with the fixed 2D positional encoding of [30]. The decoder follows the original design of the auto-regressive transformer decoder [36], including the fixed 1D positional encoding. It generates tokens from a dataset-dependent dictionary $\mathcal{D} = \mathcal{A} \cup \mathcal{S} \cup \{\langle eot \rangle\}$, where $\mathcal{A}$ is the set of characters, $\mathcal{S}$ is a set of semantic markers and $\langle eot \rangle$ is a special "end of transcription" token. The use of markers from $\mathcal{S}$ allows DAN to go beyond plain HTR, by segmenting the text output into layout elements or named-entities. It even allows increased flexibility: by adapting the ground-truth annotations with custom semantic markers and not necessarily including full transcriptions, one can train DAN to directly address other downstream natural language understanding tasks than those showcased by the authors, such as named entity recognition or key information extraction.

The model we use for fine-tuning has been trained in two phases. First, the network of the encoder and the prediction layer of the decoder have their weights initialized by transfer learning: a simple custom model is built using these modules for the task of text line transcription, and trained on a synthetic dataset of printed lines with the CTC loss. In the second phase, DAN is trained with the cross-entropy loss and teacher forcing on a combination of a real handwritten dataset and of synthetically generated printed documents, with the proportion of synthetic data gradually decreasing. The original authors ran the training on several datasets separately. In each case, synthetic documents are generated with the same layout classes as the dataset, with consistent positioning of the layout elements.

In the present work, we use the model trained in [11] on the RIMES 2009 [14] dataset. This dataset comprises grayscale images of French handwritten letters, with ground-truth annotation consisting in text regions with their layout class (one of "sender", "recipient", "date and location", "subject", "opening", "body" and "PS and attachment") and their text transcription. In order to train DAN, these annotations were automatically translated into sequences of tokens from $\mathcal{D}_{\text{RIMES}} = \mathcal{A}_{\text{RIMES}} \cup \mathcal{S}_{\text{RIMES}} \cup \{\langle \text{eot} \rangle\}$, with consistent rules for reading order determination. For our experiments, we use the open-source implementation of DAN [1].

For our fine-tuning of this model, we need to use a token dictionary $\mathcal{D}_{\text{FT}} = \mathcal{A}_{\text{FT}} \cup \mathcal{S}_{\text{FT}} \cup \{\langle \text{eot} \rangle\}$ which is different from $\mathcal{D}_{\text{RIMES}}$. Indeed, our dataset is not in the same language (Spanish instead of French) thus, while the intersection of character sets $\mathcal{A}_{\text{FT}} \cap \mathcal{A}_{\text{RIMES}}$ is significant, both $\mathcal{A}_{\text{FT}} \setminus \mathcal{A}_{\text{RIMES}}$ and $\mathcal{A}_{\text{RIMES}} \setminus \mathcal{A}_{\text{FT}}$ are non-empty. Furthermore, our fine-tuning task of information extraction is encoded by a distinct set of markers than those for layout segmentation in RIMES, so $\mathcal{S}_{\text{FT}} \cap \mathcal{S}_{\text{RIMES}} = \emptyset$. As a consequence, we need to adapt the token embedding and probability prediction layers in DAN's decoder. Since their weight matrices have a clear interpretation, with individual tokens corresponding to their columns and rows respectively, this does not pose a significant challenge: we keep the trained weights corresponding to tokens in $\mathcal{D}_{\text{FT}} \cap \mathcal{D}_{\text{RIMES}}$, remove those corresponding to $\mathcal{D}_{\text{RIMES}} \setminus \mathcal{D}_{\text{FT}}$ and create new weights for $\mathcal{D}_{\text{FT}} \setminus \mathcal{D}_{\text{RIMES}}$.

4 Datasets

For this work, we had access to a dataset composed of 201 birth certificates from 4 different years (2008, 2012, 2014 and 2016), handwritten in more than 12 writing styles (as can be seen in Fig. 3), with their corresponding transcribed text as kept in the computer system used in the local civil registry of Tacuarembó, Uruguay. Images are binarized and saved in tiff format, with a resolution of 200 dpi. The transcription of 7 fields present in the documents was provided in a CSV file. These fields are: "document number", "enrollee's full name", "year of the enrollment", "jurisdiction", "1st parent's full name", "date of birth", and "2nd parent's full name". The field "department" was added to the dataset as a means of having one more annotation with zero annotation cost given that it is constant in all 201 documents. From this original dataset, we created two different sets of annotations.

The first one was obtained by ordering the CSV files containing the original annotations (as given by the civil registry of Tacuarembó) in the document's reading order, and converting the date, originally provided in the format "YYYY-MM-DD", into "date field 1" and "date field 2". The former contains the day of birth written in letters, while the latter contains the month written in letters followed by the word "de" followed by the year written in numbers (YYYY). (For instance, the date "2014-05-31" is annotated as "treinta y uno" in "date field 1" and "mayo de 2014" in the "date field 2".) This decision was

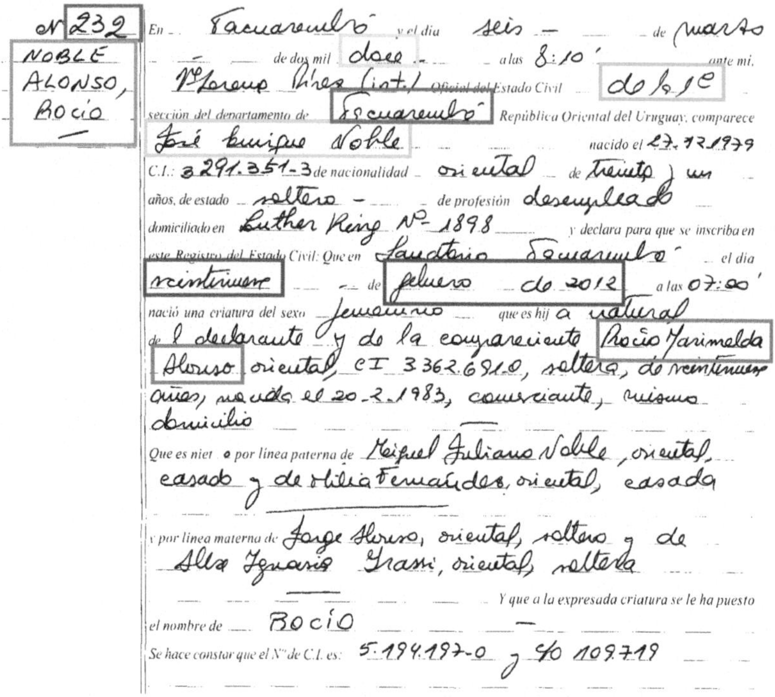
Nº 232
NOBLE ALONSO, ROCIO

En Tacuarembó y el día seis de marzo de dos mil doce a las 8:10 ante mí, Mª Lorena Pires (int.) Oficial del Estado Civil de la 1ª sección del departamento de Tacuarembó República Oriental del Uruguay, comparece José Enrique Noble nacido el 27.12.1979 C.I.: 3.291.351-3 de nacionalidad oriental de treinta y un años, de estado soltero de profesión desempleado domiciliado en Luther King Nº 1898 y declara para que se inscriba en este Registro del Estado Civil: Que en Sanatorio Tacuarembó el día veintinueve de febrero de 2012 a las 07:00 nació una criatura del sexo femenino que es hija natural del declarante y de la compareciente Rocío Marimelda Alonso oriental, CI 3362.691-0, soltera, de veintinueve años, nacida el 20.2.1983, comerciante, mismo domicilio

Que es nieta por línea paterna de Miguel Juliano Noble, oriental, casado y de Hilda Fernández, oriental, casada

y por línea materna de Jorge Alonso, oriental, soltero y de Alba Ignacia Grassi, oriental, soltera

Y que a la expresada criatura se le ha puesto el nombre de ROCÍO

Se hace constar que el Nº de C.I. es: 5.194.197-0 y s/o 109719

Field	Normalized DAN	Diplomatic DAN
document number	`232`	`232`
enrollee's full name	`Noble Alonso Rocio`	`NOBLE ALONSO ROCÍO`
year of enrollment	`doce`	`doce`
jurisdiction	`Primera`	`de la 1a`
department	`Tacuarembó`	`Tacuarembó`
1st parent's name	`José E. Noble`	`José Enrique Noble`
date field 1	`veintinueve`	`veintinueve`
date field 2	`febrero de 2012`	`febrero de 2012`
2nd parent's name	`Rocio M. Alonso`	`Rocío Marimelda Alonso`

Fig. 2. Crop of the upper part of a birth certificate created by copying fields from several real birth certificates originally handwritten by the same public servant (and belonging to the dataset used in experiments), followed by a table showing two mentioned ways of annotating data. Annotated fields are manually highlighted in colors for visualization purposes only: document number (red), enrollee's full name (orange), year of enrollment (yellow), jurisdiction (light green), department (dark green), 1st parent's name (light blue), date field 1 (dark blue), date field 2 (violet), 2nd parent's name (pink) (Color figure online).

made in order to avoid having to manually annotate the date in each of the documents. However, note that not all handwritten dates are written this way, as Table 1 shows.

Table 1. Birth certificate crop of a line where the date is handwritten in a way different from the most frequent one, together with annotations for "date field 1" and "date field 2" for both datasets. The inferred text by both models (Diplomatic DAN and Normalized DAN) is also included.

	date field 1	date field 2
Normalized annotation	veintiocho	marzo de 2016
Normalized DAN inference	veintiocho	marzo de 2016
Diplomatic annotation	Veintiocho de Marzo	dos mil dieciseis
Diplomatic DAN inference	Veintiocho de Marzo	dos mil dieciseis

In the provided CSV files, names are also transcribed in (possibly inconsistent) shortened ways. For example, "Alberto Carlos Bustos" can be transcribed as "Alberto C. Bustos" or as "A. Carlos Bustos". As a consequence, data kept in databases are not necessarily verbatim transcriptions of the handwritten text present in the documents. Although it could be ideal if this additional formatting was learned by the model to generate a formatted output (if desired), we decided to create another set of annotations, based on a verbatim transcription of the text (also known as "diplomatic annotation").

This implied creating a new annotation set by: 1) transcribing all accents, when present; 2) respecting upper and lower case; 3) writing each of the two entities corresponding to the birth date in the way they appear in each document (dates can appear as "veintiocho de marzo" and "dos mil dieciseis" in some documents, as in the example shown in Table 1, or as "veintiocho" and "marzo de 2016", which is the most usual way and the one used in the normalized annotation); 4) transcribing names and last names in full, as they appear in the documents (e.g. not abbreviating "Maria Juana González" as "Maria J. González"). Generating the diplomatic transcription of all fields for the 201 documents took about 10 h of manual work.

Having these two sets of annotations (normalized and diplomatic) enabled us to perform two different experiments, which are described in the next section.

5 Experiments

The main objective of the experiments was to understand the capabilities of DAN for transfer learning, in particular how good this process works in the context of structured birth certificates written in Spanish. We randomly divided the original dataset into train, validation and test in 80% (161 images), 10% (20 images) and 10% (20 images), respectively, and used this same partition for all experiments.

Fine-tuning was done using images of a fixed height of 1900 pixels, automatically resized in such a way that they preserved their aspect ratio. Original images were no smaller than 1800 pixels wide (therefore, they were, at the most, shrunk but never enlarged). The only pre-processing that images went through, before being used in the fine-tuning processes, was a subtle rotation in such a way that the side margin present in all images becomes vertical.

The model features an encoder with 5 convolutional layers and a decoder with 8 transformer layers, each of size 256 with 4 attention heads, employing dropout, data augmentation, and label noise for enhanced performance. The maximum number of epochs is set to 2000, with a batch size of 5 and a learning rate of 0.0001.

We shall now describe our two experiments. The first one consisted in finetuning DAN using annotations that required very little pre-processing. We name the generated model **Normalized DAN**. The second one consisted in finetuning the same model using diplomatic (verbatim, character by character) annotations. We name the generated model **Diplomatic DAN**.

The reason why we started by training a first model using the annotation dataset obtained directly from the original CSV files (computer-based annotations) is that the use of this kind of annotations is convenient for two reasons. First of all, there is no need for manual transcriptions, which is a costly and time-consuming process. Secondly, the desired result was that the model would output information in a normalized format that can be directly used by a computer system, without post-processing (as it is already formatted with the expected output).

It is worth noting that in this first experiment the model must not only learn how to identify and extract the correct information but also how to format it to the desired output which is, a priori, a more complex task.

The second experiment was designed in order to test how well DAN would perform if fine-tuned with data that reflected exactly what is handwritten in the documents for the same set of fields (i.e. by using a diplomatic annotation).

6 Results

Table 2 shows the Character Error Rate (CER) and the Word Error Rate (WER) for each field and each experiment. We have included two different values for each error measure: regular and normalized text. The first one is considering an exact, case and accent-sensitive comparison between text. For the normalized metric, we have considered uppercase and lowercase to be the same, and also accents were removed. We have included this analysis because, in several occasions, the differentiation of the casing or the presence of accents is not relevant.

A first observation of the quantitative results shown in Table 2 is that both CER and WER values are similar to those obtained in [11], even though the context is different. Comparing the results of the two different models shows that

Normalized vs Diplomatic Annotation   49

Table 2. Character Error Rate and Word Error Rate for each field of birth certificates in the test set (composed of 20 birth certificate scans). Both metrics are computed on regular text (reg.) and normalized text (norm.), in which characters are changed to lowercase and accents are removed.

Field	Normalized DAN				Diplomatic DAN			
	CER (%)		WER (%)		CER (%)		WER (%)	
	reg.	norm.	reg.	norm.	reg.	norm.	reg.	norm.
document number	3.51	-	10.00	-	3.51	-	10.00	-
year of enrollment	0.00	**0.00**	0.00	**0.00**	3.31	3.31	5.00	5.00
jurisdiction	0.00	**0.00**	0.00	**0.00**	0.92	0.92	3.33	3.33
department	0.00	**0.00**	0.00	**0.00**	4.50	0.00	5.00	0.00
date field 1	0.00	**0.00**	0.00	**0.00**	0.61	0.61	3.85	3.85
date field 2	0.00	**0.00**	0.00	**0.00**	0.35	0.00	1.67	0.00
enrollee's full name	2.25	2.25	9.64	9.64	2.94	**1.73**	14.46	**6.02**
1st parent's name	10.72	10.43	36.36	36.36	2.83	**2.02**	12.86	**10.00**
2nd parent's name	15.81	15.81	35.19	35.19	7.17	**5.70**	28.36	**20.90**
global	4.36	4.32	13.92	13.92	3.05	**2.04**	11.27	**7.51**

Table 3. Errors made by Diplomatic DAN on normalizable fields. The only errors made by Normalized DAN in all normalizable fields are also shown in the table (they are the ones made in document number). Errors made when an uppercase letter instead of its corresponding lowercase letter (or vice-versa) are marked in blue. Errors made when actually substituting one character for another one or omitting a character are marked in red. No insertion errors were detected in these fields.

Field	Annotation	Inference	Crops from documents
document number	90	20	
	1154	1156	
year of enrollment	catorce	cocer	
jurisdiction	de la 1o	de la 1a	
	de la 1ra	de la 1r_	
department	TACUAREMBÓ	Tacuarembó	
date field 1	dieciocho	diescho	
date field 2	catorce	Catorce	

Diplomatic DAN globally performs slightly better than **Normalized DAN**. Yet, a closer look at the results obtained for each field reveals that **Normalized DAN** performs better than **Diplomatic DAN** in most fields ("year of enrollment", "jurisdiction", "department", "date field 1" and "date field 2"), while **Diplomatic DAN** performs considerably better in fields that contain names and last names ("enrolee's full name", "1st parent's name" and "2nd parent's name"). We analyze the outputs based on this division of fields.

Fields that do not contain names and last names Six out of the nine fields do not contain names or last names. In these cases, the results suggest that giving the model a normalized annotation to learn (which is what was used for Normalized DAN), improves the model's learning capabilities. The number of errors that Diplomatic DAN makes in all normalizable fields is shown in Table 3. Even if the results obtained by the two models can be comparable, training the **Normalized model** is less costly in terms of annotation effort and it also generates a formatted output, which is usually preferred when saving data into information systems.

Fields that contain names and last names As mentioned before, there are three fields in the dataset that contain names and last names: "enrollee's full name", "1st parent's name" and "2nd parent's name". In the three cases, the CER and WER, if compared ignoring cases and accents, are better with **Diplomatic DAN**. It is interesting to see how this improvement is different in the tree mentioned fields.

The field "enrollee's full name" is only slightly better for Diplomatic DAN than for Normalized DAN. It is worth remembering that, in this specific field, where names and last names typically appear in several lines, in the margin of the certificate, a full annotation is provided both in the normalized dataset and in the diplomatic one. This probably explains the very similar results that are obtained by both models.

The main difference between the "1st parent's name" and "2nd parent's name" fields is that the former is surrounded by preprinted text while the latter is always surrounded by handwritten text. This probably explains why the behavior of Diplomatic DAN in the "2nd parent's name" field is almost 3 times better than Normalized DAN, while the improvement obtained by Diplomatic DAN on the "1st parent's field" is more than 5 times better (when evaluated ignoring cases and accents).

One of the possible reasons why Diplomatic DAN works better than Normalized DAN on these three fields was already explained in the description of the dataset: annotations provided by the civil registry included some abbreviation criteria on both parents' names fields where, very frequently, the second name was abbreviated to its initial (e.g., "Alberto Carlos Bustos" was transcribed as "Alberto C. Bustos"). Somehow, this was learned by the model but with mixed results. In cases where the enrollee has a second name and the model was able to understand which letter to use, Normalized DAN produced a correct output (i.e. correctly abbreviated the middle name). In other cases, it made a mistake when abbreviating the middle name. In cases where no middle name was handwritten in the document, the model just invented one, presumably to agree with the format of the data it was trained on.

7 Conclusions and Future Work

In this work we explored the capabilities of fine-tuning DAN using two annotation strategies to adapt it to a different context than the one in which it was trained. We observed that, even with very little data (181 documents), the model was able to obtain very similar results to the ones shown in the original paper.

We also explored two annotation strategies, normalized and diplomatic, and saw that the former worked better in fields that can be standardized while the latter performed better in fields where character by character transcription is needed. The aforementioned results on fine-tuning DAN for transcribing handwritten documents open several questions.

The first one is whether using a "hybrid" annotated dataset (meaning, normalized for normalizable fields and diplomatic for fields containing names and last names) would result in a model that is capable of both transcribing "normalizable" fields in a normalized way (hence, eliminating the need of post-transcription normalization) while being extremely accurate when transcribing non-normalizable fields (such as, in our case, names and last names).

A second natural line is exploring how far we can go in reducing the number of birth certificates needed to fine-tune a model without losing accuracy.

A third line we will explore is the model's generalization capabilities. For instance, when dealing with different document layouts. Related to this, it is worth testing whether using a different annotation order for birth certificates that have the margin on the left or on the right will improve accuracy.

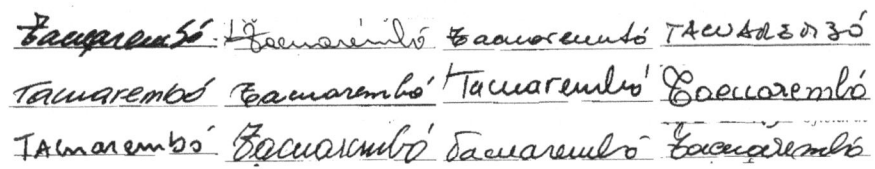

Fig. 3. Images of the word "Tacuarembó" as written by 12 different writers.

Further, one can explore whether exploiting document redundancy by annotating redundant information will result in higher accuracy. As can be seen from Fig. 2, birth certificates generally have some data redundancy such as the enrolled person's name is written both in the margin and in the body of the document. Lastly, since handwriting styles can be so different among one another, we would like to explore whether training a model on a specific writer can lead to a better understanding of the writer's specific handwriting style and, therefore, a higher transcription accuracy.

Acknowledgments. The research that originated the results presented in this publication was partly supported by the Agencia Nacional de Investigación e Innovación (ANII) and the France 2030 CollabNext project.

References

1. Dan implementation repository by TEKLIA. https://gitlab.teklia.com/atr/dan, release: 0.2.0rc6
2. Abadie, N., Carlinet, E., Chazalon, J., Duménieu, B.: A benchmark of named entity recognition approaches in historical documents application to 19th century French directories. In: Uchida, S., Barney, E., Eglin, V. (eds) Document Analysis Systems. DAS 2022. Lecture Notes in Computer Science, vol 13237, pp. 445–460. Springer, Cham (2022). https://doi.org/10.1007/978-3-031-06555-2_30
3. Akbik, A., Bergmann, T., Blythe, D., Rasul, K., Schweter, S., Vollgraf, R.: FLAIR: an easy-to-use framework for state-of-the-art NLP. In: Proceedings of the 2019 Conference of the North American Chapter of the Association for Computational Linguistics (Demonstrations), pp. 54–59 (2019)
4. Arora, A., et al.: Using ASR methods for OCR. In: 2019 International Conference on Document Analysis and Recognition (ICDAR), pp. 663–668. IEEE (2019)
5. Bluche, T., Louradour, J., Messina, R.: Scan, attend and read: end-to-end handwritten paragraph recognition with MDLSTM attention. In: 2017 14th IAPR International Conference on Document Analysis and Recognition (ICDAR), vol. 1, pp. 1050–1055. IEEE (2017)
6. Boillet, M., Tarride, S., Schneider, Y., Abadie, B., Kesztenbaum, L., Kermorvant, C.: The Socface project: large-scale collection, processing, and analysis of a century of French censuses (2024)
7. Cheplygina, V., Varoquaux, G.: Artificial intelligence in science: lessons from shortcomings in machine learning for medical imaging. In: Artificial Intelligence in Science: Challenges, Opportunities and the Future of Research. Organization for Economic Co-operation and Development (OECD) (2023)
8. Clérice, T., et al.: CATMuS medieval: a multilingual large-scale cross-century dataset in Latin script for handwritten text recognition and beyond (2024)
9. Constum, T. et al.: Recognition and information extraction in historical handwritten tables: toward understanding early 20th century Paris census. In: Uchida, S., Barney, E., Eglin, V. (eds) Document Analysis Systems. DAS 2022. LNCS, vol 13237, pp. 143–157 Springer, Cham (2022). https://doi.org/10.1007/978-3-031-06555-2_10
10. Coquenet, D., Chatelain, C., Paquet, T.: End-to-end handwritten paragraph text recognition using a vertical attention network. IEEE Trans. Pattern Anal. Mach. Intell. 45(1), 508–524 (2022)
11. Coquenet, D., Chatelain, C., Paquet, T.: DAN: a segmentation-free document attention network for handwritten document recognition. IEEE Trans. Pattern Anal. Mach. Intell. 45(7), 8227–8243 (2023)
12. Devlin, J., Chang, M.W., Lee, K., Toutanova, K.: BERT: pre-training of deep bidirectional transformers for language understanding. arXiv preprint arXiv:1810.04805 (2018)
13. Graves, A., Schmidhuber, J.: Offline handwriting recognition with multidimensional recurrent neural networks. In: Proceedings of the 21st International Conference on Neural Information Processing Systems, NIPS 2008, pp. 545–552. Curran Associates Inc., Red Hook, NY, USA (2008)

14. Grosicki, E., Carré, M., Brodin, J.M., Geoffrois, E.: Results of the RIMES evaluation campaign for handwritten mail processing. In: 2009 10th International Conference on Document Analysis and Recognition, pp. 941–945. IEEE (2009)
15. Huang, Y., Lv, T., Cui, L., Lu, Y., Wei, F.: LayoutLMv3: pre-training for document AI with unified text and image masking. In: Proceedings of the 30th ACM International Conference on Multimedia, MM 2022, pp. 4083–4091. ACM, New York, NY, USA (2022). https://doi.org/10.1145/3503161.3548112
16. Kim, G., et al.: OCR-free document understanding transformer. In: Avidan, S., Brostow, G., Cissé, M., Farinella, G.M., Hassner, T. (eds.) Comput. Vision - ECCV 2022, pp. 498–517. Springer Nature Switzerland, Cham (2022)
17. Krizhevsky, A., Sutskever, I., Hinton, G.E.: ImageNet classification with deep convolutional neural networks. Adv. Neural Inf. Proc. Syst . **25** (2012)
18. Li, M., et al.: TrOCR: transformer-based optical character recognition with pre-trained models. In: Proceedings of the AAAI Conference on Artificial Intelligence, vol. 37, pp. 13094–13102 (2023)
19. Mikolov, T., Sutskever, I., Chen, K., Corrado, G.S., Dean, J.: Distributed representations of words and phrases and their compositionality. Adv. Neural Inf. Process. Syst. **26** (2013)
20. Mikolov, T., Sutskever, I., Chen, K., Corrado, G.S., Dean, J.: Distributed representations of words and phrases and their compositionality. Adv. Neural Inf. Process. Syst. **26** (2013)
21. Monroc, C.B., Miret, B., Bonhomme, M.-L., Kermorvant, C.: A comprehensive study of open-source libraries for named entity recognition on handwritten historical documents. In: Uchida, S., Barney, E., Eglin, V. (eds.) Document Analysis Systems: 15th IAPR International Workshop, DAS 2022, La Rochelle, France, May 22–25, 2022, Proceedings, pp. 429–444. Springer International Publishing, Cham (2022). https://doi.org/10.1007/978-3-031-06555-2_29
22. Nion, T., et al.: Handwritten information extraction from historical census documents. In: 2013 12th International Conference on Document Analysis and Recognition, pp. 822–826. IEEE (2013)
23. Oliveira, S.A., Seguin, B., Kaplan, F.: dhSegment: a generic deep-learning approach for document segmentation. In: 2018 16th International Conference on Frontiers in Handwriting Recognition (ICFHR), pp. 7–12. IEEE (2018)
24. Peng, Q., et al.: ERNIE-layout: layout knowledge enhanced pre-training for visually-rich document understanding. In: Goldberg, Y., Kozareva, Z., Zhang, Y. (eds.) Findings of the Association for Computational Linguistics: EMNLP 2022, pp. 3744–3756. Association for Computational Linguistics, Abu Dhabi, United Arab Emirates (Dec 2022). https://doi.org/10.18653/v1/2022.findings-emnlp.274, https://aclanthology.org/2022.findings-emnlp.274
25. Petitpierre, R., Kramer, M., Rappo, L.: An end-to-end pipeline for historical censuses processing. Int. J. Doc. Anal. Recogn. (IJDAR) **26**(4), 419–432 (2023)
26. Puigcerver, J.: Are multidimensional recurrent layers really necessary for handwritten text recognition? In: 2017 14th IAPR International Conference on Document Analysis and Recognition (ICDAR), vol. 1, pp. 67–72. IEEE (2017)
27. Puigcerver, J.: Are multidimensional recurrent layers really necessary for handwritten text recognition? In: 2017 14th IAPR International Conference on Document Analysis and Recognition (ICDAR), vol. 1, pp. 67–72. IEEE (2017)
28. Romero, V., et al.: The ESPOSALLES database: an ancient marriage license corpus for off-line handwriting recognition. Pattern Recogn. **46**(6), 1658–1669 (2013). https://doi.org/10.1016/j.patcog.2012.11.024, https://www.sciencedirect.com/science/article/pii/S0031320312005080

29. Ronneberger, O., Fischer, P., Brox, T.: U-Net: convolutional networks for biomedical image segmentation. In: Navab, N., Hornegger, J., Wells, W.M., Frangi, A.F. (eds.) MICCAI 2015. LNCS, vol. 9351, pp. 234–241. Springer, Cham (2015). https://doi.org/10.1007/978-3-319-24574-4_28
30. Singh, S.S., Karayev, S.: Full page handwriting recognition via image to sequence extraction. In: Lladós, J., Lopresti, D., Uchida, S. (eds.) ICDAR 2021. LNCS, vol. 12823, pp. 55–69. Springer, Cham (2021). https://doi.org/10.1007/978-3-030-86334-0_4
31. Tarride, S., Boillet, M., Kermorvant, C.: Key-Value Information Extraction from Full Handwritten Pages. In: Fink, G.A., Jain, R., Kise, K., Zanibbi, R. (eds) Document Analysis and Recognition - ICDAR 2023. ICDAR 2023. LNCS, vol 14188, pp. 185–204 Springer, Cham (2023). https://doi.org/10.1007/978-3-031-41679-8_11
32. Tarride, S., Boillet, M., Moufflet, J.-F., Kermorvant, C.: SIMARA: a database for key-value information extraction from full-page handwritten documents. In: Fink, G.A., Jain, R., Kise, K., Zanibbi, R. (eds.) Document Analysis and Recognition - ICDAR 2023: 17th International Conference, San José, CA, USA, August 21–26, 2023, Proceedings, Part III, pp. 421–437. Springer Nature Switzerland, Cham (2023). https://doi.org/10.1007/978-3-031-41682-8_26
33. Tarride, S., Lemaitre, A., Coüasnon, B., Tardivel, S.: A comparative study of information extraction strategies using an attention-based neural network. In: Uchida, S., Barney, E., Eglin, V. (eds.) Document Analysis Systems: 15th IAPR International Workshop, DAS 2022, La Rochelle, France, May 22–25, 2022, Proceedings, pp. 644–658. Springer International Publishing, Cham (2022). https://doi.org/10.1007/978-3-031-06555-2_43
34. Tarride, S., et al.: Large-scale genealogical information extraction from handwritten Quebec parish records. Int. J. Doc. Anal. Recogn. (IJDAR) **26**(3), 255–272 (2023). https://doi.org/10.1007/s10032-023-00427-w
35. Tu, Y., Guo, Y., Chen, H., Tang, J.: LayoutMask: enhance text-layout interaction in multi-modal pre-training for document understanding. In: Annual Meeting of the Association for Computational Linguistics (2023). https://api.semanticscholar.org/CorpusID:258967524
36. Vaswani, A., et al.: Attention is all you need. In: Advances in Neural Information Processing System, vol. 30 (2017)
37. Wigington, C., Tensmeyer, C., Davis, B., Barrett, W., Price, B., Cohen, S.: Start, follow, read: end-to-end full-page handwriting recognition. In: Ferrari, V., Hebert, M., Sminchisescu, C., Weiss, Y. (eds) ECCV 2018. LNCS, vol. 11210, pp. 372–388. Springer, Cham (2018). https://doi.org/10.1007/978-3-030-01231-1_23

ARPC

ARPC 2024 Preface

A workshop on "Advanced Analysis and Recognition of Parliamentary Corpora" (ARPC) was organized in the framework of the International Conference on Document Analysis and Recognition (ICDAR 2024), which took place in Athens, Greece. The ARPC workshop was held on the morning of Saturday, 31 August 2024. It included four scientific papers, which underwent a single-blind peer review process by two reviewers each, a keynote talk, and a structured discussion. This workshop was designed for parliamentary researchers, data scientists, and policymakers aiming to extract nuanced insights from vast, yet unexplored legislative archives.

Scope and Motivation

Parliamentary archives house a wealth of contemporary and historical legislative and administrative documents. The study of these corpora presents remarkable opportunities, enabling the convergence of previously disparate fields like history, political science, and linguistics. This again opens novel horizons in comprehending parliamentary data and discourse. The workshop offered a state-of-the-art platform for the scholarly discussion of innovative methods for the recognition and analysis of these corpora. In doing so, it facilitated the exploration of advanced techniques that enhance our understanding of parliamentary materials, foster interdisciplinary connections, and propel research in this dynamic field.

Workshop Description

Data-driven insights from archives have the potential to steer academic research in a variety of fields. This workshop addressed the growing importance of employing advanced recognition and analytical methods and tools to decode the complexities within legislative and administrative documents of parliamentary origin. Hence, it was placed under ICDAR, the premier international event for scientists and practitioners involved in document analysis and recognition. The workshop dived deeply into cutting-edge OCR techniques for parliamentary corpora. Further attention was given to recognizing patterns, extracting meaningful insights and understanding the intricate dimensions of contemporary and historical parliamentary discourse. The relevance of this topic lies in its potential to bridge previously isolated domains of research, fostering interdisciplinary collaboration. By connecting history, political science, and linguistics, participants will unlock a richer understanding of legislative evolution, political trends, and linguistic nuances embedded in parliamentary proceedings.

Due to the synergy of perspectives from diverse stakeholders, scientific discussions during the workshop yielded outcomes that extend beyond individual disciplines. Envisioned outcomes include novel methodologies, identification of trends, and the establishment of a collaborative network that transcends traditional academic silos. The workshop was supported and promoted by the Hellenic OCR Team, a global network for analyzing parliamentary data. Established back in 2017, the Team represents the first scientific crowdsourcing initiative that aims exclusively at the processing and study of parliamentary textual data.

The topics of interest to the workshop included but were not limited to:

- The recognition of polytonic Greek fonts
- Recognition of mixed text (printed and handwritten)
- Parliamentary discourse analysis
- Historical trends in parliamentary language use
- Integration of linguistic and political science methodologies in OCR
- Cross-lingual OCR challenges in parliamentary texts
- Machine learning approaches for semantic analysis of parliamentary proceedings
- Ethical considerations in the digitization and analysis of parliamentary records
- Developing standardized formats for parliamentary data preservation
- The role of OCR technology in enhancing public access to parliamentary archives
- Comparative analysis of parliamentary rhetoric across different eras
- The impact of digital humanities tools on legislative studies
- Application of natural language processing (NLP) techniques in political discourse analysis
- Automated categorization and indexing of parliamentary documents
- Challenges and solutions in digitizing non-standard parliamentary texts.

August 2024

Fotios Fitsilis
George Mikros

Diminutives in Political Discourse – The Case of Serbian and Slovenian

Milena Oparnica(✉)

The Institute for Artificial Intelligence of Serbia, Fruskogorska 1, 21000 Novi Sad, Serbia
milena.oparnica@ivi.ac.rs

Abstract. Diminutives are well acknowledged for their contribution to the expressiveness of language, often adding layers of nuance and attitude to discourse. This paper draws comparisons between the use of diminutives in the ParlaMint-RS 4.0 (Serbian parliament) and ParlaMint-SI 4.0 (Slovenian parliament) corpora [1]. Our findings reveal a distinctive pattern within political discussions: the employment of diminutives, particularly when referring to entities other than the speaker (i.e., not in the first person), is almost invariably associated with a negative connotation or the intention to convey irony. This paper aims to underscore the significance of such linguistic nuances, highlighting how diminutives can subtly influence the tone and perceived intent of political discourse. Through the examination of verbal diminutives, we contribute to a deeper understanding of the use of language in political contexts.

Keywords: Verbal Diminutives · Political Discourse · ParlaMint · Serbian · Slovenian

1 Introduction

The exploration of verbal diminutives in the Serbian language was advanced by one of the pioneering studies, conducted by Irene Grickat in 1955 [2]. In her work on verbal diminutives across Slavic languages, Grickat highlighted the unique position of Serbo-Croatian. She noted that among the Slavic languages, Serbo-Croatian stands out for its extensive use and development of diminutive verbs. This language employs both prefixes and suffixes to form diminutive verbs, many of which are highly productive.

Dragićević [3] also conducted an analysis of diminutive verbs to determine whether they conveyed a positive or negative attitude towards the actions described. She concluded that, among all six verbs she analyzed, their use in the first person signified a positive stance towards the action expressed. In contrast, the usage of diminutive verbs in the third person was associated with a negative attitude towards the action. For instance, the following two sentences have different sentiments. The first sentence (*a.*) would probably be regarded as neutral denoting the action of writing that is not important. The second one (*b.*), related to the quality of someone else's writing would be marked as negative and it is used pejoratively.

a. Piskaram   nešto.
   Write-DIM-1.PERS something
b. On piskara   nešto.
   He write-DIM-3.PERS something

The meaning of diminutives largely depends on the context. Therefore, the type of discourse becomes crucial for inferring the meaning of diminutives. For example, talking about children and saying that children are *smiling a bit*, i.e. using a diminutive form of smile, usually does not carry a negative connotation. However, the use of the same verb in political discourse almost always conveys a negative undertone. *Contempt is particularly pronounced if the verb is used in the third person singular or plural. If used in the first person, it introduces information about the non-mandatory nature of the action and about performing it carelessly. In any case, there's almost no positive connotation* [3, p. 79]. Dragićević [3] notes that the low frequency of verbal diminutives is recorded in the Serbian language electronic corpus SrpKor [4], within the literary-artistic functional style.

Therefore, the goal of this study is to examine diminutive verbs within the parliamentary corpus in Serbian and Slovenian, aiming to gain insights into the prevailing attitude, whether positive or negative, most commonly expressed through diminutives in this particular type of discourse. In corpus linguistics, the term "semantic prosody" is frequently employed for investigating the attitudinal meanings of words by analysing their collocates (see e.g. [5, 6]). Due to the scope of the work, it was not possible to examine the collocates of all verbal diminutives. The collocates of two lexemes were analyzed in more detail, the most frequent lexeme *smeškati se* "smile-DIM" and its Slovenian equivalent *smehljati se*.

This exploration seeks to unveil how diminutives function within political language, highlighting their power to subtly influence the interpretation of verbal communication.

2 Verbal Diminutives in Serbian and Slovene

2.1 Serbian Verbal Diminutives

Grickat's analysis reveals that diminutive verbs in Serbo-Croatian can convey a range of nuanced meanings. These include the indication of an action occurring over a brief period, the frequency of an action that is both repetitive and short-lived, and the diminished significance of the action, among other interpretations. For instance, examining the suffixes containing [r], Grickat elucidates that these suffixes alter the meaning of the verb in a way that emphasizes a lack of care or the triviality of the action denoted by the root verb, i.e. *životariti*, "subsiting in a minimal". The diminution can also be achieved through prefixation, and verbs can have both a prefix and a suffix, for example, *pročačkati* "to rummage through".

In Serbian, the verbal diminutives are most commonly formed by adding the element *-k* immediately after the word's root. Grickat also tackles different suffixes for verbal diminution in Serbo-Croatian, such as *-k, -ak, -uk, -ik, -čk, -ar, -kar, -rk, -šk, -ušk, -c, -uc, -ic, -ck, -et*. The most frequent verbal diminutive suffixes are *-ka, -kara, -ucka, -uca, -uka, -uši, -uška* [3]. From more frequent verbs, diminutive verbs can be derived with various suffixes, for example, *piskati, piskarati, pisuckati* are diminutive verbs derived from the verb *pisati* "to write".

2.2 Slovenian Verbal Diminutives

As Sicherl [7] concludes, unlike nominal diminutives, verbal diminutives are less common in the Slovenian language. Verbal diminutives in Slovenian are formed by adding suffixes and prefixes to a neutral verbal base. Toporišič [8] and Vidovič Muha [9] identify a range of suffixes for diminutive formation: *-k-, -čk-, -c-, -ic-, -inc-, -lj-, -ik-* and *-uck-* (the last one added by Vidovič Muha).

In Slovene, as in Serbian, verbal diminutives can exhibit multiple levels of diminution, much like nominal diminutives, though they are less common. This multiplicity in verbal forms is achieved through either double suffixation or the addition of a prefix and a diminutive suffix, where the prefix connotes a "small quantity". Sicherl [7, p. 148] notes derivations such as *stopati* to *stopicati* to *stopicljati* ("to step", "stepping lightly", "stepping very lightly") and *vohati* to *vohljati* to *ovohljati/povohljati* ("to smell", "smell-DIM", "smell-DIMDIM"). Multiple diminutiveness in verbs is possible, though less frequent than in nouns. Verbal diminutiveness can also be expressed analytically or periphrastically, involving an adverbial component like "a little", "a bit", or "lovely", "nicely" to convey a quantitative or positive evaluation of the verbal action.

3 Methodology and Data Sources

Diminutive verbs are investigated using comparable corpora from Serbian and Slovenian parliamentary discourse – ParlaMint 4.0, accessible via the Clarin platform [1]. ParlaMint 4.0 encompasses a set of comparable corpora, consisting of transcriptions of parliamentary debates. The Serbian segment of ParlaMint 4.0 includes transcripts spanning from 1997 to 2022, totaling 83,266,206 words, while its Slovenian counterpart covers the period from 2000 to 2022, amassing a total of 69,032,700 words.

Guided by the suffix classifications [2, 3], we employed regular expressions to isolate potential diminutive forms in ParlaMint-RS 4.0. After compiling lists of these diminutives, we manually reviewed them to ensure accuracy in classification. After identification and extraction of diminutive verbs for Serbian, we started seeking equivalent translations of these diminutive verbs into Slovenian and subsequently searched these Slovenian diminutive forms within the corpus. It is crucial to highlight that determining whether a verb qualifies as a diminutive was not always straightforward, as some instances were ambiguously positioned within a continuum of diminution.

This study excludes diminutive forms that involve altering a vowel within the verb stem, as these are not readily identifiable through our methodology. For example, the transformation from *sip-a-ti* "to pour" to its diminutive form *sip-i-ti* "to pour a bit" is omitted from our analysis, given the complexity of accurately detecting such variations.

4 Results

4.1 Verbal Diminutives Within Serbian Parliamentary Discourse

Table 1 presents a lemmatized list of diminutive verbs with a frequency greater than 10, extracted from the given corpus. Although, as we observed from Dragićević [3], diminutive verbs are almost nonexistent in the corpus of literary-artistic style, associating them more with conversational language, it is interesting to note that these verbs are frequent in political discourse.

Table 1. Verbal diminutive forms extracted from ParlaMint-RS 4.0 (Serbian parliament)

Verb	Absolute frequency	Freq. Per million	English
smeškati	381	3.94	smile-DIM
kraduckati	73	0.75	steal-DIM
Čačkati	68	0.70	poke-DIM
Tužakati	38	0.39	accuse-DIM
zveckati	33	0.34	clang/cling-DIM
Šetkati	31	0.32	stroll-DIM
Seckati	28	0.29	chop-DIM
Spiskati	28	0.29	spend-DIM
trčkarati	26	0.27	run-DIM
Krckati	25	0.26	knap-DIM
Koknuti	24	0.25	knock-DIM
Brčkati	24	0.25	bath-DIM
Krčkati	23	0.24	cook-DIM
švrćkati	23	0.24	stroll-DIM
skoknuti	22	0.23	jump-DIM
Čačnuti	22	0.23	poke-DIM
vozikati	21	0.22	drive-DIM
Šuškati	21	0.22	rustle-DIM
mućkati	21	0.22	shake-DIM
smejuljiti	20	0.21	smile-DIM
cerekati	20	0.21	grin-DIM
moljakati	18	0.19	pray-DIM
začeprkati	18	0.19	rummage-DIM
čeprkati	16	0.17	rummage-DIM

(*continued*)

Table 1. (continued)

Verb	Absolute frequency	Freq. Per million	English
pročačkati	14	0.14	touch-DIM
začepkati	13	0.13	touch-DIM
Cepkati	13	0.13	split-DIM
iščačkati	12	0.12	rummage out-DIM
mućnuti	12	0.12	stir-DIM
Iseckati	11	0.11	chop-DIM
zamumuljiti	11	0.11	muddy-DIM
Kuckati	11	0.11	knock-DIM
Lupkati	10	0.10	tap-DIM
Merkati	10	0.10	measure-DIM
pijuckati	10	0.10	drink-DIM

Similar to the meanings associated with nominal diminutives in political discourse (as discussed by Oparnica & Panić Cerovski [10]), verbal diminutives are almost invariably used with a negative or ironic connotation in this type of discourse. Given the extensive presence of the reflexive verb *smeškati se* "smile-DIM" in the corpus, we carried out a further analysis of the collocations in order to determine their semantic prosody. We analyzed the appearance of words occurring within three words on either side of the verb *smeškati se* (cf. e.g. [11]). Table 2, in addition to the number of co-occurrences and the total number of occurrences of the collocate (candidates), displays the values of the LogDice score, a measure of collocational strength that adapts to varying corpus sizes (cf. [12]) As illustrated in Table 2, apart from "se", a reflexive marker, it frequently collocates with the words *vi* "you.2PL" and *samo* "just", specifically in the construction *samo se ti/vi smeškaj/smeškajte* "just keep smiling" (see examples c. and d. below).

Examples below are related to the collocate *samo* "just", indicating criticism, encapsulating the speaker's critical view of the others' responses to serious situations.

c. *Dalje, ona vaša kesa, zaboravili ste jednu stvar – to je nekada bila čarapa, pa ste se sada sofisticirali i mislite da sa kesom možete da završite posao.* **Samo se vi smeškajte**. *Ponovo ćete da izađete za ovu govornicu, a bogami, ovih dana ćete poprilično da izlazite da se* **smeškate** *i da dajete poruke građanima o novoj realnosti. Da, građani Srbije, nova realnost – zadesila nas je elementarna nepogoda.*

"Furthermore, that bag of yours, you've forgotten one thing – it used to be a sock, and now you've sophisticated yourselves thinking that with a bag you can finish the job. **Just keep smiling.** You will stand at this podium again, and indeed, these days you will be coming out quite a bit to **smile** and to deliver messages to the

Table 2. Collocates of the verb *smeškati se* "smile-DIM" within Parlamint-RS 4.0

Word	Cooccurrences	Candidates	LogDice	English
zadovoljno	5	86	8.45	contentedly
Kiselo	3	32	7.89	sourly
smeškali	3	46	7.85	smile-DIM.PST.PL
Mile	3	191	7.43	dear.F.GEN.SG
Glavom	3	1707	5.56	head.INSTR.SG
Samo	22	20739	5.09	Just
Možete	3	3120	4.81	can-2PL
Lepo	9	13838	4.37	nicely
Vi	76	165755	3.91	you.2PL
nemojte	6	13105	3.87	do.2PL.NEG
Nemojte	3	9813	3.27	Do.2PL.NEG
Ti	9	31281	3.22	you.2SG
Hoćete	3	11423	3.06	want.2PL
Vidim	4	15400	3.05	see.1SG
Vi	9	40017	2.87	You.2SG.FORMAL /2PL
Se	355	1,594,918	2.87	REFL
samo	37	189,602	2.67	just

citizens about the new reality. Yes, citizens of Serbia, a new reality – a natural disaster has struck us."

d. ...*Bojana i ove druge, imaju oni uzora svog. Samo se ti smeškaj, samo se ti smeškaj. Smeškajte se i vi što prodajete Telekom. To su vaši epohalni rezultati. Mi tražimo odgovor na pitanja – kad ćete da prekinete*

"...Bojan and the others, they have their role models. Just keep smiling, just keep smiling. You who are selling Telecom, keep smiling as well. Those are your epochal results. We are asking for an answer to the question – when will you stop"

Another collocate is the negation "not" – *nemojte da se smeškate* "do not laugh - DIM", which from the speaker's point of view refers to inappropriate behavior of some participants in the meeting and serves to remind the listener or the audience to behave seriously. Adjective *kiselo* "sour" in *kiselo smeškanje* "sour smile" connotes skepticism. Interestingly, the terms *lepo* "nice" and *zadovoljno* "contendly" in the context of smiling, though primarily positive, signify a form of mocking acceptance of what the speaker finds unacceptable, hinting at irony. As noted by Louw [5], a prosodic clash indicates irony. This effect is achieved by deliberately creating a clash between the expected associations of words and their actual use, which often subtly reveals the speaker's true attitudes. In the example below, *vi se lepo smeškate* "you smile nicely" seems to be used ironically and critically, pointing out that while the person maintains a pleasant demeanor, serious underlying issues are being overlooked or mishandled. A "nice smile" in a situation where contentment is unexpected or inappropriate could serve to heighten the ironic effect.

e. *Kako je moguće da najveći evropski giganti konkurišu za neki posao, a kažete da nemate projektnu dokumentaciju? Prema tome, samo se vi smeškajte (**vi se lepo smeškate**, imate lep osmeh, ovako, sa gospođom Čomić kada razmenite, i to je vrlo fino), ali ja vam kažem da vi sad uvodite ovu državu u velike probleme. Počinjete da radite ponovo bez para, bez obezbeđenih sredstava, bez zatvorenih finansijskih konstrukcija,*

"How is it possible that the largest European giants are competing for a job, and yet you say that you don't have project documentation? So, just keep smiling (you do have a **nice smile**, like this, when you exchange looks with Mrs. Čomić, and that's very fine), but I'm telling you that you are now leading this country into great trouble. You begin to work again without money, without secured funds, without closed financial structures."

The verb *kraduckati*, implying petty theft or stealing on a small scale, was among the three most frequently used verbal diminutives, because many politicians quoted V. Ilić, the former Serbian Minister, V. Ilić, who once made a remark similar to the following: "I understand that you steal a little, but you're stealing too much." The following example illustrates such use. Here, the adverb *malo* "little" used along with the diminutive verb emphasizes the diminutiveness associated with the verb.

f. *Liči na Veljinu izreku – pustio sam vas da **malo kraduckate**, ali, ljudi, vi kradete.*

"It resembles Velja's saying – I let you **steal a little**, but, folks, you are stealing."

The verb *vozikati se* does not appear in the first person in the ParlaMint-RS 4.0 corpus. It is interesting to compare this with the register of conversational discourse. For example, in a corpus containing texts from *.rs* domains, such as PDRS [13], it appears in the first person with a frequency of 0.02 per million tokens. The meaning is associated with casual driving (e.g., "usually I fix up the bike and then I go for a leisurely ride"). Here is an example of the use of this verb in parliamentary discourse (*g.*). The use of the verb *vozikati se* in political discourse conveys contempt. Even without the use of a diminutive, the content would be negative. However, the employment of a diminutive verb makes the statement even more expressive and further amplifies its negativity.

g. *Znači, kupujete avione, vozikate se avionima i baš vas briga kako je građanima Srbije i od čega će da žive ovi ljudi koji su na ivici egzistencije, milion nezaposlenih.*

"So, you're buying airplanes, flying around in them, and you couldn't care less about the citizens of Serbia and how they're going to live, these people who are on the brink of existence, a million unemployed."

4.2 Verbal Diminutives Within Slovenian Parliamentary Discourse

After extracting verbal diminutives from the ParlaMint-RS 4.0 corpus, translations for diminutives with a frequency of ten or more were sought. Subsequently, we examined their occurrence frequency within the ParlaMint-SI 4.0 corpus. Translations were primarily searched through the *Fran.si* [14] and Amebis.si [15]. In Table 3, the translation equivalents of Serbian verbal diminutives with a frequency of ten or greater, which are also diminutives in the Slovenian language, are provided. Lexemes that could be translated with the same word, depending on the context, are bolded. It should be emphasized that these translated verbs may not perfectly align with the Serbian diminutive forms. Sometimes their applicability varies, being usable in a greater or lesser number of contexts. For instance, verbs such as *brskati* and *pobrskati* – "to search-DIM", are notably common in Slovenian. However, these verbs do not have direct equivalents in Serbian diminutive forms such as *čeprkati*, *začeprkati*, and *pročačkati*, but are sometimes equal to *tražiti/pretraživati* "to search". Contextual analysis revealed that verbs like *brskati* and *pobrskati* meaning "to rummage", "to dig into the past" or "to browse (the internet)", sometimes carry an expressive meaning in the parliamentary corpus, and sometimes they do not. Specifically, when it comes to browsing the internet (14 tokens in total), they do not bear an expressive meaning at all, and such uses were excluded from the absolute frequency calculation.

Furthermore, it was observed that the use of this verb in the first person usually does not carry a negative meaning (examples *h.* and *i.*), except when the verb is negated (example *j.*).

h. ***Brskam*** *in iščem in tega zakonika ne najdem. Ustavnopravnega zakonika ne najdem.*

"I'm searching and searching, and I can't find this code. I can't find the constitutional law code."

The example below (*i.*) suggests a more involved or meticulous process of searching or investigating, especially when used in reference to looking for information or trying to understand a situation better. The addition of an adverb *malce* "a bit" emphasizes the speaker's effort in a nuanced way, indicating a deliberate but not overly extensive search.

i. *Spraševal sem se, zakaj Ministrstvo za finance tako hiti zdaj, kar naenkrat, da hoče sprejeti ta zakon.* *Malce sem še **brskal**, pa sem ugotovil, da poteka še en spor pred Evropskim sodiščem za človekove pravice, in sicer zaradi umanjkanja pravnega sredstva pritožbe.*

"I wondered why the Ministry of Finance is in such a hurry now, all of a sudden, to pass this law. I **did a bit** more **digging** and found out that there is another dispute before the European Court of Human Rights, due to the lack of a legal remedy for appeal."

However, when this verb is negated in the first person, it typically conveys a negative attitude, such as in the example below.

Diminutives in Political Discourse 67

Table 3. Verbal diminutive equivalents in Slovenian with their frequency within ParlaMint-SI 4.0

Original Verb	Frequency	Slovenian Translation	Frequency (absolute)	Freq. per million	English
smeškati	381	smehljati	51	0.62	smile-DIM
tužakati	38	tožariti	45	0.55	accuse-DIM
zveckati	33	cingljati	2	0.02	clink-DIM
Seckati	28	rezljati	0	-	cut-DIM
spiskati	28	razfračkati	0	-	blow (money, resources)-DIM
trčkarati	26	tekati	44	0.54	run-DIM
Krckati	25	hrstljati, hrskati, hrstati	0	-	crack-DIM
Brčkati	24	čofotati	0	-	paddle-DIM
skoknuti	22	skakljati	2	0.02	jump-DIM
Šuškati	21	šušljati	15	0.18	rustle-DIM
smejuljiti	20	smejčkati	2	0.02	smile-DIM
začeprkati	18	**pobrskati**	86	1.25	**search-DIM**
čeprkati	16	**brskati, bezati**	140, 2	2.02, 0.02	**search-DIM**
nahuškati	14	nahujskati	4	0.05	incite-DIM
pročačkati	14	**brskati, pobrskati**	140	2.02	**search-DIM**
iščačkati	12	**zbezljati**	15	0.18	**search-DIM**
Iseckati	11	zrezljati	0	-	chop-DIM
Kuckati	11	trkati	4	0.05	tap-DIM
Lupkati	10	tapljati	3	0.04	tap-DIM
pijuckati	10	srkati	9	0.11	sip-DIM
smuljati	10	zmutiti	3	0.04	confuse-DIM

j. *Z vsem spoštovanjem do predlagatelja, kolega Koprivca, ne verjamem temu. Pa **ne bom** tukaj zdaj **brskal** po magnetogramih, kaj vse ste, predvsem Socialni demokrati, v času opozicije vedno govorili, ampak ko je prišla pa koalicija, ki nekaj naredi, imate pa probleme.*

"With all due respect to the proposer, colleague Koprivc, I do not believe this. And I **won't start rummaging** through the transcripts now, about everything you, especially the Social Democrats, always said while in opposition, but when a coalition that actually accomplishes something comes along, you have problems."

Due to the frequent occurrence of *brskati* and *pobrskati*, a more detailed contextual analysis of these verbs is required, which falls outside the scope of this paper.

5 Comparative Analysis

Based on the results outlined in Tables 1, 2, and 3, it can be inferred that the use of diminutive verbs is more varied and more frequent in Serbian than in Slovenian parliamentary discourse. The following are observed similarities and differences in the use of diminutives in parliamentary discourse (Table 4).

Table 4. The comparison of the occurrences of verbal diminutives in Parliament-RS 4.0 and Parlamint-SI 4.0

Serbian		Normalized frequency (pmw)
Total number of words within corpus	83,266,206	**13.82**
Unique dim. Verbs (freq. Equal or higher than 10)	37	
Total number of dim. Verbs (freq. Equal or higher than 10)	1151	
Slovenian		
Total number of words within corpus	69,032,700	**6.24**
Unique dim. Verbs (freq. Higher than 10)	24	
Total number of dim. Verbs (freq. Equal or higher than 10)	431	

Low Frequent Diminutive Verbs in Slovenian. It is interesting to analyze the frequency comparison between the pairs *smeškati* – *smehljati* "to smile-DIM", as well as *smejuljiti* and its Slovenian equivalent *smejčkati* "smile-DIM". The pair *smeškati* – *smehljati* "to smile-DIM" has a higher degree of lexicalization, indicating that it is more commonly used and recognized. The pair *smejuljiti* – *smejčkati* is noted for being more expressive. Both Serbian counterparts appear significantly more often within parliamentary corpora (see Tables 1 and 3). Sicherl [7] mentions Slovenian *smehljati* as a lexicalized diminutive. Notably, within the parliamentary corpus, the semantic prosody is predominantly negative as evidenced by its collocations with words such as *posmehljivo* "mockingly", *cinično* "cynically" and *sadistično* "sadistically", *grobu* "grave.DAT" as well as alluding to Orwell (see Table 5). The example below is an example of the positive collocate *lepo* "nicely". It can be noted that the entire statement is ironic, and can be interpreted similarly to the example e. above.

> k. V tem, da se na koncu podjetniki sami organizirajo in preko odvetniške družbe Grilc Vouk Škof vložijo direktne tožbe zaradi tega, ker vi tega niste naredili, ker morate skrbeti, da ste prijatelji, da greste skupaj na večerje in se**lepo smehljate** , ob tem pa oni žagajo ekonomski temelj Slovenije.

> "In this, the entrepreneurs end up organizing themselves and through the law firm Grilc Vouk Škof file direct lawsuits because you did not do this, because you need to make sure you are friends, that you go out to dinners together and **smile nicely**, while they are undermining the economic foundation of Slovenia."

Table 5. Collocates of the verb *smehljati se* in Parlamint-SI 4.0

Word	Cooccurrences	Candidates	LogDice	English
naslovnic	2	10	10.07	cover.GEN.PL
stanovskimi	2	16	9.93	resident.GEN.PL
Orwell	2	20	9.85	Orwell
sadistično	1	2	9.27	sadistically
časopisja	2	62	9.18	newspapers.GEN
obtoženci	1	7	9.14	accused.NOM.PL
Ajde	1	16	8.93	Come on
luštno	1	31	8.64	Cute
užitkom	1	32	8.62	pleasure.INSTR
posmehljivo	1	35	8.57	mockingly
naslonil	1	61	8.19	leaned
grobu	1	63	8.17	grave.DAT
levičarske	1	67	8.12	leftist
ekranov	1	69	8.09	screens.GEN
cinično	2	206	7.99	cynically

More Frequent Diminutive Verbs in Slovenian. As already observed above, there are diminutive verbs that are more frequent in Slovenian. *Brskati* and *pobrskati* were most frequent verbs among the examined equivalents (see Table 3). These verbs correspond to various equivalents in Serbian, as illustrated in Table 3. Given their frequency, these verbs require further contextual analysis, as already noted above. Additionally, there is also *tekati* (with frequency per million equal to 0.54 compared to Serbian *trčkarati* 0.27). However, it should be mentioned that in Serbian other similar forms were used in the corpus with a frequency less than 10, for example: *trčkati, istrčkavati, trknuti*.

l. ...*imate petnaest minuta, nemojte da idete na kafu,* **trknite** *do Pravnog fakulteta, uzmite diplomu.*

"...you have fifteen minutes, don't go for a coffee, dash to the Faculty of Law, and grab your diploma."

m. *Ako bude ovako stalno* **istrčkavao** *i ulazio, znate, gubimo na kontinuitetu obraćanja i on bi trebalo nešto da pribeleži i eventualno da nam da posle odgovore.*

If he keeps running out and coming back in like this, you know, we lose the continuity of the speech, and he should take some notes and possibly provide us with answers afterward.

Verbs that are Not Diminutive in Slovenian. In searching for equivalent translations of Serbian diminutive verbs with a frequency equal to or higher than ten in Slovenian, we came across some lexemes for which there is a translation, but they are not diminutive in Slovenian (see Table 6). Generally, like the diminutives, they carry a negative meaning within the corpus itself. In an example below, an equivalent to *čačkati* in Serbian, *bezati* "to poke/to pick at somebody" is positioned within a broader critique of governmental and parliamentary inaction, and failure to address significant socio-economic issues.

n. *Sem rekel, poglejte, vam bom odkrito povedal, da s to in takšno vlado, s to in takšno koalicijo, s temi in takšnimi pogledi na razvoj in Slovenijo bodo penzije šle kvečjemu dol. Han se je odločil, da bo **bezal** v Združeno levico. Legalno, legitimno. Velikega predvolilnega pol leta se bo odvrtelo, bistvena razlika med tovrstnimi našim ubesedovanji je, ne le da Trček govori o penezih, ampak da tudi razmišlja makroekonomsko. Ampak v tem državnem zboru se ne razmišlja.*

"I said, look, I'm going to be frank with you, with this kind of government, this kind of coalition, with these sorts of views on development and Slovenia, pensions are only going to decrease. Han has decided he will **pick at** the United Left. Legally, legitimately. A significant pre-election half-year will unfold, the fundamental difference between such expressions of ours is, not only does Trček talk about money, but he also thinks macroeconomically. However, in this national assembly, thinking does not occur."

Mućkati and *mućnuti* are highlighted in italics because, even in the Serbian parliamentary corpus, they are most often used in a lexicalized form in the expression *mućnuti/mućkati/promućkati glavom* ("put on one's thinking cap"), which is frequently used in informal discourse. The following examples show different uses and meanings of this verb in parliamentary discourse. In the example *o.* speaker accuses the president of offending members of the parliament by using an informal phrase *mućnuti glavom*. The repetition of the phrase emphasizes the negative attitude of the speaker towards the statement. Using *mućkati* "tampering" in the example *p.* implies concerns about manipulation or improper actions within the process.

o. *Gospodine predsedniče, vi ste povredili Poslovnik tako što ste rekli da svi poslanici treba da **mućnu glavom**. Da **mućkaju glavom**. Koji je to član?*

"Mr. President, you have violated the Rules of Procedure by stating that all members should **put on their thinking cap**. **To put on their thinking cap**. Which article is that?"

p. *Pitaće ministar – ne predlažete rešenje kako ubrzati privatizaciju? Vratiti ove stimulativne mere, promeniti, dopuniti; uvesti i treći metod – zašto se beži od dokapitalizacije? Navodno, tu će da se **mućka**. Ne može tu da se **mućka**, ali nemam vremena da obrazložim zašto.*

"The minister will ask – don't you propose a solution to accelerate privatization? To return these incentive measures, to change, to supplement; to introduce a third method – why avoid recapitalization? Supposedly, there will be **tampering**. One cannot **tamper** there, but I don't have time to explain why."

Table 6. Verbal diminutives in Parlamint-RS 4.0 and Parlamint-SI 4.0

Serbian diminutive verbs	Freq. per million	Slovenian translation	Freq. per million	English
čačkati	0.70	bezati	0.02	touch-DIM
čačnuti	0.23	bezati	0.02	touch-DIM
vozikati	0.22	vozariti se	-	drive-DIM
mućkati	*0.22*	*mešati, premisliti, misliti*	-	*stir/nod/think-DIM*
cerekati	0.21	režati	0.70	grin-DIM
moljakati	0.19	moledovati, prosjačiti	-	pray-DIM
cepkati	0.13	trgati	-	split-DIM
smaknuti	0.13	umakniti	-	remove-DIM
mućnuti	0.12	*premisliti, misliti*	-	*stir/nod/think-DIM*
merkati	0.10	pogledovati	-	look-DIM

Verbs that are Not Translatable into Slovenian. Finally, there are verbs such as *kraduckati* "steal-DIM" or *šetkati se* and *šećkati se* "walk-DIM" that are not translatable to Slovenian. We would have to use a phrase as a translation instead of a single lexeme. E.g. the Serbian diminutive verb *šetkati se* "walk-DIM" would be translated similay to "po malem se sprehajati".

6 Conclusion

In this paper, diminutive verbs extracted from the Serbian parliamentary corpus, ParlaMint-RS 4.0, were presented. Translation equivalents in the Slovenian language were then sought for these verbs, and subsequently, the frequency of the translated diminutive verbs was examined. As we have observed, the meanings of diminutive verbs greatly depend on the context, making it very difficult to determine their exact equivalents.

It is noted that diminutive verbs are more common and varied in Serbian than in Slovenian parliamentary discourse. This may be due to the generally higher usage and productivity of diminutive verbs in conversational discourse [2, 3]. Interestingly, in this type of discourse they are almost always used in a negative connotation, often with an ironic meaning, except when they are used in the first person. Collocations that include the Serbian *smeškati se* "smile-DIM" and the Slovenian *smehljati se* "smile-DIM" were presented to determine and compare their semantic prosodies. Certainly, a more detailed analysis of each of the verbs listed in this work together with their collocates could contribute to a more complete picture of diminutive verbs as a means of achieving various pragmatic effects in political discourse. Additionally, it would be interesting to compare them with their originating verbs to perform more detailed analyses.

Finally, This study confirmed what Grickat [2] wrote about verbal diminutives, namely that they are more productive in Serbo-Croatian than in any other Slavic language. Also, as observed by Grickat [2] and Dragićević [3], the context plays a significant role in determining the meaning of diminutives. Through our exploration of the usage of diminutive verbs, we enhance the comprehension of how language is strategically utilized in political environments, illuminating the complex roles diminutives play as expressive devices, frequently imbued with a layer of criticism or irony.

References

1. Erjavec, T., et al.: Linguistically Annotated Multilingual Comparable Corpora of Parliamentary Debates ParlaMint.ana 4.0. Slovenian Language Resource Repository CLARIN.SI (2023). ISSN 2820-4042. http://hdl.handle.net/11356/1860
2. Грицкат, И.: Деминутивни глаголи у српскохрватском језику. Јужнословенски филолог, 21, 45–96 (1955)
3. Драгићевић, Р.: Глаголски деминутиви између творбеног и употребног значења. У: Kowalski, P. (ed.) Słowotwórstwo w przestrzeni komunikacyjnej, pp. 61–84. Instytut Slawistyki Polskiej Akademii Nauk, Warszawa (2021)
4. SrpKor. Corpus of Contemporary Serbian Language (2013). http://www.korpus.matf.bg.ac.rs/
5. Louw, B.: Irony in the text or insincerity in the writer?—The diagnostic potential of semantic prosodies. In: Baker, M., Francis, G., Tognini-Bonelli, E. (eds.) Text and Technology: In Jonor of John Sinclair, pp. 157–176. Benjamins, Amsterdam (1993)
6. Hunston, S.: Semantic prosody revisited. Int. J. Corpus Linguist. **12**(2), 249–268 (2007). https://doi.org/10.1075/ijcl.12.2.09hun
7. Sicherl, E.: Diminutive nouns and verbs in Slovene compared to their English equivalents. In: Slovene Linguistic Studies (2013). https://hdl.handle.net/1808/11435
8. Toporišič, J.: Slovenska slovnica. Založba Obzorja, Maribor (1991)
9. Vidovič Muha, A.: Slovensko skladenjsko besedotvorje ob primerih zloženk. ZIFF, Partizanska knjiga, Ljubljana (1988)
10. Oparnica, M., Panić Cerovski, N.: The use of diminutives in everyday communication. In: Belgrade Linguistics Days (BeLiDa), vol. 1, pp. 327–345 (2022). https://doi.org/10.18485/belida.2022.1.ch13
11. Milicevic, M.: Semi-automatic construction of comparable genre-oriented Corpora of Serbian in Cyrillic and Latin scripts (2015)
12. Heaton, D., Nichele, E., Clos, J., et al.: "The pingdemic has been a greater challenge than Covid itself": semantic prosodies in UK newspaper articles during the pandemic. SN Soc. Sci. **3**, 146 (2023). https://doi.org/10.1007/s43545-023-00740-5
13. Wasserscheidt, P.: Serbian Web Corpus PDRS 1.0. Slovenian Language Resource Repository CLARIN.SI (2023). ISSN 2820-4042. https://hdl.handle.net/11356/1752
14. Slovar slovenskega knjižnega jezika: SSKJ 2, 2nd edn., Cankarjeva založba (2014). Accessed via Fran.si platform
15. Amebis.si. 2023. https://www.amebis.si/

Loghi: An End-to-End Framework for Making Historical Documents Machine-Readable

Rutger van Koert[1](✉) [iD], Stefan Klut[1] [iD], Tim Koornstra[2] [iD], Martijn Maas[1] [iD], and Luke Peters[2] [iD]

[1] Humanities Cluster KNAW, Oudezijds Achterburgwal 185, 1012 DK Amsterdam, The Netherlands
{rutger.van.koert,stefan.klut,martijn.maas}@di.huc.knaw.nl
[2] Nationaal Archief, Prins Willem-Alexanderhof 20, 2595 BE Den Haag, The Netherlands
{tim.koornstra,luke.peters}@nationaalarchief.nl

Abstract. Loghi is a novel framework and suite of tools for the layout analysis and text recognition of historical documents. Scans are processed in a modular pipeline, with the option to use alternative tools in most stages. Layout analysis and text recognition can be trained on example images with PageXML ground truth. The framework is intended to convert scanned documents to machine-readable PageXML. Additional tooling is provided for the creation of synthetic ground truth. A visualiser for troubleshooting the text recognition training is also made available. The result is a framework for end-to-end text recognition, which works from initial layout analysis on the scanned documents, and includes text line detection, text recognition, reading order detection and language detection.

The Loghi pipeline has been used successfully in several projects. We achieve good results on the layout analysis and text recognition of both the handwritten and printed archives of the Dutch States General on resolutions spanning the 17th and 18th century. The CER on handwritten 17th century material is below 3%. Loghi is open source and free to use.

Keywords: Handwritten Text Recognition · Layout Analysis · PageXML

1 Introduction

Historical documents in archives are a true treasure trove for researchers, however often these archives are not digitally accessible and if they are, it is often in the form of just the scan without an accompanying transcription. Millions of pages have however been scanned and are now waiting to be processed further, so their contents can live in the digital world and be searched and used further by both machines and humans. In order to extract machine-readable text from historical documents, the scanned documents are processed using software that detects and recognises text lines. Depending on the type of scanned document, it is often referred to as either Optical Character Recognition (OCR) for machine printed text or Handwritten Text Recognition (HTR) for handwritten documents. In this paper we will use the term HTR for both OCR and HTR as our method can process both separately and together when they appear on the same text line. A typical processing pipeline for HTR consists of trainable baseline detection, text line segmentation and trainable HTR, resulting in an output format of e.g. PageXML [22] or hOCR [3].

Although other OCR/HTR engines exist, we present Loghi as a new framework for historical document recognition. The Loghi framework consists of a set of tools that enables users to create their own pipeline for HTR processing. Example scripts are provided to make starting your own pipeline as easy as possible.

The Loghi framework is fully open source using an MIT Licence[1] and built for production environments. The intermediate stages produce PageXML making it interoperable with other frameworks that use PageXML. The software can be found at https://github.com/knaw-huc/loghi

2 Related Work

HTR and the related field layout analysis are active research areas that have produced other solutions as well. Transkribus [10] by READ Coop is probably the most well known software package that focuses on HTR for historical documents. Some of the Transkribus software is made available as open source, and we gratefully make use of their Expert client to produce ground truth.

Kraken [13] is a well known open source HTR engine, which integrates nicely with eScriptorium [14], which provides a friendly GUI for the ground truth creation of documents.

Calamari [31] is another well known open source engine for OCR of historical documents.

We mention Tesseract [28] here because, although its focus is mainly on modern machine printed documents, it is possible to use Tesseract successfully on historical documents.

3 Method

Workflow. We follow a classic approach of divide and conquer where we first detect where the text lines in the image are and then do the actual text recognition on single text lines without knowing which lines become before or after. The text lines that are obtained by the layout analysis are cut out from the image and processed using a separate tool for the text recognition. Along with the separate layout analysis and text recognition, Loghi provides tools to do the pre- and post-processing for these core parts.

---

[1] https://opensource.org/license/mit.

Specific tools are provided that can help when training a network. These are a visualiser that can show layer activations and alternative characters during the decoding phase, a tool for generating synthetic ground truth data and an automatic baseline correction tool for the typical case where baselines are too wide. Loghi provides additional tools that are useful for enriching the PageXML it generates. These tools are primarily run as a post-processing step.

3.1 Stage One: Layout Analysis

The layout analysis in the Loghi framework is primarily done using Laypa [15], which is developed in conjunction with Loghi. It makes information stored in the PageXML format suitable for usage with segmentation networks. In the Loghi framework, these are primarily semantic segmentation networks, which predict a class for every pixel. The current model uses a ResNet [8] backbone and a feature pyramid head, for its pixel wise classifications.

The first application of Laypa is the detection of baselines, which is done by classifying each pixel as either background or baseline. Figure 1b shows the intermediate prediction for the original image seen in Fig. 1a. The next step in the Loghi pipeline converts this binary image into baselines. First, the different baselines are separated using connected components. Second, the middle pixels of the separated baselines are taken as the points of the baseline. Finally, the baselines are simplified using the Ramer-Douglas-Peucker algorithm [6,25] to remove redundant coordinates.

The second application of Laypa is optional. It can also be used to predict text regions and their classes, for example: marginalia, page number, resolution, date. An example of the Laypa predictions can be seen in Fig. 1c. This method is still done using semantic segmentation, where each class has their own pixel value. The contours of each region class are extracted [29] and simplified using the Ramer-Douglas-Peucker algorithm. Baselines that overlap with an extracted text region (by an adjustable threshold) are assigned to this text region in the PageXML. This can for example be used to distinguish the marginalia from the main running text. If no text region recognition is done, Loghi will either assign all baselines to a text region spanning the entire image, or it tries to group the baselines based on the distance between lines depending on settings (See Sect. 3.3). However, the text regions obtained using this method will all be of the generic type "Text".

While the default layout analysis is done using Laypa, it is possible to interchange parts of the Loghi pipeline for others. In this case, it is possible to use other public models, such as P2PaLA [24], as long as they support PageXML.

3.2 Stage Two: Handwritten Text Recognition

Preparation. After detecting the baselines, the text lines are extracted from the original image, as the core software, Loghi-HTR[2] requires individual text

---

[2] https://github.com/knaw-huc/loghi-htr

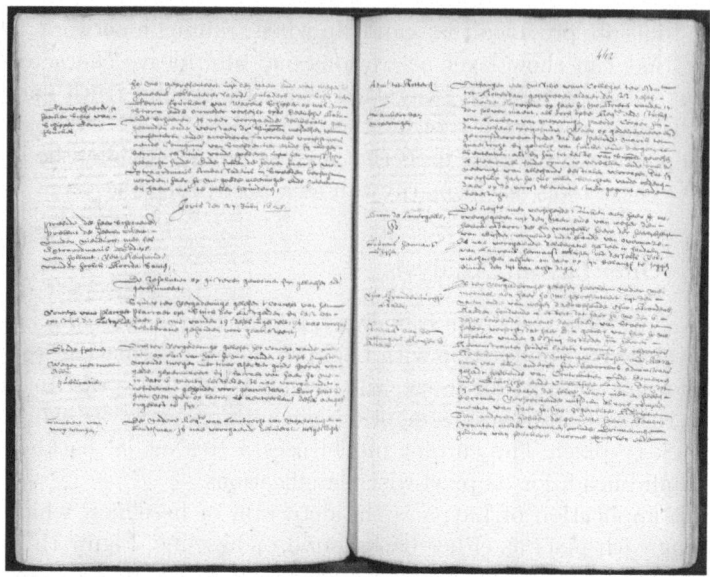

(a) Original image.

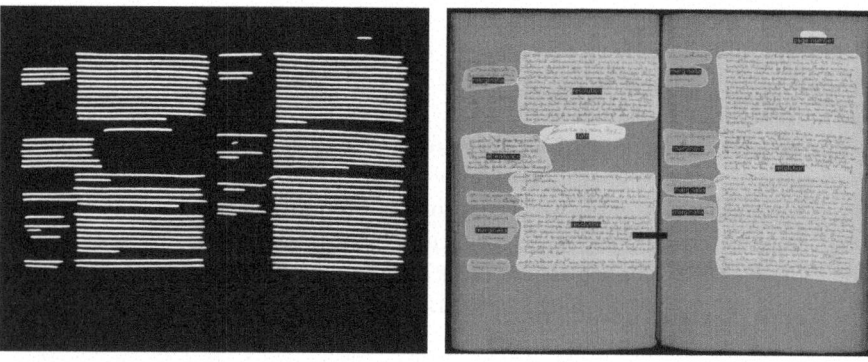

(b) Intermediate baseline output.　　　(c) Region detection output.

Fig. 1. Progression from the original image through intermediate baseline detection to the optional region detection result. Original image, Nationaal Archief, Resoluties van de Staten-Generaal, 1576-1796, number archive inventory 1.01.02, inventory number 3204. Intermediate baseline and region detection outputs using Laypa [15].

lines. When cutting the text lines from the image, either existing polygons from the PageXML can be used, or the polygons can be (re-)calculated using seam carving [2], using the extracted baseline coordinates. Seam carving is a content-aware image resizing algorithm. In Loghi, it works by identifying horizontal "seams" (connected paths that separate two text lines) of pixels that have the lowest energy. The energy is calculated by running the Sobel [11] edge detection algorithm, and Niblack binarisation [20] and combining these results. This combined result is blurred to allow for smoother path finding in the seam carving step. We carve seams for both the upper and lower boundaries of the text line polygon, and do not allow seams to cross the baseline, as this can result in the two seams joining above or below the text line. The text lines can be effectively isolated, while minimising distortion to the important textual content, using this method. For an example seam-carved text line, see Fig. 2.

Fig. 2. Seam carved text line

The resulting images of the text lines are stored as 4-channel RGBA PNG. The input for Loghi-HTR also allows images that are greyscale or plain RGB without any seam carving. The alpha channel of the RGBA PNG images designates background areas (e.g., adjacent text lines) as determined by the seam carving algorithm. Additionally, the framework also supports optional Otsu's [21] or Sauvola [26] binarisation as an image preprocessing step.

Model Specifications. The HTR module is built on top of the Keras framework [5] for training, development, and inference. The training process involves feeding the HTR network with text line image inputs, their corresponding transcriptions, and optional per-line weights that can lower or raise the learning rate for specific examples. Before being presented to the network, the text transcriptions undergo character-based tokenization. In this process, each unique character (including spaces and punctuation) is assigned a numerical token.

The core HTR neural network of Loghi typically employs a combination of several convolutional [19] layers, recurrent layers (GRU [4] or LSTM [9]) and an output layer with a softmax activation function. This final layer produces probabilities for each character token. To produce a sequence of characters from the individual probabilities, we employ Connectionist Temporal Classification [7] (CTC) decoding and loss. CTC is designed for sequence labelling tasks where the alignment between the input and the output is not explicitly known. In HTR, CTC allows us to train the model without requiring perfect character-level segmentation of the input images.

To help users get started, we provide several fixed network architectures in the training model library. Additionally, it is possible to give your own specific module architecture as a Variable-size Graph Specification Language (VGSL) spec [30]. VGSL-spec allows for a short, but powerful description of the network which can be translated by the software into a functional Keras model. Our implementation of VGSL-spec is slightly extended to also allow for ResNet-like networks using skip connections. For someone who wants to do more complex things, it is possible to write your own model architecture, as the software is open source.

Data Augmentations. Data augmentation plays a crucial role in improving the robustness and generalisation capabilities of the HTR model. By artificially expanding the training dataset, we can reduce overfitting and increase the model's ability to handle variations in real-world handwriting. The HTR module supports a diverse range of optional data augmentation techniques, including:

- **JPEG Distortion:** Simulates varying levels of JPEG compression artefacts by randomly adjusting image quality within a specified range (min quality 50, max quality 100). This helps the model become less sensitive to image compression noise.
- **Elastic Transform:** Introduces non-linear distortions to the text line images by applying a randomly generated displacement field. This augmentation aids the model in recognising handwriting with irregular spacing and character shapes.
- **Random Vertical Crop:** Randomly crops portions of the text line image along the vertical axis. By exposing the model to only partial text lines, it learns to focus on local features and becomes more robust to occlusions or incomplete text lines.
- **Random Width Adjustment:** Randomly stretches or compresses the text line image horizontally. This augmentation forces the model to adapt to variations in character width and spacing.
- **Random Shear:** Applies a random shear transformation along the x-axis, simulating slanted handwriting styles.
- **Image Blur:** Adds Gaussian blur with a randomly selected sigma value. This helps the model better handle low-resolution or blurry text lines.
- **Colour Inversion:** Inverts the text line image colours with a 50% probability. This augmentation encourages the model to focus on shape-based features, rather than solely on colour or intensity information.

Inference. The network produces a file that contains, for each input text line, the input image location, the corresponding produced transcription, and a confidence score that quantifies the reliability of the transcription. The confidence score is derived by taking the log-likelihood obtained during the CTC decoding process. The normalisation process adjusts the confidence score in proportion to the number of characters or elements in the transcription, reflecting the algorithm's certainty in its output on a per-character basis.

Formally, let P denote the log-probability of the top path as output by Tensorflow's [1] Beam Search Decoder[3], and T represent the total number of time steps in the decoding process. The confidence score C is then calculated as follows:

$$C = \exp\left(\frac{P}{T}\right) \quad (1)$$

The confidence score C lies within the range $[0, 1]$. A higher confidence score signifies a greater likelihood that the generated transcription is accurate. This

---

[3] https://www.tensorflow.org/api_docs/python/tf/nn/ctc_beam_search_decoder.

confidence metric aids in identifying potentially incorrect transcriptions, facilitating decision-making in downstream applications.

Merging Results to PageXML. The final step in this stage is converting the intermediate output of the Loghi-HTR to PageXML. Both PageXML version 2013 or 2019 are supported. Internally, 2019 is used and 2013-support is achieved by converting it using a XSLT-transformation. It is possible to swap in other HTR engines such as PyLaia [23] or Kraken in this stage. Inferencing using the Loghi framework is fast, it can process about 2 million single page scans per week using dual NVIDIA A30 GPUs.

Extra Tools. Some extra tools were created for improving the training of the Loghi-HTR software by visualising network internals and synthetic ground truth generation.

Visualisation. To enhance our understanding and troubleshoot during model training of Loghi-HTR, we have developed a visualisation module. This tool is designed to examine post-training representations learned by the model for a given sample image. Specifically, the module creates a top-3 time step prediction plot as shown in Fig. 3a, indicating at which parts of the image which characters were considered during the prediction process. Additionally, for further investigation, it generates a XLSX file of all the other top-X predictions per time step and a heat map based on the prediction values.

Moreover, the module outputs a plot displaying a random selection of layer activations when the trained convolutional layers are provided with random noise input. As demonstrated in Fig. 3 each layer shows activations of increasing complexity for the sample image. The initial layers tend to identify edges or colours, while deeper layers show more specialised activations related to specific parts of characters or shapes. These visualisations can further explain how the model's output was generated and act as a basis for determining why certain text lines are not recognised correctly.

Synthetic Ground Truth. A tool is provided which takes fonts and flat text files as input and generates images and corresponding ground truth in PageXML format. These images can then be used for (pre-) training a model. The effect of pre-training is that the amount of required real-world ground truth is significantly less than if no pre-training is done. This tool is especially helpful when creation of ground truth on real documents is expensive and/or time-consuming.

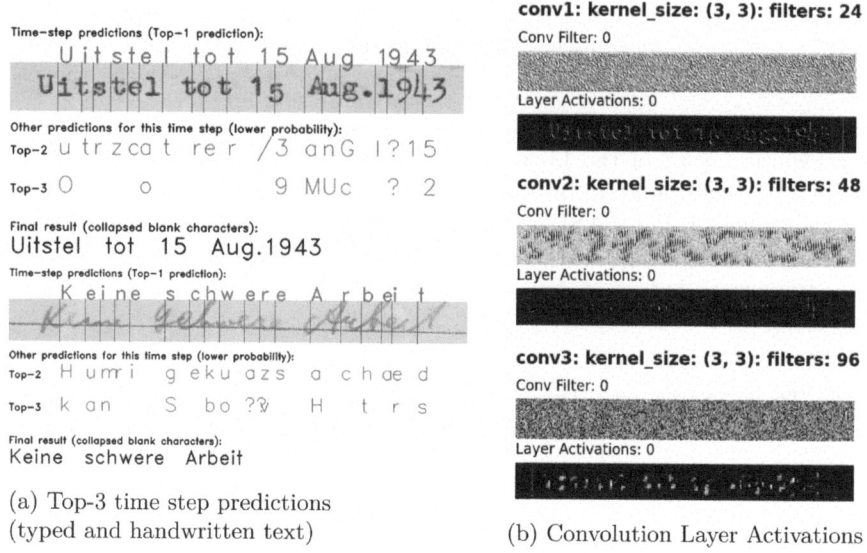

(a) Top-3 time step predictions (typed and handwritten text)

(b) Convolution Layer Activations

Fig. 3. Example visualisations

3.3 Stage Three: Post-processing

Reading Order Detection. Reading order is determined by clustering text lines into paragraphs based on their proximity, taking into account that lines in the same paragraph are typically located above each other. Proximity is determined by calculating the distance to another text line for each text line and taking the median of that distance for a scan. Every two lines that are within a distance of 1.5 times the median interline distance of each other are considered part of the same cluster. The resulting paragraphs are linked, working from the top left, finding the closest paragraph below it.

Language Detection. Language detection is based on langident[4] and works on several levels in the PageXML. It will provide results on text line, text region and page level. Although some preset training is provided, we suggest providing your own language files.

Word Splitting. Word splitting is the calculation of where each word of the text line starts and end. This is useful for highlighting purposes when results are used in a search environment. Word splitting is achieved by taking the total length of the baseline and dividing it by the number of characters including punctuation, resulting in an average width per character. The word length is then calculated by multiplying the number of characters in a word with the

---

[4] https://github.com/HuygensING/langident.

average character width. Polygons for each separate word are then calculated using the text line polygon.

4 Results

We have used this pipeline successfully on millions of scans for several projects ranging from 16th century materials to "modern" materials from World War 2, but will focus on only one here: The Republic project.

4.1 Republic

The Republic project[5] is creating digital access to the handwritten and printed resolutions of the Dutch States General, the supreme governing body of the Dutch republic from 1576 until 1796 [17]. The archive covers about 225,000 scans containing all resolutions (decisions) made by the States General in that period.

The handwritten part spans from 1576 until 1703 and the printed part from 1704 until 1796. The handwritten part has problems with bleed through and degradation or damage (Fig. 4). Specific problems of the Republic Printed set are: ligatures, the long s: f, initial characters (Fig. 6a) and in a lesser extent bleed through (Fig. 6a).

Both sets do however use very formulaic language [18], which can be used to automatically extract the logical structure of text elements, such as individual resolutions and the date and attendance list of each daily meeting. The handwritten resolutions were copied by professional writers and can be considered fairly neat for the period.

Republic Print. The initial project goal for the printed text was to reach a CER (Character Error Rate) under 10%, as this would make NER and other downstream tasks possible. Initially the layout analysis was performed by P2PaLA and the OCR by a fine-tuned version of Tesseract, where we were able to reach a CER of 2%, meaning 98% of the characters are correct. The ground truth was created by manually correcting 9 scans and then running the fine-tune training.

We expanded the training set by running the resulting Tesseract model against a large set of documents, automatically labelling lines as correct that are known common phrases such as "WAAR op geen resolutie is gevallen" (EN: "ON which no decision was made") and labelled lines with extremely high confidence scores (>0.9) also as correct. Added to these lines were some manually corrected lines with medium (>0.4 and <0.8) confidence scores and lines that provided rare characters such as the capital character Q. The decision for selecting lines with medium confidence scores was made with the thought that these would help the learning process more than lines with very low confidence – which

---

[5] https://republic.huygens.knaw.nl/index.php/en/republic-english/.

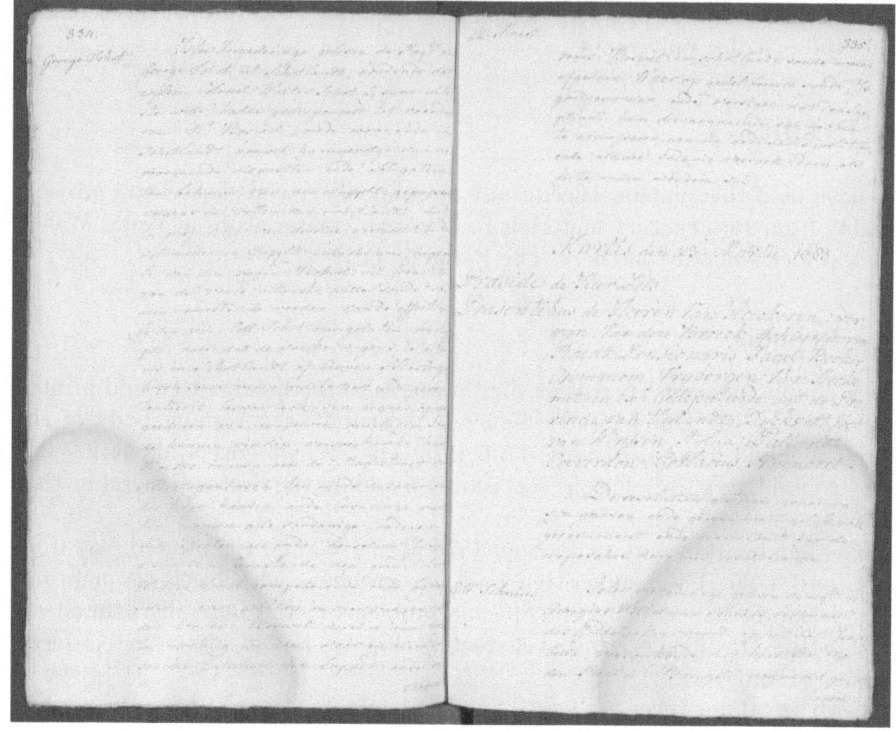

Fig. 4. Water damage

often contained damage – or lines with very high confidence, where the model does not learn much.

We then trained the model again using both the pseudo-labels and the real ground truth and repeated this process a couple of times resulting in a training set of 100,000 text lines with and a validation set of 10,000 text lines that were most likely to be correct and a very small set of actual ground truth that was mixed in the training and validation set.

The CER using Tesseract eventually reached 0.28% on the validation set, where we have to mention that a large part of the validation data contained the automatically selected lines with high confidence, thereby not being representative of the full corpus.

When manually inspecting the results on other scans, it became clear that the actual CER on the corpus was much higher, more likely close to 2%. In addition, we sometimes noticed completely different transcriptions on average text lines (see Example 5).

As an experimentation, we tried training the Loghi-HTR using an NVIDIA 2080TI on the same input data as Tesseract was trained on using settings that were used earlier to train on 17th and 18th century handwritten materials; after just a few hours of training, the CER on the validation set reached 1%. Continu-

Fig. 5. Normal line transcribed as "PNWWW" by Tesseract

ing the training eventually led to a CER on the validation set of 0.099% (99.9% correct). Again, this validation set is not fully representative of the corpus.

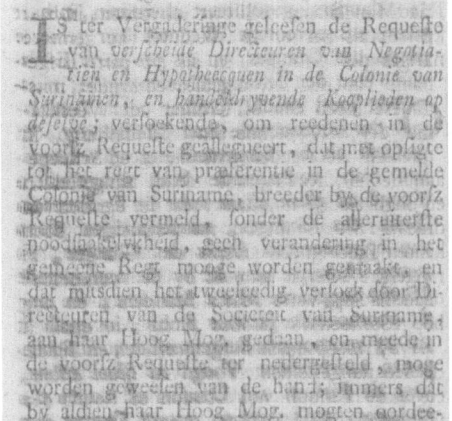

(a) Bleed trough on printed text

(b) Initial characters have wildly different font size

Fig. 6. Dataset specific problems

Republic Handwritten. Ground truth for 515 randomly selected scans from the handwritten part of the corpus were professionally created and made available on Zenodo containing both layout and transcriptions [27].

For the layout analysis, we used Laypa and trained a network to detect and recognise the different regions and the baselines in scans. For details, see [15].

For the HTR we used Loghi-HTR, a batch size of 16 and including augmentations of random shearing, horizontal stretching and elastic transformations, which lead to a CER of 2.69% (see Fig. 7) on the validation set after 194 epochs when early stopping was triggered by 20 consecutive epochs without improvement. Without augmentations, it stops after 121 epochs, leading to CER 2.78% (see Fig. 8). Another model was trained by using IJSBERG [12] data and combining this with the 515 scans ground truth of Republic Handwritten. This model was trained using horizontal stretching and random shearing and resulted in a CER of 2.50% on the Republic Handwritten validation set. All handwritten scans from the Republic project were HTR-ed and selectively made available in

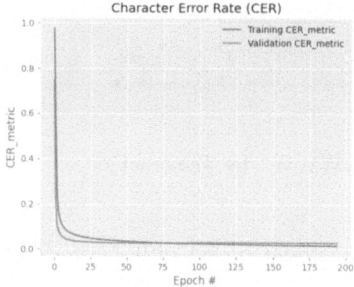
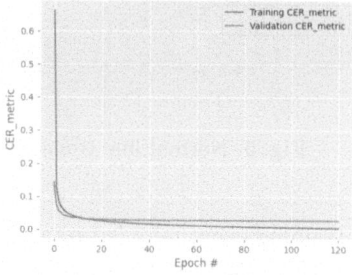

Fig. 7. CER Republic Handwritten with augmentations

Fig. 8. CER Republic Print without augmentations

chunks for corrections by citizen scientists. The results of these corrections were used again as input training data, combined with the original ground truth. This resulted in a CER of 1.99% on the validation set. We took care not to include scans from the validation set in the citizen science set.

The final HTR results were further processed by a team that focused on the textual contents, formulaic expressions and on how these books were set up as an information system of 17th and 18th century decision-making. This was done using pagexml-tools[6] and fuzzy-search,[7] which assist in (mostly algorithmically) detecting the logical structure of resolutions, attendance lists and contemporary indexes [17,18].

As is typical of archives of administrative documents, the resolutions contain many formulaic phrases that signal where certain text sections start and end [16]. This allows reorganising the text lines according to these existing logical structures, to give more fine-grained access via the search application (users can search for combinations of keywords in individual resolutions instead of in pages). Moreover, the logical structure can also be algorithmically linked via the corresponding text lines to label the text regions in the PageXML. These can then be fed back to the HTR pipeline to enhance or improve text region classification.

The final results of the Republic Project will be published in December 2024 and will include all transcriptions.

5 Conclusion

5.1 Discussion

It is clear that the Loghi framework does not solve all problems encountered when creating automatic transcriptions from historical documents. Although some of the problems of document degradation can be partially overcome by creating

---

[6] https://github.com/knaw-huc/pagexml.
[7] https://github.com/marijnkoolen/fuzzy-search.

more ground truth for training, it does not solve them completely. The creation of extra ground truth is not always feasible due to budget or time constraints.

While we are unsure of the exact cause, we believe the issues encountered by Tesseract were due to overfitting.

The training and validation sets of Republic Print are biased, and the CER can not be fully trusted to be representative of the entire corpus. We manually checked several pages and can safely assume that the CER is far below 1%. Nevertheless, we decided to change this procedure of semi-automatic ground truth collection to be more random and less biased for future projects, especially for validation sets.

The results on the Republic Handwritten set are very good, considering it spans a period of over 100 years and multiple writers. We think the use of formulaic language in these documents makes it easier for the model to learn to correctly predict the transcriptions.

Regarding the detection of reading order, we note that this is a challenge with no simple solution. All but the most simple documents can have multiple reading orders, based on how you are using the text. When reading the main text, the page numbers, headers and footers are not part of the reading order, but when you are skimming the pages, the headers are a valuable part of the text. Similarly, the handwritten resolutions of the States General have marginalia next to the resolutions containing keywords. When searching for a specific text, those are read first. When you are doing named entity extraction, the marginalia are ignored and only the resolutions are read.

Loghi shows good results on the layout analysis and text recognition of historical data, and we feel that it is a welcome addition to the existing tools for the automatic transcription of historical documents. All software is made freely available via GitHub[8]. Currently, we rely on other tools such as Transkribus for the creation of ground truth.

5.2 Conclusion

The Loghi framework has been applied successfully to several real-world projects such as Republic and has been applied on various datasets.

The Loghi framework is a nice, fully open source addition to the tools currently available for humanity researchers. It gives archives the ability to process large amounts of scans in-house without any licensing costs for the HTR.

The Loghi framework achieves high quality results with very low error rates on both layout and text recognition. In addition, it can process scans relatively fast.

5.3 Future Work

We would like to get accurate representative numbers concerning CER on the Republic Printed set by creating some representative randomly selected ground

---

[8] https://github.com/knaw-huc/loghi.

truth. It is however expensive and time-consuming, and other work will likely have our priority.

We have several different augmentations on our input data, but we do not know exactly the benefit of each augmentation. We hope to be able to further look into the benefits each augmentation brings and optimising their impact on training results by running an ablation study. Of special interest is how the effect of the augmentations is impacted by turning other augmentations on and off. We think that the effect of the augmentations with larger datasets is relatively small compared to the effect it has on tiny datasets (<100 scans).

The use of transformers has not gone unnoticed to us, and we are currently experimenting on implementing a new TrOCR based HTR-module that can be a drop-in replacement for the current HTR-module. Initial results are promising.

We hope that we can include tools that partially automate the creation of ground truth. We lack a GUI where ground truth can be created, and currently depend on other tools. In the future, we hope to be able to create an integrated GUI for the creation of ground truth. We also would like to have a more standardised transcription convention, so ground truth can benefit other projects more.

We will continue to support and further develop the framework until 2027.

Acknowledgments. This study was made possible by the REPUBLIC (https://republic.huygens.knaw.nl/) project of the Huygens Institute of the Royal Netherlands Academy of Arts and Sciences (KNAW) funded by the Dutch Research Council (NWO) grant 175.2017.024. We kindly thank NWO, KNAW and the Dutch National Archive for the time and support to make this project possible. We are very grateful for all persons who helped in the creation of ground truth, the volunteers who corrected the HTR output and all persons who make their ground truth available, so these projects are made possible. Additionally, we thank the GLOBALISE (https://globalise.huygens.knaw.nl/) project, CLARIAH (https://www.clariah.nl/) and the Digital Infrastructure department of the KNAW Humanities Cluster (https://di.huc.knaw.nl/) for the use of computing hardware. We thank Marijn Koolen for very constructive reviews and comments on the drafts of this paper. We sincerely thank the editor and reviewers for taking the time to review our manuscript and providing constructive feedback to improve our manuscript.

Disclosure of Interests. The authors have no competing interests to declare that are relevant to the content of this article.

References

1. Abadi, M., et al.: TensorFlow: large-scale machine learning on heterogeneous systems (2015). https://www.tensorflow.org/, software available from tensorflow.org
2. Avidan, S., Shamir, A.: Seam carving for content-aware image resizing. In: ACM SIGGRAPH 2007 Papers, SIGGRAPH 2007, p. 10-es. Association for Computing Machinery, New York (2007). https://doi.org/10.1145/1275808.1276390
3. Breuel, T., Kaiserslautern, U.: The hOCR microformat for OCR workflow and results. In: Ninth International Conference on Document Analysis and Recognition,

ICDAR 2007, vol. 2, pp. 1063–1067 (2007). https://doi.org/10.1109/ICDAR.2007.4377078
4. Cho, K., van Merriënboer, B., Gülçehre, Ç., Bougares, F., Schwenk, H., Bengio, Y.: Learning phrase representations using RNN encoder-decoder for statistical machine translation. CoRR **abs/1406.1078** (2014). http://arxiv.org/abs/1406.1078
5. Chollet, F., et al.: Keras (2015). https://keras.io
6. Douglas, D.H., Peukers, T.K.: Algorithms for the reduction of the number of points required to represent a digitized line or its caricature. Cartographica Int. J. Geogr. Inf. Geovis. **10**(2), 112–122 (1973). https://doi.org/10.3138/FM57-6770-U75U-7727
7. Graves, A., Fernández, S., Gomez, F., Schmidhuber, J.: Connectionist temporal classification: labelling unsegmented sequence data with recurrent neural networks. In: Proceedings of the 23rd International Conference on Machine Learning, ICML 2006, January 2006, vol. 2006, pp. 369–376 (2006). https://doi.org/10.1145/1143844.1143891
8. He, K., Zhang, X., Ren, S., Sun, J.: Deep residual learning for image recognition (2015)
9. Hochreiter, S., Schmidhuber, J.: Long short-term memory. Neural Comput. **9**, 1735–80 (1997). https://doi.org/10.1162/neco.1997.9.8.1735
10. Kahle, P., Colutto, S., Hackl, G., Mühlberger, G.: Transkribus - a service platform for transcription, recognition and retrieval of historical documents. In: 2017 14th IAPR International Conference on Document Analysis and Recognition (ICDAR), vol. 04, pp. 19–24 (2017). https://doi.org/10.1109/ICDAR.2017.307
11. Kanopoulos, N., Vasanthavada, N., Baker, R.L.: Design of an image edge detection filter using the Sobel operator. IEEE J. Solid-State Circ. **23**(2), 358–367 (1988)
12. Keijser, L.: 6000 ground truth of VOC and notarial deeds 3.000.000 HTR of VOC, WIC and notarial deeds, October 2020. https://doi.org/10.5281/zenodo.4159268
13. Kiessling, B.: Kraken-an universal text recognizer for the humanities. In: ADHO, Éd., Actes de Digital Humanities Conference (2019)
14. Kiessling, B., Tissot, R., Stokes, P., Stökl Ben Ezra, D.: eScriptorium: an open source platform for historical document analysis. In: 2019 International Conference on Document Analysis and Recognition Workshops (ICDARW), vol. 2, p. 19 (2019). https://doi.org/10.1109/ICDARW.2019.10032
15. Klut, S., van Koert, R., Sluijter, R.: Laypa: a novel framework for applying segmentation networks to historical documents. In: Proceedings of the 7th International Workshop on Historical Document Imaging and Processing, HIP 2023, pp. 67–72. Association for Computing Machinery, New York, NY, USA (2023). https://doi.org/10.1145/3604951.3605520
16. Koolen, M., Hoekstra, F.: Detecting formulaic language use in historical administrative corpora. In: Proceedings of the Computational Humanities Research Conference 2022, pp. 127–151 (2022)
17. Koolen, M., et al.: The value of preexisting structures for digital access: modelling the resolutions of the Dutch states general. ACM J. Comput. Cult. Herit. **16**(1), 1–24 (2023)
18. Koolen, M., Hoekstra, R., Sluijter, R., Oddens, J.: Formulas and decision-making: the case of the states general of the Dutch Republic. In: Proceedings. http://ceur-ws.org (2023). ISSN 1613-0073
19. LeCun, Y., et al.: Backpropagation applied to handwritten zip code recognition. Neural Comput. **1**(4), 541–551 (1989). https://doi.org/10.1162/neco.1989.1.4.541

20. Niblack, W.: An Introduction to Digital Image Processing. Delaware Symposia on Language Studies5. Prentice-Hall International (1986). https://books.google.nl/books?id=XOxRAAAAMAAJ
21. Otsu, N.: A threshold selection method from gray-level histograms. IEEE Trans. Syst. Man Cybern. **9**(1), 62–66 (1979). https://doi.org/10.1109/TSMC.1979.4310076
22. Pletschacher, S., Antonacopoulos, A.: The page (page analysis and ground-truth elements) format framework. In: 2010 20th International Conference on Pattern Recognition, pp. 257–260 (2010). https://doi.org/10.1109/ICPR.2010.72
23. Puigcerver, J., Mocholí, C.: PyLaia (2018). https://github.com/jpuigcerver/PyLaia
24. Quirós, L.: Multi-task handwritten document layout analysis. arXiv e-prints arXiv:1806.08852, June 2018. https://doi.org/10.48550/arXiv.1806.08852
25. Ramer, U.: An iterative procedure for the polygonal approximation of plane curves. Comput. Graph. Image Process. **1**(3), 244–256 (1972). https://doi.org/10.1016/S0146-664X(72)80017-0. https://www.sciencedirect.com/science/article/pii/S0146664X72800170
26. Sauvola, J., Seppänen, T., Haapakoski, S., Pietikäinen, M.: Adaptive document binarization. Pattern Recogn. **33**, 147–152 (1997). https://doi.org/10.1109/ICDAR.1997.619831
27. Sluijter, R., et al.: REPUBLIC PageXML ground truth handwritten resolutions States General (2023). https://doi.org/10.5281/zenodo.7695131
28. Smith, R.: An overview of the tesseract OCR engine. In: Ninth International Conference on Document Analysis and Recognition, ICDAR 2007, vol. 2, pp. 629–633. IEEE (2007)
29. Suzuki, S., be, K.: Topological structural analysis of digitized binary images by border following. Comput. Vis. Graph. Image Process. **30**(1), 32–46 (1985). https://doi.org/10.1016/0734-189X(85)90016-7. https://www.sciencedirect.com/science/article/pii/0734189X85900167
30. VGSL Specs - rapid prototyping of mixed conv/LSTM networks for images (2024). https://tesseract-ocr.github.io/tessdoc/tess4/VGSLSpecs.html
31. Wick, C., Reul, C., Puppe, F.: Calamari - a high-performance TensorFlow-based deep learning package for optical character recognition. Digit. Humanit. Q. **14**(1) (2020)

Open Parliamentary Data as a Tool for Linguistic Research: Exploring the 'Greek Language Question' in the *Journal of Parliamentary Debates*

Maria Kamilaki[✉]

Hellenic Parliament, Directorate-General for e-Administration, Library and Publications, Athens, Greece
m.kamilaki@parliament.gr

Abstract. Parliamentary Libraries currently face the challenge of shifting from being gate-keepers of a Parliament's archival and contemporary "treasures" to functioning as dynamic information and knowledge hubs, through the production, management and availability of open diachronic and synchronic data; in this light, this paper presents the Hellenic Parliament Library experience, focusing on a subcategory of historical parliamentary data, included in the Hellenic Parliament Library Repository under construction. In the first part, more technically-oriented, an outline of the text detection and recognition process is provided, in order to render the materials in question machine-readable, while in the second part the potential for linguistic research is highlighted, through a case-study exploring aspects of the 'Greek language question', as discussed in the parliamentary context, within the wider framework of language policy making.

Keywords: parliamentary data · Greek language question · text line detection & recognition · deep neural networks · corpus linguistics · collocations · lexical profile

1 Introduction

The decisive role of Parliaments in shaping contemporary democracies has been increasingly highlighted in an unprecedented way through internet and communication technologies, which in turn has, to a large extent, modified institutional relations with the public and the mass media [2, 16]: Parliament Houses and their legislative procedures are no longer reserved to elite social classes or accessible only through press coverage and mediation; there are a number of different tools and channels that allow practically all interested stakeholders to have full access to data produced in them (e.g. in most Western countries, parliamentary debates are publicly available in various formats -text, video, image-, plenary sessions and committees are transcribed and annotated etc.).

In this light, Parliamentary Libraries and Research Services contribute significantly to the digital transformation of Parliaments as representative institutions, through the

© The Author(s), under exclusive license to Springer Nature Switzerland AG 2024
H. Mouchère and A. Zhu (Eds.): ICDAR 2024 Workshops, LNCS 14935, pp. 89–102, 2024.
https://doi.org/10.1007/978-3-031-70645-5_7

production, management and availability of open data, fostering core values of parliamentarism, such as openness, transparency and accountability. The process of producing open data involves both synchronically-generated materials, relevant to law-making and oversight, and historical/archival data, providing a holistic and comprehensive outlook on parliamentary history. Therefore, parliamentary libraries, functioning as an interface between Parliaments as high prestige institutions and potential users, can serve as a mediation point for the vernacularization of knowledge and information pertaining to data of parliamentary interest [5, 19, 20].

Among the priorities set by the relevant institutions is to promote production and analysis of synchronic and diachronic parliamentary data in an interdisciplinary approach, a venture which presents mind-blowing potential in an array of cognitive domains, like history, political & social sciences, economics, linguistics, namely every aspect of public policy-making. If, additionally, one reflects upon fields of application of AI-based technologies [21][1] within the parliamentary environment, the opportunities seem not only endless, but also meaningful for wider audiences within contemporary democratic contexts. Moreover, shifting from local to global, the development of integrative analytical tools that create comparable, transnational corpora offers a better universal understanding of parliamentary discourse [8, 9, 27] and its wider societal impact (e.g. with respect to diverse or vulnerable parts of society like women, minorities and marginalized groups).

In this context, the aim of this paper is to present the Hellenic Parliament Library (hence: HPL) experience, focusing on a subcategory of historical parliamentary data included on the HPL Repository currently under construction, also exploring potential for linguistic research through a case-study, regarding parliamentary discussions on the controversial 'Greek language question'.

2 The HPL in the Digital Sphere: Some Background Information

In the case of the Hellenic Parliament, the long-standing (since 1845) and emblematic presence of the Library as a gate-keeper of national parliamentary tradition, has led to the existence of a vast archival treasure, offering valuable insight and supporting the scarce existing analyses on Greek parliamentary data (e.g. see [12–14]).[2] However, all this wealth of information had been digitized during past decades using once pioneer, but now obsolete techniques, being accessible only in image format (jpeg). It soon became evident for the Library administration and the various stakeholders involved that, in order to facilitate research into these documents, converting their content into

---

[1] Although numerous technologies are assigned to AI today and several parliaments seem to recognize the need to introduce AI, parliamentary institutions still seem slow in assimilating the technological advances, often viewing them with scepticism, with only a few examples of actual implementation of such technologies in the parliamentary workspace (for a detailed account see [11]).

[2] It is worth mentioning that the Hellenic Parliament, via its Research Service, has been participating in forward-looking research projects for more than a decade (e.g. see participation in the ManyLaws project (https://www.manylaws.eu), through which an Akoma Ntoso-based codified version (e-code) of a collection of Greek, Austrian and EU legislation was created, see https://www.manylaws.eu/about/discover-manylaws).

digital form and comprehending their textual and non-textual information through an automated procedure was a conditio sine qua non.

These assumptions functioned as a springboard for the creation of the project *Digitization of the holdings of the Hellenic Parliament Library*, an EU-funded program of digitization, digital image processing with optical character recognition (OCR) and documentation, which began in June 2022 and will be completed in October 2024. The program is managed by the European Programs Implementation Service (E.P.I.S.) of the Hellenic Parliament and is run, through outsourcing, by the company COSMOTE as an external partner,[3] with the collaboration and monitoring of the HPL staff; it develops along two main axes: 1) digitization of printed materials (e.g. archives of parliamentary interest, some of which were unknown to research, like the archives from the period of the Junta (Military Dictatorship, 1971–1973), collections of emblematic political figures, but also manuscript codices, historical maps etc.), 2) digital reprocessing of already digitized materials (e.g. the historical parliamentary archives 1843–1967: parliamentary debates, introductory bills, law proposals, constitutional revisions etc., amounting to 552.598 pages, coupled with a vast Press Collection of 18.000.000 digitized pages). The materials included are versatile, in order to highlight the quantitative and qualitative cultural capital of the HPL, since the main goal of this Project was to produce a uniformly encoded, interoperable set of subcorpora, providing a representative outlook on the HPL holdings.

Among the materials described above, in the following paragraphs emphasis is placed on the *Journal of Parliamentary Debates* corpus, forming a subcategory of open parliamentary data.

3 The Journal of Parliamentary Debates as a Corpus

Official proceedings/transcripts of parliamentary debates are transcriptions of formal spoken language, produced in a controlled and regulated context. Their content, structure and discourse forms a unique text-type of political discourse, with a wide range of theoretical and applied implications in disciplines like political science, sociology, history, linguistics etc. (indicatively see [10, 30, 31]).

The corpus used for the present study relies on the official transcripts of the plenary debates, typed in Greek polytonic fonts (see [15]) and included in the *Journal of Parliamentary Debates* (hence: JPD), recording the full version of the parliamentary sessions between 1863–1967, while the Minutes of Parliamentary Debates present a condensed, summarized form.

There are currently two available sources for accessing the JPD a) the HPL Repository under construction, in which the respective subcorpus includes 167 volumes (1863–1967), offering the respective data in pdf, epub and txt format (still in process), b) the Centre for Greek Language "David Antoniou Archive:[4] Journal of Parliamentary Debates

---

[3] The company was evaluated and selected through international public competition, based on the national legal framework for public procurements.

[4] The David Antoniou Archive is an important archive for the study of the history of the Greek language. It was put together by the scholar David Antoniou, who later endowed it to the Centre for the Greek Language (as cited in: https://inventory.clarin.gr/corpus/1084).

subcorpus (1877–1934)", available through CLARIN-EL,[5] comprising excerpts of parliamentary discourse (selected fragments of parliamentary speeches, debates, plenary sessions, publications of the Hellenic Parliament etc.), referring to legislative and other matters having to do with the Greek Language. The size of the corpus amounts 264.285 words and the data format is text/plain (character encoding: UTF-8).

The two above mentioned sources have a different orientation and purpose, the first one being designed to meet the needs of a library and its users, while the second one adheres to a corpus linguistics methodology [3]. Still, both can be used in a complementary basis, to cross-check hypotheses and findings, and to cross-fertilize conclusions.

Drawing on these two sources, in the first, technically-oriented part of this paper, focus is placed on the HPL Repository under construction, more specifically on transforming the respective documents to machine-readable form through text line detection and recognition, in order to make them readily accessible for research in the near future. The second part, more linguistically-oriented, explores the JPD offered in the David Antoniou Archive, which is linguistically annotated, rendering it suitable for exploring potential research applications, among others, in the field of historical Sociolinguistics.

3.1 Preparing the Data: Document Image Analysis in the HPL Repository

Recently, research in the field of document image analysis has investigated utilizing deep learning for the segmentation and recognition of typewritten and handwritten text. In the case of the HPL Repository all materials subjected to document image analysis were typewritten. The JPD materials, more specifically, had already been scanned in jpeg form (76.942 jpeg) – thus material preparation for digital processing and remastering included only quality and condition assessment; organization and classification of documents followed the layout of the pre-existing 167 leather-bound volumes. Given the often poor quality of the original scans, pre-processing of the scanned images was performed, in order to enhance their quality, including noise reduction, skew correction, de-speckling and binarization.

The next two key steps, which are crucial for a successful outcome, as Kaddas et al. point out in [24], are text line detection (also see indicatively [4]) and text line recognition. Text line detection is a technique used in document processing to detect and extract text lines from a scanned document, often being applied as a pre-processing step for Optical Character Recognition (OCR) and Layout Analysis. Recognizing the text in documents has been frequently considered as a subsequent step to the text line detection task.

In preparing the data for the HPL Repository, a variation of the YOLOv5 [23, 29] Deep Neural Network model (YOLOv5-OBB1) has been used, providing speed and efficiency (of primary importance in this case, given the huge amount of materials being processed within a very strict time-plan), accuracy, versatility (allowing for customization and adaptation to different domains), ease of use and user-friendliness, scalability (enabling users to choose the appropriate model size, proposing modifications during

---

[5] See: https://inventory.clarin.gr/.

training, based on the specific requirements of each project, coupled with resource constraints), state-of-the-art performance and, last but not least, active development and community support. According to Kaddas et al. ([24], pp. 215–216), YOLOv5 has a major change in the Neck of the architecture, when compared to its predecessors: as it is shown in Fig. 1, a variant of Spatial Pyramid Pooling layer feeds a Path Aggregation Network (PANet), in which the CSP strategy has been incorporated via the BottleNeckCSP layer. Finally, the Head of the network is composed from 3 convolution layers for predicting the location of bounding boxes, as long as the object classes with their scores and the angle estimation.

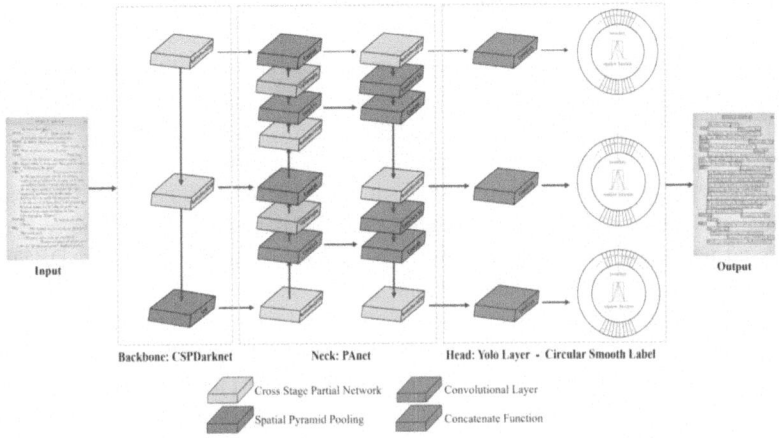

Fig. 1. Architecture of the proposed YOLOv5-OBB model

Text line detection procedure can be summarized as follows: at a first step, the automatic text line detection results to a set of polygons that surround each text line of the document, as shown in Image 2. Then, the user can correct the segmentation results, using the following system functions: • To correct a polygon by moving the desired points in the right position. • To add a new polygon before or after a polygon. • To delete a polygon. • To connect a polygon with another polygon.

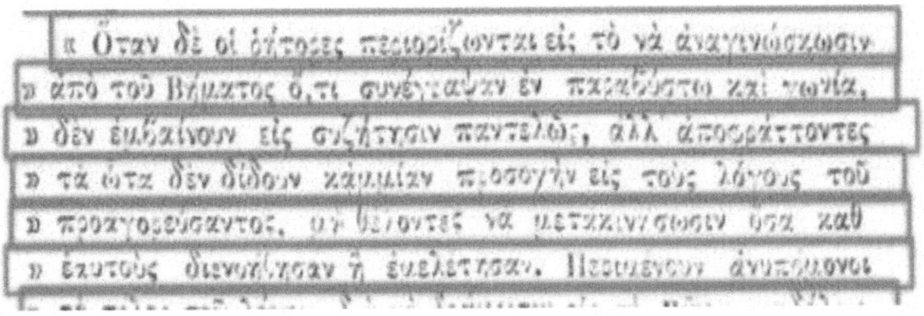

Image 2. Excerpt from the YOLOv5 output

In order to localize text lines with orientation, the Circular Smooth Label (CSL) technique (see [33]) was used to classify the rotation angle of the object's bounding polygon. With the CSL method, the angles are encoded in a circular/repeating pattern, while windowing functions (pulse, rectangular, triangle and Gaussian) can be used to exterminate boundary problems, in order to obtain better prediction of the angle's class (see [24], p. 216).

Having observed that, in many cases, ground-truth text lines are strictly annotated and the polygonal edges almost overlap with the characters, by loosening the enclosing polygon over the horizontal axis, both line detection and recognition systems improve. This technique is called 'Extended Polygonal Labeling'. When the procedure of text line detection and correction is completed, the system creates text line images with the image parts inside each polygon.

These images will be then used as input for the text line recognition module (also cf. Kaddas et al. in [24], p. 370), relying on the open-source, TensorFlow-based Calamari-OCR engine,[6] that uses advanced deep neural network. The best configuration for the recognition engine is selected after experimentation and training, using the text lines from the wider dataset of 552.598 jpeg tokens included in the corpus of the HPL scanned parliamentary materials. In order to construct the full page OCR from the predictions, the detected text lines were sorted using Density-based spatial clustering (DBSCAN) and the coordinates of the extracted clusters.

Due to the low image quality of the original PDF document, sometimes the process had to be repeated several times, with most of the time consumed by the data validation and quality control step. As evaluation metrics, the Character Error Rate (CER) and Word Error Rate (WER) have been considered, in order to validate the robustness of the text line recognition module.

A major drawback in the whole process was the fact that the Project's technical specifications predicted that, due to the vast quantity of the materials to be processed, the result of the OCR procedure was to be delivered in raw form – therefore, post-processing of OCR processed texts packages to correct errors and improve accuracy was not performed, irreparably reducing optimization. Techniques such as spell-checking, dictionary look-up, and pattern recognition may be applied in the future, to correct misrecognized characters or words. Human proofreaders may also review the OCR output to manually ensure the accuracy of the transcribed text.

After the transcription of the text, the next step was indexing with metadata: following international standards (Dublin Core Element Set),[7] a schema of parliamentary-oriented metadata was created, such as title, speaker(s), date, and keywords, to facilitate search and retrieval of the digitized archival material. Again, the historical specificities of the materials posed various obstacles: e.g. the names of the speakers were often not mentioned, distinguishing Members of Parliament (MPs) from guest speakers and/or regular speakers from the session chairs, who speak on procedural matters, was sometimes hard, official names of administrative units had changed many times over the decades etc. Further obstacles anticipated in the near future may include: data security concerns,

---

[6] See: https://github.com/Calamari-OCR/calamari.

[7] See: https://www.dublincore.org/.

resistance to change on behalf of the parliamentary staff, user engagement and training, issues of digital curation and long-term sustainability.

At the current stage, although documentation of the Parliamentary Collection, including the JPD, has finished, quality control of metadata is still in process, expected to be completed by June 2024, in order for the Repository to enter its pilot testing. Taking into consideration the limitations and drawbacks of the work presented, evaluation and feedback from users and various stakeholders involved will allow for continuous improvement, once the Repository enters its production phase.

Upon completion of the scheduled works, the corpus of the *Journal of Parliamentary Debates*, produced within the wider framework of the HPL Repository, will be delivered to the public, as already mentioned, in various formats (txt, pdf, epub). Given the linguistic orientation of this paper, among these, txt lends itself ideally for processing and analysis with Corpus Linguistics tools and advanced text mining methods. Therefore, although, strictly speaking, the materials included in the HPL Repository do not follow the methodology of Corpus Linguistics, still the compilation of such data load and the possibility to extract/download corpora and subcorpora, readily available for analysis using online tools (like AntConc, Sketch Engine etc.), offers tremendous research opportunities in exploring various aspects of diachronic and synchronic language use, like morphological change, lexical features (e.g. words with ideological connotations), sociolinguistic case-studies concerning gender, status, political affiliation etc.

For the time being, however, given that the HPL Repository is still under construction, making any research practically impossible, in the following section we will limit ourselves to the JPD corpus offered by the David Antoniou Archive, focusing on parliamentary debates on language issues.

3.2 Exploring the Data: A Case-Study Based on the David Antoniou Archive

Considering parliamentary debates as par excellence nexuses of societal power relations and ideologies, expressed through the use of language alongside with other forms of physical action [22, 30, 31], in the second part of this paper we will exemplify the research opportunities offered by open parliamentary data, by experimenting with two metalinguistic terms that have defined the notorious Greek language question[8]: *katharevousa*, standing for the high variety vested with overt sociolinguistic prestige and used in formal settings (education, administration, religion), and the *demotic*, representing the low variety, having covert prestige and being activated in everyday discourse and literature.

---

[8] The 'Greek language question' has been one of the most heated issues in Greece's intellectual, socio-political and policymaking history. The term refers to the diverse and multi-layered disputes, first appearing by mid-eighteenth century within the framework of the Greek Enlightenment, over which form of the Greek language was most suitable for educational and scholarly writing and, after the formation of the independent Greek state (1830), as the official language [18]: the archaic *katharevousa* (literally 'of a pure form'), an early nineteenth-century construction, articulated by the scholar Adamantios Korais and based on the idea of a compromise between ancient Greek and the vernaculars spoken at the time, or the *demotic*, a form of spoken language derived from different dialects, exhibiting thus extensive variation.

Since the Greek language question has become an integral part of the process of nationhood and statehood for Greece, each variety being indeed associated with a distinct ideology about the nation and its historical destiny [28], parliamentary discourse constitutes an ideal field for foregrounding the symbolic value of language as a carrier of ideas and mentalities, via the close inspection of the two metalinguistic labels. If, additionally, we take into consideration the fact that parliamentary debates are the context in which language attitudes [7] of MPs are manifested and where language policy takes place, this kind of approach stresses the importance of using parliamentary corpora as a text-type of political discourse alongside others [10, 17], in order to draw meaningful conclusions on language use and policy.

Using the Sketch Engine tools [26], by uploading the David Antoniou corpus of the Journal of Parliamentary Debates, we can get valuable insight into the two words' actual usage by the MPs, across a very large time span (1877–1934). To begin with, using the wordcount tool (nouns), we see that the two terms appear very high in rank with regards to frequency of occurrence: among 10.607 words included in the corpus, *katharevousa* appears at the 18th position (743 occurrences) and the *demotic* at the 30th position (576 occurrences), which in turn demonstrates their prevalence in the discussions about language.

Implementing the powerful 'word sketch' tool,[9] processing a word's collocates and other word combinations in its surroundings and offering shorthand knowledge of its lexical profile, we observe that katharevousa (see Fig. 3) collocates e.g. with cognitive verbs (μαθαίνω [learn], εννοώ [mean]), which comes as no surprise, but also with verbs of emotional turbulence (σιχαίνομαι [loathe], υβρίζω [swear]), actually not expected to appear in a parliamentary discussion on linguistic variation. Looking at its modifiers, we find designations which range from descriptive terms, referring to mode of discourse (γραφομένη [written] vs λαλουμένη [spoken]), way of learning (διδασκομένη [taught at school], as opposed to 'naturally acquired') and temporal indexing (σημερινή [contemporary]), to features relevant to form (απλή [simple]) and internal structure (άκρα [extreme], αμιγής [pure]), or even purely evaluative terms with expressive meaning (κακολογουμένη [badly reputed], αποσκορακισθησομένη [to be expelled], ατημέλητος [careless]). All these attributes evoke the different versions of the "high" variety that were actually in use over time, covering the continuum from relatively simple/plain (cf. απλή) to archaic and obsolete (cf. άκρα) variants.

The same holds for the demotic (see Fig. 4): the modifiers include many adjectives with an emphasis on its "genuine" character (ακατέργαστος [unprocessed], γνησία [genuine], καθαρά [clean], αμιγής [pure]), relating this variety to non-standard, vernacular ones, having evolved through natural language change and being spontaneously used by people in their everyday discourse. On the contrary, negative attributes include μαλλιαρή [hairy] and χυδαϊστική [vulgar], capturing the association of this variety with connotations of trivialization and vulgarization of the (perceived as) "glorious" linguistic past, a conviction deeply rooted in the history of Greek diglossia. The demotic also co-occurs with terms of adherence, like θιασώτης, οπαδός [follower], and επικράτηση [prevalence], pertaining to issues of power relations and prestige distribution; we also find terms denoting evolution (εξέλιξη) and progress (πρόοδος),

---

[9] See: https://www.sketchengine.eu/guide/word-sketch-collocations-and-word-combinations/.

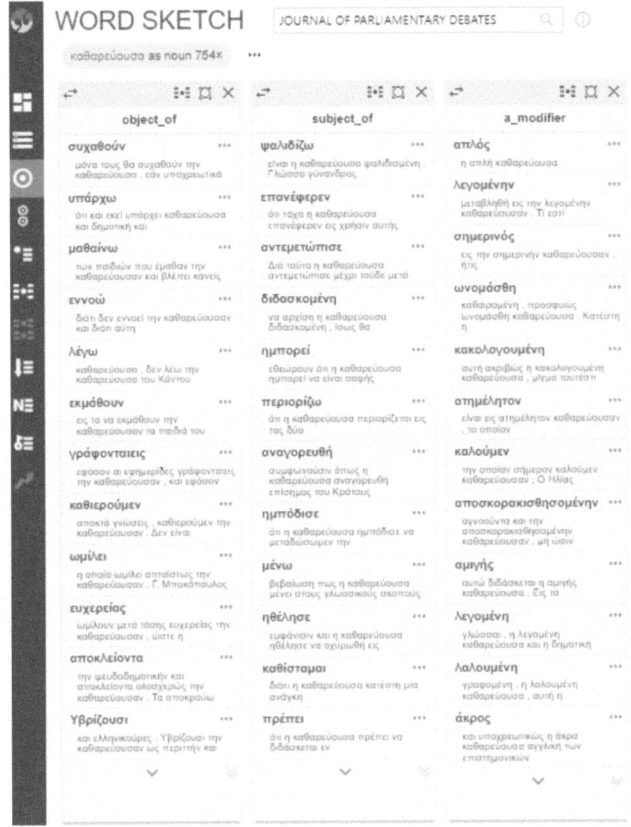

Fig. 3. The lexical profile of *katharevousa* using the word sketch tool

recalling widespread debates all around Europe about language change as progress or decay, a popular dilemma with opposing views [1]; terms like εισαγωγή [introduction], κατοχύρωση [consolidation] and διδασκαλία [teaching] reveal the longterm struggle of the demoticist movement to establish the low variety as the official language of education and administration. Finally, the participle λεγομένη [so-called], used for both katharevousa and the demotic, demonstrates their non-standardized, not unanimously accepted character, open to debate between the two sociolinguistic "camps".

Closer inspection of the concordances, in terms of historical period and specific MP, provides further insight into the sociocultural and political context of the language debate across time: e.g. searching the collocation δημοτική γλώσσα [demotic language] as a Key Word In Context (KWIC), we can reconstruct the various opposing representations (positive vs negative) of the variety in question, arising from the argumentation posed by MPs during various parliamentary debates. As it is shown in Fig. 5, supporters of the demotic put forward its use as the language of education (ex. 1); the demotic is linked to Greek oral, folk poetry (ex. 2, 3) and to the soul and spirit of the Greek nation (ex. 10, 13); on the contrary, the "sworn enemies" of the variety associate it with the

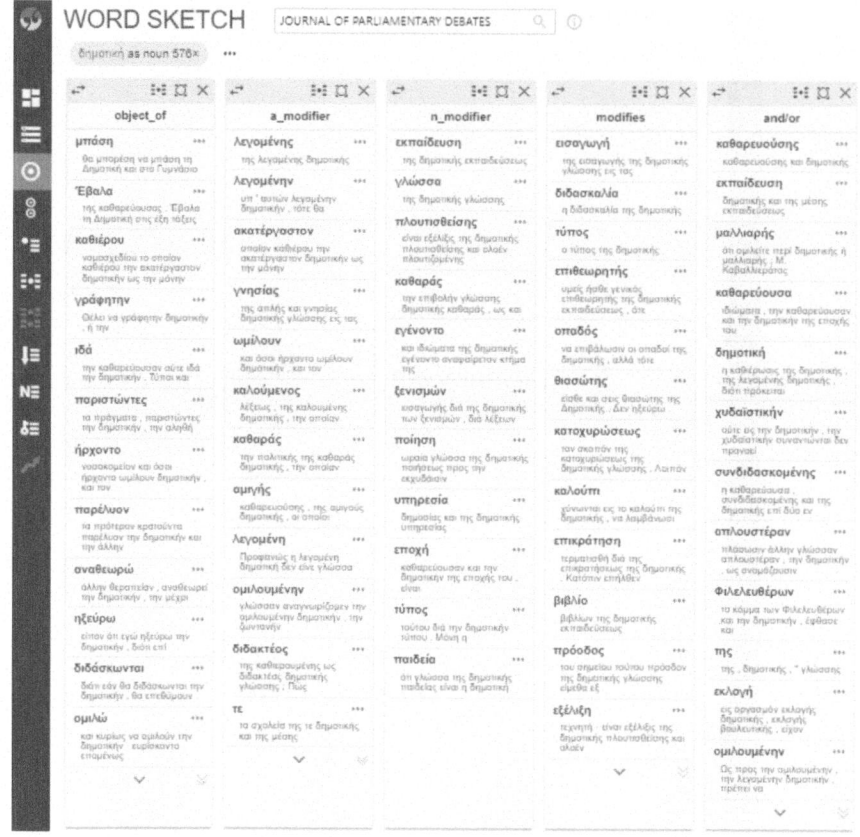

Fig. 4. The lexical profile of the *demotic* using the word sketch tool

extreme, radical μαλλιαρή [hairy] (ex. 4, 11); it is also argued that the demotic is a product of long-term slavery (ex. 5), that it lacks rules (ex. 6), given that it had been exempted from public discourse (ex. 9); reference is made to "entitled opinions", like the linguist G. Hatzidakis, a famous proponent of katharevousa (ex. 7, 8), etc. Based on these argumentative chunks, one can then analyse them in many different ways, e.g. in correlation with the speakers' political stance and partisan adherence, in terms of semantic prosody, argumentation-wise using analytical tools like *topoi* (see [32]), etc., thus drawing valuable conclusions not only for Greek Linguistics, but also for recent Greek political & sociocultural history, and policymaking.

Overall, the above mentioned comments shed light upon the rivalry surrounding the parliamentary debates on language, echoing the passionate and ardent discussions of the past with respect to the language question inside the Parliament House, but also within intellectual circles and wider societal milieus (cf. [25]).

Fig. 5. Concordance of the collocation δημοτική γλώσσα (*demotic language*)

4 Conclusions

In this paper, distributed in two parts, a technical/methodological and an exploratory/linguistic one, it was attempted to stress the importance of open parliamentary data for research, outlining the various steps from rendering materials machine-readable through text line detection and recognition, to formulating research questions in the field of historical sociolinguistics through a small-scale case-study, exploring metalinguistic terms. Availability of the same parliamentary data (e.g. of a particular discussion or of a specific period, recorded in the JPD) through different channels, like the HPL Repository and the David Antoniou Archive, also provides opportunities for comparative analysis, e.g. using the latter as a specialized corpus and the former as a reference corpus (to test e.g. whether katharevousa and demotic are keywords in the David Antoniou Corpus, which would in turn mark their centrality in the discussion on language and educational policies over a long period of time), cross-checking frequencies, stats etc.

Beyond the scope of Linguistics, this kind of data-driven approach places emphasis on using qualitative historical parliamentary data and corpora as a springboard for evidence-based policymaking (EBP, see [6, 25]), i.e. an approach that 'helps people make well informed decisions about policies, programs and projects by putting the best available evidence at the heart of policy development and implementation'. Extending the same line of reasoning presented in the research-oriented part of the paper, we indicatively cite some examples: studying past parliamentary debates on education can help policymakers evaluate the societal impact of past policies and readjust current educational measures; likewise, by examining the language and argumentation used by policymakers in the past with respect to gender issues or people with special needs, analysts can propose reforms that better meet the needs of vulnerable social groups; assessing past regulatory interventions in economics, as discussed in parliamentary discourse, can help better address emerging financial challenges of today; or, by exploring parliamentary discourse on foreign policy, policymakers can better understand past international relations across various geostrategic contexts, making more informed decisions for the future. The examples are limitless, all highlighting the need for interdisciplinary collaboration, among others, in the production of open parliamentary data and corpora, but also the emergency of networking between different stakeholders (e.g. public bodies, research institutions and private companies), in order to maximize the impact of the separate efforts made and to minimize potential expense, both in terms of human capital and financial resources.

Joining forces in the above suggested manner forms a safe road to institutional innovation for Parliaments, in order to really open up archival treasures and contemporary data to research, but most importantly to citizens as basic beneficiaries in a democratic society.

Acknowledgments. This paper falls within the scope of the author's post-doctoral thesis, entitled "From language attitudes to language policies: Discussing the Greek Language Question at the Hellenic Parliament", carried out under the supervision of Prof. Eleni Karantzola (University of the Aegean, Department of Mediterranean Studies).

Special thanks are due to Panagiotis Kaddas, PhD student at the National Centre of Scientific Research "Demokritos" and main responsible for OCR processing of the materials included at the HPL Repository (external partner: InDigital), for providing the technical outline of this paper.

Disclosure of Interests. The author has no competing interests to declare that are relevant to the content of this article.

References

1. Aitchison, J.: Language Change: Progress or Decay? Cambridge University Press, Cambridge (2012)
2. Andrews, P., da Silva, F.S.C.: Using parliamentary open data to improve participation. In: Proceedings of the 7th International Conference on Theory and Practice of Electronic Governance, pp. 242–249. NYC: Association for Computing Machinery (ACM) (2013). https://doi.org/10.1145/2591888.2591933
3. Biber, D., Reppen, R.: Corpus Linguistics, vol. 1. Sage Benchmarks in Language and Linguistics (2012)
4. Boillet, M., Kermorvant, C., Paquet, T.: Robust text line detection in historical documents: learning and evaluation methods. J. Doc. Anal. Recogn. (IJDAR) **25**(2), 95–114 (2022)
5. Bryant, R., Lavoie, B., Rinehart, A.K.: Building research data management capacity: case studies in strategic library collaboration. OCLC Research, Dublin, OH (2023)
6. Cairney, P.: The Politics of Evidence-Based Policy Making. Palgrave MacMillan, London/New York (2016)
7. Dragojevic, M.: Language attitudes. In: Oxford Research Encyclopedias (2017). https://oxfordre.com/view/10.1093/acrefore/9780190228613.001.0001/acrefore-9780190228613-e-437

8. Erjavec, T., et al.: Multilingual comparable corpora of parliamentary debates. ParlaMint 1.0. Slovenian language resource repository CLARIN.SI. (2020). http://hdl.handle.net/11356/1345
9. Erjavec, T., et al.: Linguistically annotated multilingual comparable corpora of parliamentary debates ParlaMint.ana, 2.1. Slovenian language resource repository CLARIN.SI. (2021). http://hdl.handle.net/11356/1431
10. Fišer, D., Lenardič, J.: CLARIN corpora for parliamentary discourse research. In: Proceedings of the LREC2018 Workshop ParlaCLARIN: Creating and Using Parliamentary Corpora. European Language Resources Association (2018). http://lrec-conf.org/workshops/lrec2018/W2/summaries/14_W2.html
11. Fitsilis, F.: Artificial Intelligence (AI) in parliaments – preliminary analysis of the Eduskunta experiment. J. Legislative Stud. **27**(4), 621–633 (2021). https://doi.org/10.1080/13572334.2021.1976947
12. Fitsilis, F., Bayiokos, V.: Implementing structured public access to the legal reports on bills and law proposals of the Scientific Service of the Hellenic Parliament, Greece. Knowl. Manage. Dev. J. **13**(2), 63–80 (2017). http://journal.km4dev.org/
13. Fitsilis, F., Koryzis, D.: Parliamentary control of governmental actions on the interaction with European organs in the Hellenic Parliament and the National Assembly of Serbia. Online Papers on Parliamentary Democracy V. (2016). https://www.pademia.eu/publications/online-papers-on-parliamentary-democracy/online-papers-on-parliamentary-democracy-v2016/
14. Fitsilis, F., Mikros, G.: Development and validation of a corpus of written parliamentary questions in the Hellenic Parliament. J. Open Humanit. Data **7**(18), 1–14 (2021). https://doi.org/10.5334/johd.45
15. Gatos, B., et al.: GRPOLY-DB: an old Greek polytonic document image database. In: 2015 13th International Conference on Document Analysis and Recognition (ICDAR), pp. 646–650 (2015)
16. Granickas, K.: Parliamentary informatics: what data should be open and how multi stakeholder efforts can help parliaments achieve it. European Public Sector Information Platform Topic Report No. 2013/05 (2013). https://www.europeandataportal.eu/sites/default/files/report/2013_parliamentary_informatics.pdf
17. Hladká, B., Kopp, M., Straňák, P.: Compiling Czech parliamentary stenographic protocols into a corpus. In: Proceedings of the LREC 2020 Workshop on Creating, Using and Linking of Parliamentary Corpora with Other Types of Political Discourse (ParlaCLARIN II), pp. 18–22. European Language Resources Association (ELRA) (2020)
18. Horrocks, G.: Greek: A History of the Language and Its Speakers. Longman, London & New York (1997)
19. IFLA: Access and opportunity for all: How Libraries contribute to the United Nations Agenda 2030 (2014). https://www.ifla.org/wp-content/uploads/2019/05/assets/hq/topics/libraries-development/documents/access-and-opportunity-for-all.pdf
20. IFLA: Guidelines for Parliamentary Libraries, 3rd edn. (2022). https://www.ifla.org/news/guidelines-for-parliamentary-libraries-3rd-edition/
21. IFLA: IFLA Statement on Libraries and Artificial Intelligence (2020). https://repository.ifla.org/bitstream/123456789/1646/1/ifla_statement_on_libraries_and_artificial_intelligence-full-text.pdf
22. Ihalainen, P., Ilie, C., Palonen, K. (eds.): Parliament and Parliamentarism. A Comparative History of a European Concept. Berghahn, New York/Oxford (2016)
23. Jocher, G., et al.: ultralytics/yolov5: v7.0 - YOLOv5 SOTA Realtime Instance Segmentation (2022). https://doi.org/10.5281/zenodo.7347

24. Kaddas, P., Gatos, V., Palaiologos, K., Christopoulou, K., Kritsis, K.: Text line detection and recognition of Greek polytonic documents. In: Coustaty, M., Fornés, A. (eds.) ICDAR 2023 Workshops. LNCS, vol. 14194, pp. 213–225. Springer, Cham (2023). https://doi.org/10.1007/978-3-031-41501-2_15
25. Kamilaki, M.: Parliamentary discourse analysis and language policymaking: the role of language ideologies as qualitative evidence. In: Fitsilis, F., Mikros, G. (eds.) Smart Parliaments: Data-driven Democracy, pp. 7–16. European Liberal Forum (2022). https://doi.org/10.53121/ELFTPS4
26. Kilgarriff, A., et al.: The sketch engine: ten years on. Lexicography **1**, 7–36 (2014)
27. Kouklakis, G., Mikros, G., Markopoulos, G., Koutsis, I.: Corpus Manager: a tool for multilingual corpus analysis. In: Matthew, D., Rayson, P., Hunston, S., Danielsson, P. (eds.) CL2007 Proceedings, pp. 27–30, July 2007, Birmingham, UK (2007). http://ucrel.lancs.ac.uk/publications/CL2007/paper/244_Paper.pdf
28. Mackridge, P.: Language and National Identity in Greece, 1766–1976. OUP, Oxford (2009)
29. Redmon, J., Divvala, S., Girshick, R., Farhadi, A.: You only look once: unified, real-time object detection. In: 2016 IEEE Conference on Computer Vision and Pattern Recognition (CVPR), pp. 779–788 (2016)
30. Reisigl, M., Wodak, R.: The Discourse Historical Approach (DHA). In: Flowerdew, J., Richardson, J.E. (eds.) The Routledge Handbook of Critical Discourse Studies, pp. 87–121. Routledge (2017)
31. van Dijk, T.A.: Political identities in parliamentary debates. In: Illie, C. (ed.) European Parliaments Under Scrutiny: Discourse Strategies and Interaction Practices, pp. 29–56, John Benjamins Publishing (2010)
32. Wodak, R., Meyer, M. (eds.): Methods of Critical Discourse Analysis. Sage, London (2006)
33. Yang, X., Yan, J.: Arbitrary-oriented object detection with circular smooth label. In: Vedaldi, A., Bischof, H., Brox, T., Frahm, J.-M. (eds.) ECCV 2020. LNCS, vol. 12353, pp. 677–694. Springer, Cham (2020). https://doi.org/10.1007/978-3-030-58598-340

Digitization of Written Parliamentary Questions from the Historical Archive (1974–1977) of the Hellenic Parliament

Fotios Fitsilis[1(✉)] , Basilis Gatos[2] , Konstantinos Palaiologos[2] ,
Panagiotis Kaddas[2], Charalambis Kyrkos[1], Maria-Eleni Georgoulea[3] ,
Yiannis Armenakis[3], Christina Tasouli[3] , George Mikros[4] , Olivier Rozenberg[5] ,
and Eleni Kiousi[1]

[1] Hellenic Parliament, Vas. Sofias Avenue 2-4, 10021 Athens, Greece
fitsilisf@parliament.gr
[2] National Center for Scientific Research "Demokritos", Patriarchou Grigoriou E' and Neapoleos 27, 15341 Agia Paraskevi, Greece
[3] Hellenic OCR Team, Aretis 5B, 15343 Agia Paraskevi, Greece
[4] Hamad Bin Khalifa University, Education City - Gate 8, Ar Rayyan, Qatar
[5] Paris, France

Abstract. This article outlines the digitization process and methodology applied to the archive of parliamentary questions from the 1st Parliamentary Term (1974–1977) in the Hellenic Parliament. A collaborative pilot project involving parliament, academia, and a research center facilitated the conversion of printed material to open data. The main tasks of the project include capturing digital images, a custom Optical Character Recognition (OCR) software solution employing machine learning, and rigorous validation for accuracy of a fragmented and of variable quality polytonic corpus in a variety of modern Greek language called Katharevousa. The article discusses the approach and challenges as well as the initial results of the digitization effort, emphasizing ongoing research steps. Overall, 1,674 images were digitally processed corresponding to 1,338 questions. Following algorithmic training, character recognition accuracy is over 98.5%. Successful implementation streamlines further similar digitalization operations in the vast parliamentary archives, while enabling in-depth studies on parliamentary control in the turbulent period of the immediate post-junta era in Greece. A preliminary comparative analysis with a corpus of newer parliamentary questions (2009–2019) provides insights and incentives for the further study of the characteristics and evolution of the Greek language.

Keywords: Hellenic Parliament · written questions · machine learning · OCR · parliamentary control · polytonic corpus

---

O. Rozenberg—Independent researcher.

© The Author(s), under exclusive license to Springer Nature Switzerland AG 2024
H. Mouchère and A. Zhu (Eds.): ICDAR 2024 Workshops, LNCS 14935, pp. 103–117, 2024.
https://doi.org/10.1007/978-3-031-70645-5_8

1 Introduction

Parliamentary control or -in a broader sense- oversight involves the Parliament's ability to examine the Government's activities, ensuring accountability through institutional mechanisms and processes [1, 2]. A key instrument is the submission of written questions by Members of Parliament (MPs). These questions serve as a fundamental means of parliamentary control, constituting formal requests for information from the pertinent Ministers, addressing issues of public concern [3]. Essentially, the purpose of these inquiries is to obtain information regarding the accuracy of particular incidents and compel the Government to outline steps for addressing the raised concerns [4]. In the context of permanently archiving, preserving, and analyzing written question data of previous parliamentary terms from various perspectives, it is imperative to accurately convert the historical documents from paper into digital objects with full-text access.

The article presents the digitization of parliamentary archive documents from the 1st Parliamentary Term (17 November 1974–22 October 1977) at the Hellenic Parliament. These documents consist exclusively of written parliamentary questions from the aforementioned period in handwritten and typed forms. This ongoing initiative serves as a pilot project led by the Department of Scientific Documentation and Supervision in the Scientific Service of the Hellenic Parliament. Collaborators involve the Department of Parliamentary History and the Directorate of Parliamentary Control of the Hellenic Parliament, in conjunction with the Document Image Analysis Group[1] of the Institute of Informatics and Telecommunications of the National Center for Scientific Research "Demokritos". The latter provided the tools and expertise for the recognition process essential for project success. The digitization process unfolds in three distinct stages: capturing digital images of the entire corpus, initial line detection and correction, and subsequent digital character recognition and correction via a custom designed OCR platform. The Machine Learning (ML) technology used adapts and improves character recognition while trained on already validated subsets of documents, thus gradually minimizing human intervention during the correction stage. In addition, detailed metadata, encompassing the qualitative elements of each question, are extracted. The comprehensive process gradually leads to the creation of a validated digital corpus with significant value for subsequent multi-level analyses.

The parliamentary corpus encompasses written questions and questions combined with applications to submit documents (see Art. 126 and 133 parliamentary Rules of Procedure [RoP],[2] respectively) from the archives of the Hellenic Parliament. Given their extended storage duration under non-optimal conditions, there were major concerns of deterioration or permanent damage of the target material. These concerns were unfortunately confirmed during the operations undertaken by the project team. The utilization of state-of-the-art OCR technology along with a fine-tuned validation process preserves and provides access to documents of parliamentary and historical significance. Given the significance of parliamentary control documents, especially those from the 1st Parliamentary Term, their digitization becomes imperative due to the risk of loss or full damage owing to their age. Their digitization serves a dual purpose: preserving the files

---

[1] https://users.iit.demokritos.gr/~bgat/DIA.htm.
[2] https://www.hellenicparliament.gr/en/Vouli-ton-Ellinon/Kanonismos-tis-Voulis/.

Digitization of Written Parliamentary Questions   105

electronically, replacing their traditional physical storage under non-optimal conditions, and facilitating a comprehensive study on their content for the respective parliamentary period. This article concentrates on the digitization stage, specifically extracting the text and metadata from a set of 1,338 parliamentary questions from the 1st Parliamentary Term into an editable electronic format. It commences with a brief analysis of the pivotal role of parliamentary control, specifically via written questions (Sect. 2). The next part outlines the utilized OCR technology and the related ML algorithms for increasing recognition accuracy, while presenting the underlying software platform (Sect. 3). Subsequently, the structure and digitization methodology of the pilot project is detailed, elucidating the rationale behind envisaging the 1st Parliamentary Term for digitization (Sect. 4). The report then thoroughly examines the preliminary project results and discusses the most significant outcomes (Sect. 5). A preliminary analysis assesses the potential of the digital corpus (Sect. 6). Ultimately, the article outlines the research conclusions and puts forth recommendations for the progression and potential expansion of the project (Sect. 7).

2 Written Questions and the Significance of Parliamentary Control

Parliamentary control is characterized by the Parliament's capacity to scrutinize the actions of the Government, holding it accountable through institutional tools and processes, while taking necessary measures to safeguard public interests. A diversity of tools is usually available in view of overseeing the government - especially within parliamentary regimes [5], ranging from votes of no confidence to committees of investigations [6]. Among them, parliamentary questions have been identified as particularly crucial given their capacity to follow topical issues and their decentralized procedural feature. Hence, they should be distinguished given their different political saliency [7]. Written questions typically constitute a personal instrument that individual legislators freely decide to use without much political constraints from their leaders. For this reason, they are considered by political scientists and historians as unique tools enabling them to identify where political attention goes. More generally, written questions are considered to fill a diversity of key parliamentary functions [8]. Representation, oversight and information have especially been distinguished with written questions aiming at expressing local issues, assessing the government activity or filling the general information gap faced by legislatures.

Typically, studies assess the impact of local considerations or topical events on the issues addressed by individual legislators through parliamentary questions. In particular, the capacity of questions to bring electoral advantages to their constituents is discussed in the literature [9]. Whatever their electoral impact, the local focus of written questions is a general feature that has been observed throughout parliaments. A way of coding this dimension has also been suggested [10]. Political scientists and sociologists are also interested in the profile of questioning legislators, including through the gender dimension [11] or seniority. Lastly, written questions have been considerably investigated within the European Parliament [12–16].

The above literature indicates that, although being rather politically marginal, written questions constitute discrete signaling tools used by legislators to inform their peers and third actors (especially group leaders) of their concerns. Less strategically, they also constitute an instrument of self-promotion.

In the Constitution of the Hellenic Republic, parliamentary control is explicitly enshrined in Art. 70 para. 6,[3] according to which the Plenum has the general competence to exercise parliamentary control. Simultaneously, the legislator is authorized to delegate this task to other formations, such as the Section enshrined in Art. 71 and the standing parliamentary committees. More specifically, in the Hellenic Parliament, written questions are defined as a means of parliamentary control in Art. 126 RoP. Since the 1st Parliamentary Term under investigation, effective control by the Hellenic Parliament over the Government has been a crucial mechanism for holding the executive power accountable and delineating parliamentary and political responsibilities. The process of parliamentary control, often yielding specialized information, serves as a valuable source for historical, geographical, legal, and linguistic data. This data proves particularly beneficial for modern multidisciplinary scholars, including legal experts, political scientists, and historians, offering insights into the evolution of the functions of the Greek State, governance structures, and parliamentary behavior.

The study of parliamentary documents has already commenced on a pan-European scale [17–20], as well as in Greece. The first digitization project on parliamentary questions from the 16th Parliamentary Term (5.2.2015–28.8.2015) was conducted in 2016–2017 [21] acting as a catalyst for the present research on the 1st Parliamentary Term. Notably, the basic structure of questions, as it appears throughout the 1st Parliamentary Term, has persisted to the present day. This original initiative also sparked the creation of the Hellenic OCR Team [22, 23], which is a scientific crowdsourcing initiative for the processing and analysis of parliamentary data. Facilitating access to parliamentary control records is crucial for scholars, including political scientists, lawyers, and historians, enabling a comprehensive examination of the function's effectiveness across parliamentary periods and under changing socio-political conditions.

3 OCR Technology and Platform

The recognition of Greek polytonic texts of the corpus (mainly typewritten but also handwritten) is accomplished using the platform described in [24, 25] and comprises two main modules, one for text line segmentation and one for text line recognition. In this section, we will briefly describe these two modules and present an overview of the provided functionalities of the recognition platform. The functionality of the platform is discussed in Sect. 4.3 and Sect. 4.4.

Text line segmentation is a necessary procedure to obtain precise and accurate recognition results. Automatic text line segmentation is carried out using a variation of the well-known YOLOv5 [26] Deep Neural Network model utilized for the detection of oriented quadrilateral polygons [27] (YOLOv5-OBB). The detected text line polygons are extended since extended quadrilaterals yield better results and assist significantly the text line recognition task [25]. To be able to automatically edit detection results acquired from YOLOv5-OBB and to efficiently apply OCR, detected polygons are sorted using

---

[3] https://www.hellenicparliament.gr/en/Vouli-ton-Ellinon/To-Politevma/Syntagma/.

Density-based spatial clustering (DBSCAN) to preserve the correct reading order of the text lines.

The text line recognition module is based on the open-source Calamari-OCR engine [28] that uses advanced deep neural networks. We follow a CONV1-> MAXPOOL-> CONV2 -> MAXPOOL -> BiLSTM scheme, where: CONV1 is Convolutional Layer with 40 filters, stride 1 and 3 × 3 receptive field, followed by ReLU activation. CONV2 is similar to CONV1 but with 60 filters. MAXPOOL is a 2 × 2 max pooling layer and BiLSTM is a Bidirectional Long Short-term Memory layer with 200 hidden nodes. After the BiLSTM layer, Dropout is applied with a skip ratio of 0.5. Data augmentation including padding, distortions, blobs, and multiscale noise is also involved during the training phase.

4 Digitization Methodology

4.1 Preliminary and Field Research

The 1st Parliamentary Term (1974–1977) was selected for investigation through the pilot project. The written questions in paper form from this period are located in a dedicated archive in the premises of the Hellenic Parliament. Following the identification of the material from the reference period, further research was undertaken to evaluate its condition and determine the optimal approach for digitization. Having secured access to the archive and during in situ research, it was observed that only a limited number of questions have been preserved in this physical form over the past decades. These were placed in six folders that displayed significant signs of dampness and mold altering the physical medium, though not homogeneously distributed across the material.

The elevated risk of further degradation prompted an immediate initiation of digitization processes. To avoid cross-contamination and protect the target material, protective gloves and masks were used by the project team. The actual number of the original folders and the questions therein was significantly higher. This was confirmed both via interviews with personnel that dealt with the inventory in the archives that was conducted in 2019 and through comparison with the protocol books of the 1st Parliamentary Term that were found to be located in a different archivation space. The parliamentary term in question comprised three regular sessions and three summer recess sessions. According to the protocol books, a total of 8,724 questions were recorded during this period. Preliminary research identified 1,338 extant written questions, and digitizing these was the main objective of the project. It is evident that only a small fraction (15.3%) of the written questions have been fully preserved.

Ultimately, the decision was made to digitize the entire available corpus of written questions rather than utilizing sampling methods. The identified documents consist of predominantly printed questions using a polygraph/typewriter. Handwritten questions were also identified, constituting 3.8% of the total number. The considerably smaller, poorly defined, and incomplete corpus of corresponding responses was excluded from the current research phase.

4.2 Capturing Digital Images

The initial phase of digitization involved capturing images of the written questions to generate .jpg files (see Fig. 1). For practical reasons and to efficiently structure the digitization process, the documents were clustered in batches. The batches typically contain between 100 and 150 pages, equivalent to the respective number of image files. Overall, 10 batches have been formed.

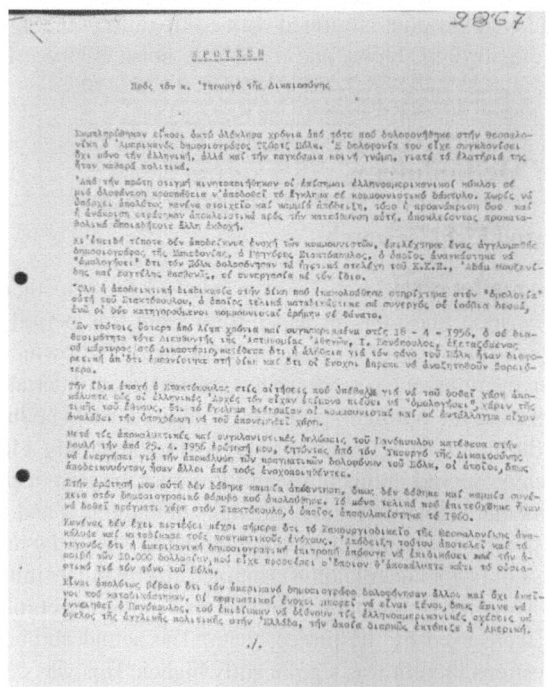

Fig. 1. Question with protocol number 2867, captured in a jpg file.

Image capturing commenced in December 2022 and concluded in March 2023, with an average session duration of 1 to 1.5 h per batch, depending on the volume of documents and their quality. Statistical analysis reveals that, on average, each question corresponds to 1.2 pages of text, while in rare cases there are multi-page questions to be found extending up to 10 pages. A high-resolution fixed-position digital camera was utilized, capable of capturing the entire width of the documents measuring 21 × 29.7 cm (A4 paper size). The photographs have a fixed resolution of 3024 × 4032 pixels and 24-bit color depth.

The questions were photographed as they were found in the folders. Consequently, it is possible that they were not processed sequentially. Any potential shifts in the order of submission were addressed during the metadata extraction process (see Sect. 4.5). After processing each batch, the digital images are exported in dedicated folders as .jpg files to be used for further processing.

Digitization of Written Parliamentary Questions 109

4.3 Line Correction

The next phase of digitization involves the recognition platform, equipped with functionality to predict and correct both segmentation and recognition results. After the initial automatic text line detection, the platform offers the user the following editing functionalities (see Fig. 2): a) Correcting a polygon by adjusting the desired points to their correct positions, b) Adding a new polygon before or after an existing one, c) Deleting a polygon, and d) Merging two polygons.

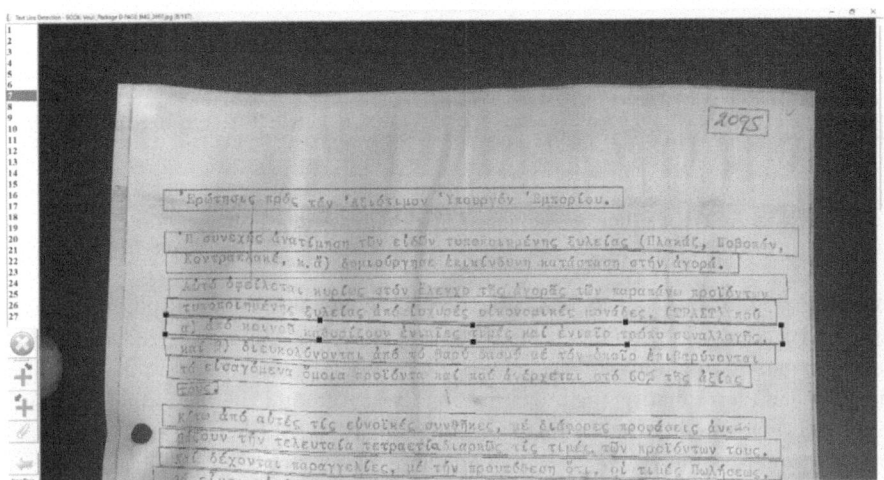

Fig. 2. The text line detection and correction module of the recognition platform.

Line correction is important for improving recognition accuracy. In order to keep track of the reading order, each predicted line is assigned a number, starting from the top-left corner of the document, and progressing sequentially from top to bottom. In some cases, user intervention is needed to edit problematic text-containing areas within each document. Using a cursor, polygon points can be adjusted or corrected, as text lines may contain extraneous artifacts such as smudges, stains, holes, and ink stamps that need to be excluded from the text line region. The corrected lines are automatically saved after any editing function.

4.4 OCR Process

Early attempts to perform OCR using a commercial ABBYY Finereader 14 software (standard edition) yielded rather low character recognition rates. Therefore, a different OCR engine was deemed necessary to capture the unique aspects of the material, particularly typewritten Greek polytonic text.

Following the text line correction phase, the platform uses the text line polygon coordinates to predict text for each line. The selected line is marked within the main image (see Fig. 3 [A]), placing it into context in relation to the surrounding text. A magnified

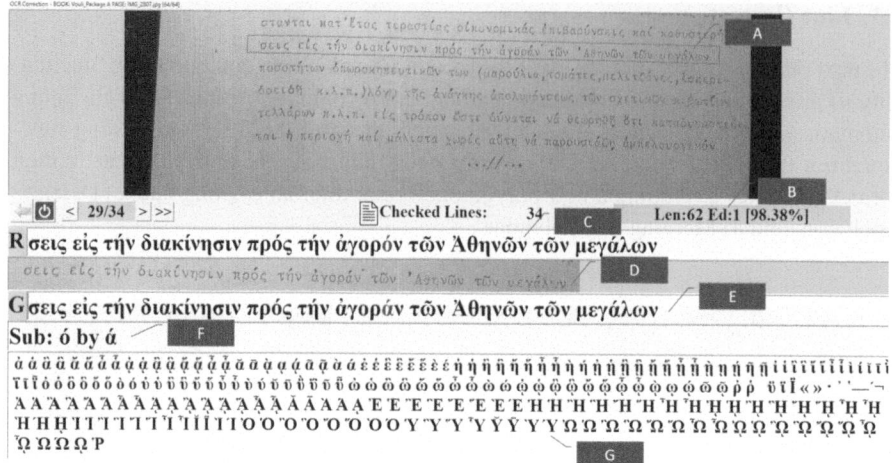

Fig. 3. The text line recognition and correction module of the platform.

image (see Fig. 3 [D]) appears below the text box containing the OCR prediction (see Fig. 3 [C]), aiding users in identifying and correcting erroneous recognized characters. Users can make corrections with a cursor (see Fig. 3 [E]) by selecting the corresponding character and then either typing the correct one, or choosing from the virtual keyboard below (see Fig. 3 [G]), which contains a comprehensive list of Greek polytonic characters. Throughout the verification and correction process, users can navigate between lines using dedicated buttons. Users are also presented with information, including the total number of lines, progress indicators denoting checked lines, character length, total corrections made, and OCR success rate (see Fig. 3 [B]). Additionally, the edit function performed by the user is presented below (see Fig. 3 [F] where a substitution is detected). These metrics collectively assist users in evaluating in situ the accuracy and efficiency of the OCR workflow.

4.5 Metadata Extraction

The data derived from the OCR process can be exported in .txt format, from which various written question elements can be identified. As an intermediate step towards producing open data, key parameters (metadata) include: a) protocol number, b) submission date of the question, c) document type, d) subject, e) session and parliamentary term, f) political party affiliation of the inquiring Member of Parliament (MP), g) name of the inquiring MP, h) the Ministry or Ministries addressed, i) name of the corresponding Minister, and, finally j) serial number of the image file corresponding to the question are place into an .xls sheet. The sheets containing question data are categorized into the three specific years of the period, namely, 1975, 1976, and 1977. It is noteworthy that in most questions from the 1st Parliamentary Term, certain details such as the subject matter or the name of the Minister or Deputy Minister that the question is addressed to are not specifically indicated.

5 Digitization Results

In this paper, we present recognition results using a dataset containing 1338 written parliamentary questions. The dataset consists of 1674 pages and 41626 text lines, 33298 used for training, 4163 for validation and 4165 for test. Text lines were produced automatically by our system (see Sect. 3) and verified and corrected by a user as it is described in Sect. 4.3. Also, the corresponding transcription was edited when needed following the procedure described in Sect. 4.4 to produce the correct transcription (Ground Truth). The test set contains a total of 192951 characters while the reported error characters are 2452. This corresponds to a Character Error Rate (CER) of 1.27% and Word Error Rate (WER) of 5.93% (Character Recognition Accuracy = 98.73%, Word Recognition Accuracy = 94.07%), which is satisfactory considering the problems in quality that are discussed below.

Following training via ML, the algorithm makes overall few mistakes in character recognition. However, the main reason for the high average OCR recognition rate exceeding 98% is that most questions are typed and not handwritten and that the number of the ones that are severely worn or altered appears to be low.

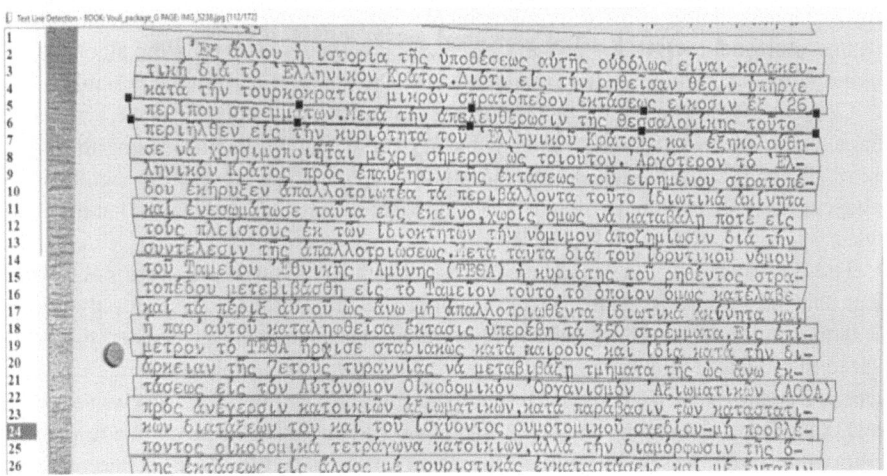

Fig. 4. Example of a document with dense, yet distinct lines.

As mentioned in Sect. 4.3, the process of line verification and correction sometimes requires user intervention. When encountering challenges due to dense text, it becomes crucial to capture the distinctive features of characters (see Fig. 4). For instance, omitting the small line beneath the circle in the Greek letter "ρ" may result in confusion with "ο". Moreover, ensuring the correct segmentation of each line is essential to avoid misinterpretation. For instance, the line in the Greek letter "μ" could be mistaken for an accent in a letter from the line below. Text line ordering also serves a critical role in maintaining text continuity. This is necessary also in cases of text columns where text line ordering has to be set per column (see Fig. 5).

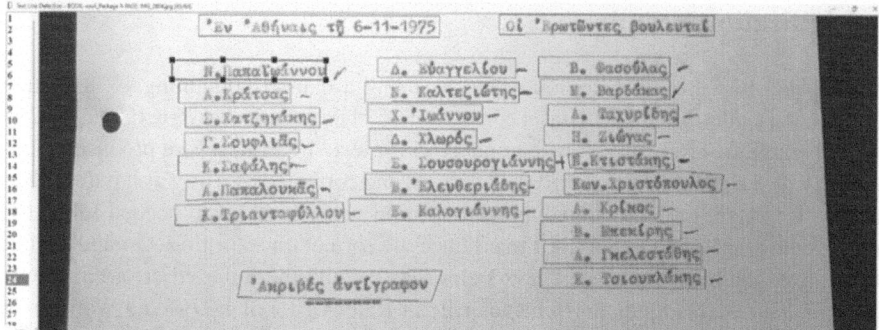

Fig. 5. Example of a document with text arranged in columns.

Validating the characters recognized through the OCR process is a fundamental element of the digitization process. The platform user holds a pivotal role in reducing, or ideally eliminating, OCR prediction errors to achieve the utmost similarity between the original text and its digital twin. Throughout the validation and correction process, meticulous adherence to the original text of the document is paramount. Any spelling errors, syntactic ambiguities, or typographical mistakes by the author were intentionally left uncorrected as one of the primary research objectives was to train the algorithm to autonomously recognize digital documents with the maximum accuracy possible.

Consistency during the correction process is necessary. Using the same symbol or character for identical instances establishes a pattern and uniformity, aiding the training of the prediction model. Achieving consistency in character editing significantly enhances text recognition, particularly in handwritten questions with varied handwriting styles.

The documents in the corpus of Greek parliamentary questions were not always clear and fully readable. Some typed questions are in poor condition, not only due to the deterioration of the paper material, but mainly because the ink has faded to such a degree that the text is almost invisible today. In other instances, letters and characters were corrected either by retyping the correct character on the same typewriter or by hand, namely by placing the correct character with a pen above the error. In a few cases, errors were simply erased. Introducing these corrections is preferred only if the corrected characters are clearly legible, and the prediction includes them, even if the identification was unsuccessful.

Despite striving for uniformity, the most challenging task for users validating and correcting the text lies in identifying letter characters in handwritten documents (see Fig. 6). Despite their official parliamentary status, the authors of some questions hastily wrote them with their unique handwriting, proving a challenge even for the human eye to decipher.

Success rates for the OCR model on these documents were generally low, averaging below 85%. In such cases, maximum fidelity to the original text is sought during validation, alongside efforts to train the algorithm on writing styles it initially struggled to recognize.

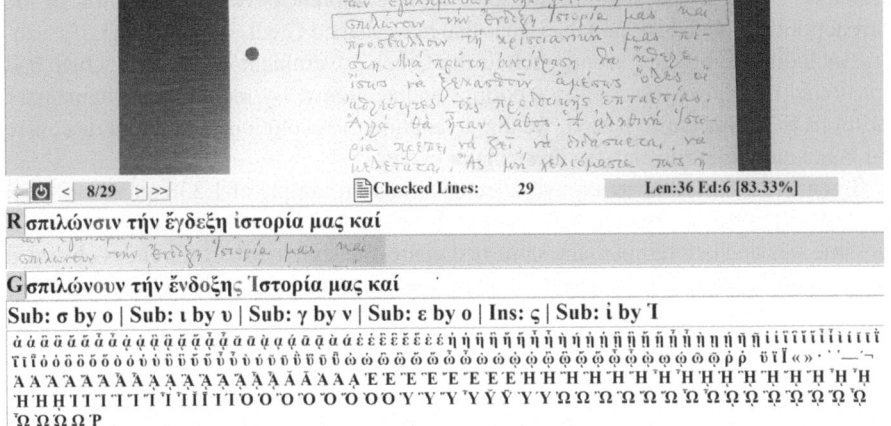

Fig. 6. Handwriting where OCR scored a low recognition rate.

6 A First Comparative Approach to Analyzing the Corpus and Its Language Variety

In this section, we present a preliminary analysis of the corpus of Greek parliamentary questions from the period 1975–1977 using text mining techniques. This corpus provides a unique opportunity to gain insights into the political landscape and the key issues that concerned Greek society during this crucial transitional period. Moreover, this corpus is particularly unique as it reflects a variety of the Greek language called Katharevousa. Katharevousa, which literally translates to "purified," was a form of Modern Greek developed in the late 18th century as a compromise between Ancient Greek and the more vernacular Demotic Greek. It was conceived as a cleansed version of the Greek language, one that was closer to Classical Greek while still being accessible for contemporary use [29].

Katharevousa was the official language of the Greek state from its establishment in 1832 until 1976, and it was widely used in official documents, literature, and public discourse during this period. The parliamentary questions in this corpus, spanning from 1975 to 1977, capture the language at a critical juncture, just before the official transition to Demotic Greek in 1976.

Analyzing this corpus not only provides insights into the political issues of the time but also offers a valuable linguistic resource for studying the characteristics and evolution of Katharevousa. By examining the vocabulary, syntax, and style employed in these parliamentary questions, we can gain a better understanding of how this particular form of the Greek language was used in official contexts and how it may have influenced the political discourse of the period.

To illustrate the extent of changes in the Greek language, we conducted a stylometric comparison between the corpus of parliamentary questions from 1975–1977 and those from the more recent period of 2009–2019 [22]. The 2009–2019 period was specifically chosen to capture the contemporary use of the Greek language in parliamentary discourse. This 11-year span allows for a more comprehensive representation of the current political landscape and the issues that have shaped Greek society in the 21st century, including significant events such as the Greek government-debt crisis, which had a profound impact on the country's politics and economy. By including parliamentary questions from this entire period, we aim to capture the evolution of political discourse and language use in response to these major events.

To ensure a fair comparison, we selected a random sample of 1,338 questions from the 2009–2019 corpus, matching the size of the 1975–1977 corpus. By using corpora of the same size and belonging to the same text genre, the stylometric differences observed more accurately reflect the impact of language variation and the evolving ways in which MPs express their discourse.

The descriptive statistics of the analyzed corpus can be seen in Table 1.

Table 1. Stylometric comparison of the Greek parliamentary questions in two different time periods (1974–1977) and (2009–2019).

Stylometric features	Parliamentary questions	
	1974–1977	2009–2019
Corpus size (in texts)	1,338	1,338
Total Words in the corpus	222,218	440,686
Mean text size (in words)	165	329
SD Text Size	82	253
Min (words)	4	12
Max (words)	559	2,416
StTTR (TTR base = 100 words)	76.95%	69.29%
SD StTTR	4.42	12.97
Mean Word Length (in characters)	5.67	5.70
SD Mean Word Length	0.33	0.29
Mean Sentence Length (in words)	14.12	16.82
SD Sentence Length	10.19	8.49

This comparison of the above basic stylometric features between Greek parliamentary questions from two different periods, 1975–1977 and 2009–2019, reveals significant insights into the evolution of political discourse over time. More specifically, there is a significant rise in mean text size in the more recent corpus, which has nearly doubled, indicating that MPs are now posing longer and more elaborate questions. Additionally, the greater standard deviation in text size reveals higher variability in the length of questions posed, potentially signaling a broader range of topics or varying levels of specificity in MPs' inquiries. This is reinforced by the broader range between minimum and maximum word counts in the recent period, where the minimum and maximum values are both higher than those from 1975–1977.

As for the vocabulary diversity, the standardized type-token ratio (StTTR), a measure of lexical diversity per 100 words [30], is lower in the later period, pointing to a reduction in vocabulary diversity. This decline could suggest that the vocabulary has become more standardized or specialized, potentially due to the greater emphasis on specific topics or increased use of common terminology in parliamentary discourse. However, the considerably higher standard deviation of the StTTR in the recent period indicates substantial variability in vocabulary diversity across different texts, possibly attributable to the broader range of issues covered or the varying styles of MPs.

Furthermore, the stability in the mean word length and its standard deviation across both periods implies consistent levels of linguistic complexity in word construction. Similarly, the increase in mean sentence length in the more recent corpus may suggest a shift towards more sophisticated sentence structures, possibly reflecting MPs' efforts to articulate questions more comprehensively. Conversely, the decrease in the standard deviation of sentence lengths in the recent period points to a more uniform sentence construction, which might be due to changes in format or guidelines for parliamentary questions.

7 Conclusions and Next Steps

A collaborative digitization project using an ML-enhanced OCR process for written parliamentary questions from the 1st Parliamentary Term (1974–1977) in the Hellenic Parliament has been successfully concluded for the available corpus of 1,338 written questions. The digitization of these historical parliamentary documents posed significant challenges related to their poor condition, diversity in format (typewritten or handwritten), and the use of a polytonic system.

These nuances initially led to low-quality recognition through OCR, necessitating substantial human intervention for validation and correction. Through training, the OCR algorithm progressively yielded improved results. A stable methodology enables the seamless integration of different users/operators, without variation in the quality of the results. This feature is considered important in large digitization projects, where the project team is likely to change over time. The core system, encompassing both line correction and OCR capabilities, takes the form of an online platform.

Ongoing efforts involve further training of the OCR algorithm, particularly in the areas of handwritten and heavily damaged documents. The results indicate that the possibility should be examined to extend this initiative to the entirety of the parliament's records printed in polytonic Greek. Beyond the modernization of the parliamentary archive, exporting question text in a digital format with a polytonic system could pave the way for the development of digital tools with even more robust OCR algorithms. In practice, the condition of the original documents poses limitations on the accuracy of character recognition, leading to errors in the extracted text that require correction through manual quality control (validation), thereby restricting the overall process efficiency.

A preliminary comparative analysis with a newer already studied corpus from 2009–2019 was attempted to reveal the potential of the digitized corpus. The comparison suggests that the parliamentary questions from the 2009–2019 period are generally longer, more detailed, and use slightly more complex language compared to those from the 1975–1977 period. However, there seems to be a decrease in lexical diversity, indicating a more repetitive use of vocabulary in the more recent questions. These observations could be related to changes in political discourse, communication styles, or the complexity of issues addressed over time. Further analysis would be needed to explore the factors contributing to these differences and their potential implications.

Overall, the results from applying the text recognition process to the 1st Parliamentary Term's questions are promising. Considering the substantial improvement through training in automatic recognition, even for handwritten text, the process is thought to have entered a productive stage. Hence, in terms of productivity and processing speed, it might be considered appropriate for the full digitization of the entire polytonic parliamentary question archive that remains in physical format (1982–1997). Scaling up the process, however, would necessitate the engagement and training of multiple operators.

Acknowledgments. The authors would like to thank Angeliki Karapanou, Head of the Department of Parliamentary Archives, Hellenic Parliament, for her expert support during the final stages of the study. The research concerning the recognition platform has been partially co-financed by the European Union and Greek national funds through the Operational Program Attica 2014–2020, under the call "RESEARCH AND INNOVATION PARTNERSHIPS IN THE REGION OF ATTICA", project reBook (Digital platform for re-publishing Historical Greek Books, project code: ATTP4-0331172).

Disclosure of Interests. The authors have no competing interests to declare that are relevant to the content of this article.

References

1. Saalfeld, T.: Members of parliament and governments in Western Europe: agency relations and problems of oversight. Eur. J. Polit. Res. **37**(3), 353–376 (2000)
2. Mavrias, K.: Syntagmatikó Díkaio [Constitutional Law], 6th edn. Ekdoseis P.N. Sakkoulas, Athens (2022)
3. Martin, S., Rozenberg, O. (eds.): Roles & Functions of Parliamentary Questions. Routledge, Abingdon (2012)
4. Kaliviotou, M.: Koinovouleftikós Élenchos – Syntagmatikó plaísio kai ória [Parliamentary Scrutiny – Constitutional framework and limits]. Ekdoseis Sakkoula, Athens (2017)
5. Lauvaux, P.: Le contrôle, source du régime parlementaire, priorité du régime présidentiel. Pouvoirs **134**, 23–36 (2010)
6. Griglio, E.: Parliamentary Oversight of the Executives. Tools and Procedures in Europe. Hart, Oxford (2020)
7. Rozenberg, O., Martin, S.: Questioning parliamentary questions. J. Legislative Stud. **17**(3), 394–404 (2011)
8. Raunio, T.: Parliamentary questions in the European Parliament: representation, information and control. J. Legislative Stud. **2**(4), 356–382 (1996)
9. Lazardeux, S.: Une Question Ecrite, Pour Quoi Faire? The causes of the production of written questions in the French Assemblée Nationale. French Politics **3**, 258–281 (2005)
10. Martin, S.: Parliamentary questions, the behavior of legislators, and the function of legislatures: an Introduction. J. Legislative Stud. **17**(3), 259–270 (2011)

11. Cornacchione, T., Tuning, R.: Women behaving differently: anti-establishment party membership and female parliamentary activity. J. Women Polit. Policy **41**(4), 457–476 (2020)
12. Brack, N., Costa, O.: Parliamentary questions and representation of territorial interests in the EP. In: Costa, O. (ed.) The European Parliament in Times of EU Crisis: Dynamics and Transformations, pp. 225–254. Palgrave Macmillan, Cham (2019)
13. Kaniok, P., Kominkova, M.: Parliamentary questions: expressions of opposition(s) within the European Parliament? Baltic J. Eur. Stud. **9**(1), 34–56 (2019)
14. Navarro, J.: Il n'y a pas de question idiote? Les questions des deputés européens à la Commission européenne et au Conseil depuis 1979. Parliaments, Estates and Representation **39**(2), 236–256 (2019)
15. Proksch, S.O., Slapin, J.B.: Parliamentary questions and oversight in the European Union. Eur. J. Polit. Res. **50**(1), 53–79 (2010)
16. Jensen, C.B., Proksch, S.O., Slapin, J.B.: Parliamentary questions, oversight, and national opposition status in the European Parliament. Legis. Stud. Q. **38**(2), 259–282 (2013)
17. Marx, M., Schuth, A.: DutchParl: a corpus of parliamentary documents in Dutch. In: Proceedings of the 10th Dutch-Belgian Information Retrieval Workshop, pp. 82–83, Nijmegen, Netherlands (2010)
18. Drobac, S., Sinikallio, L., Hyvönen, E: An OCR pipeline for transforming parliamentary debates into linked data: case ParliamentSampo-Parliament of Finland on the semantic web. Digit. Humanit. Nordic Baltic Countries Publ. **5**(1), 287–296 (2023)
19. Ogrodniczuk, M.: Polish parliamentary corpus. In: Proceedings of the LREC 2018 Workshop ParlaCLARIN: Creating and Using Parliamentary Corpora, pp. 15–19. European Language Resources Association (2018)
20. Steingrímsson, S., Barkarson, S., Örnólfsson, G.T.: IGC-Parl: Icelandic Corpus of parliamentary proceedings. In: Proceedings of the Second ParlaCLARIN Workshop, pp. 11–17. European Language Resources Association (2020)
21. Fitsilis, F., Schwemmer, C., Saalfeld, T.: Content reconstruction of parliamentary questions - combining metadata with an OCR process. In: Proceedings of the 5th International Virtual Conference on Advanced Scientific Results, pp. 107–112 (2017)
22. Fitsilis, F., Mikros, G.: Development and validation of a corpus of written parliamentary questions in the Hellenic Parliament. J. Open Humanit. Data **7**, 18 (2021)
23. Hellenic OCR Team. https://hellenicOCRteam.gr. Accessed 29 May 2024
24. Kaddas, P., Palaiologos, K., Gatos, B., Katsouros, V., Christopoulou, K.: A system for processing and recognition of Greek Byzantine and Post-Byzantine documents. In: Fink, G.A., Jain, R., Kise, K., Zanibbi, R. (eds.) Document Analysis and Recognition - ICDAR 2023, LNCS, vol. 14190, pp. 366–376. Springer, Cham (2023)
25. Kaddas, P., Gatos, B., Palaiologos, K., Christopoulou, K., Kritsis, K.: Text line detection and recognition of Greek polytonic documents. In: Coustaty, M., Fornés, A. (eds.) Document Analysis and Recognition – ICDAR 2023, International Workshop on Machine Learning (4th edition), LNCS, vol. 14194, pp. 213–225. Springer, Cham (2023)
26. Jocher, G., et al.: ultralytics/yolov5: v7.0 - YOLOv5 SOTA Realtime Instance Segmentation. https://doi.org/10.5281/zenodo.7347926. Accessed 29 May 2024
27. Yolov5 for Oriented Object Detection. https://github.com/hukaixuan19970627/yolov5_obb. Accessed 29 May 2024
28. Calamari OCR. https://github.com/Calamari-OCR/calamari. Accessed 29 May 2024
29. Mackridge, P.: The Modern Greek Language. Oxford University Press, Oxford (1985)
30. Malvern, D., Richards, B.: Investigating accommodation in language proficiency interviews using a new measure of lexical diversity. Lang. Test. **19**(1), 85–104 (2002)

MANPU

MANPU 2024 Preface

The International Workshop on coMics ANalysis, Processing and Understanding (MANPU) is the main workshop related to comics. It gathers mainly researchers in the field of computer science, but also some researchers in the field of human sciences. Comics is a medium constituted of images combined with text and graphic information in order to narrate a story. Nowadays, comic books are a widespread cultural expression all over the world. The market of comics continues to grow, and especially the market of digital comics. For example, the market in Japan is about 5.5 billion USD, 1.2 million USD in the US and 500 million € in France. The digital part of the market has reached respectively 2 billion USD, 200 million USD and 7 million USD in these countries.

From a research point of view, comics images are attractive targets because the structure of a comics page includes various elements (such as panels, speech balloons, captions, leading characters and so on), the drawing of which depends of the style of the author and presents large variability. Therefore, comics image analysis is not a trivial problem and is still immature compared with other kinds of image analysis. Moreover, digital comics such as webtoons introduce new challenges in terms of analysis and indexing.

In 2016, we held the 1st MANPU in Cancun, Mexico in conjunction with ICPR 2016. In 2017, we held the 2nd MANPU in Kyoto, Japan in conjunction with ICDAR 2017. As a characteristic point, the latter part of MANPU 2017 was held in Kyoto International Manga Museum near to the conference venue. In 2019, we held the 3rd MANPU in Thessaloniki, Greece in conjunction with MMM 2019. In 2020, MANPU 2020 was impacted by the CoVid-19 crisis and was organized as a virtual event in conjunction with ICPR 2020. In 2022, we held the 5th MANPU in Montreal, Canada in conjunction with ICPR 2022. This edition was held in Athens, Greece in conjunction with ICDAR 2024.

This year, we received 11 submissions for review, from authors from 11 distinct countries. After an accurate and thorough single-blind peer-review process, with 3 reviewers assigned to each paper, we selected 8 papers for presentation at MANPU, 2024 involving researchers from eight countries. In this edition, MANPU 2024 consisted of 3 sessions dealing with "Comic Understanding", "Text Detection, Recognition and Analysis" and, "Benchmarking and Utilization". The workshop started with an invited talk. The speaker was Neil Cohn, from Tilburg University, Netherlands. He presented his research work on The Patterns of Comics: the Visual Languages of Comics from Asia, Europe, and North America.

Last but not least, we would like to thank the MANPU 2024 Program Committee, whose members made the workshop possible with their precise and prompt reviews. We would also like to thank the ICDAR 2024 organizers for hosting the workshop, and the ICDAR workshop/publication chairs for their valuable help and support.

August 2024

Jean-Christophe Burie
Motoi Iwata
Yusuke Matsui
Rita Hartel
Tien-Tsin Wong
Ryosuke Yamanishi

Organization

General Co-chairs

Jean-Christophe Burie	University of La Rochelle, France
Motoi Iwata	Osaka Metropolitan University, Japan
Yusuke Matsui	University of Tokyo, Japan

Program Co-chairs

Rita Hartel	Paderborn University, Germany
Tien-Tsin Wong	Chinese University of Hong Kong, China
Ryosuke Yamanishi	Kansai University, Japan

Advisory Board

Kiyoharu Aizawa	University of Tokyo, Japan
Koichi Kise	Osaka Prefecture University, Japan
Jean-Marc Ogier	University of La Rochelle, France
Toshihiko Yamasaki	University of Tokyo, Japan

Program Committee

Olivier Augereau	École nationale d'ingénieurs de Brest, France
John Bateman	University of Bremen, Germany
Ying Cao	City University of Hong Kong, China
Wei-Ta Chu	National Chung Cheng University, Taiwan
Alexander Dunst	Paderborn University, Germany
Felix Giesa	Goethe University Frankfurt, Germany
Seiji Hotta	Tokyo University of Agriculture and Technology, Japan
Rynson W. H. Lau	City University of Hong Kong, China
Jochen Laubrock	University of Potsdam, Germany
Tong-Yee Lee	National Cheng-Kung University, Taiwan
Chengze Li	St. Francis University, China
Xueting Liu	Chinese University of Hong Kong, China
Muhammad Muzzamil Luqman	University of La Rochelle, France
Mitsunori Matsushita	Kansai University, Japan
Naoki Mori	Osaka Metropolitan University, Japan

Mitsuharu Nagamori	University of Tsukuba, Japan
Satoshi Nakamura	Meiji University, Japan
Nhu Van Nguyen	University of La Rochelle, France
Frédéric Rayar	University of Tours, France
Christophe Rigaud	University of La Rochelle, France
Yasuyuki Sumi	Future University Hakodate, Japan
Miki Ueno	Kyoto College of Graduate Studies for Informatics, Japan
Emanuele Vivoli	CVC, Autonomous University of Barcelona, Spain
John Walsh	Indiana University, USA
Minshan Xie	Chinese University of Hong Kong, China
Lvmin Zhang	Stanford University, USA

Retrieving and Analyzing Translations of American Newspaper Comics with Visual Evidence

Jacob Murel[(✉)] and David A. Smith

Northeastern University, Boston, MA, USA
j.murel@northeastern.edu, dasmith@ccs.neu.edu

Abstract. Research on image classification and text translation for comics have transpired largely independent of one another. Machine translation tools focus on comics' text features, thereby largely ignoring comics' heavily visual dimension. Image classification applications for comics focus primarily on genre and artist attribution. This paper bridges the gap between these areas by investigating image classification model accuracy for identifying translations of American newspaper comic strips. How might machine learning algorithms leverage comics' distinguishing visual features in order to identify pre-existing translations? To what extent do textual differences affect classification model accuracy in identifying otherwise identical comics? Using a dataset of 18,000 English and Spanish comics, we generate embeddings from three CNNs and a Vision Transformer. We generate additional embeddings from binarized images and images with text redacted using an OCR model. We compute the cosine distance between given pairs of comics and evaluate its accuracy at retrieving translations. The best models rank the true translation first for 97% of queries, falling to 94% when the language is not known.

Keywords: Information Retrieval · Translation · CNNs · OCR

1 Introduction

Machine translation tools for comics utilize optical character recognition (OCR) and other text-based approaches. These have shown promising results. Many textual features in comics are simultaneously visual however. One notable example is onomatopoeia. In many comics, the size, shape, and color of words—i.e., the visual components of text—convey important semantic information. Machine translation for comics, then, may need to integrate image-based approaches for translation. Our paper makes initial steps in this direction.

The interplay of text and image is a distinguishing feature of comics. In attempting to define their craft, cartoonists frequently highlight comics' visual features as the medium's defining quality [8,17,19]. Additionally, comics historians criticize tendencies to focus on comics' textual-linguistic elements and neglect comics' formal visual qualities, wherein they claim much of a given comics' meaning resides [18]. Coinciding with this, machine learning research notes the need for integrating text and image-based analyses and has begun to make steps in this direction [1,20].

With the larger aim of developing a more holistic approach to comics translation, this paper explores image classification methods for identifying translations of comics images. Specifically, we employ pre-trained image classification models to generate embeddings for several thousand comic strips and their pre-existing translations. We then measure machine performance in identifying translation pairs based solely on visual properties. We institute several modifications to examine how text affects purely image-based retrieval accuracy. In this way, we here lay groundwork for a larger project that analyzes how comics' visual and textual features interact in translation.

2 Related Work

Research on text translation and image classification for comics have transpired largely independent of one another. After reviewing machine translation for comics (Sect. 2.1) and the use of image classification for entire strips (Sect. 2.2), we also note prior work on identifying text within panels (Sect. 2.3), as our experiments explore text's effect on translation retrieval performance.

2.1 Machine Translation

Open-access machine translation tools for comics-such comic-translate[1], ImageTrans[2], and manga-image-translator[3]-all operate using OCR engines to extract text and generate translations thereof. The language learning platform Manga Vocabulometer uses Google Translate to source translations of manga text [13]. Current comics translation tools thus use exclusively textual information to generate translations.

One recent study explores context-aware, multimodal translation for manga text. This uses Faster R-CNN and illustrate2vec in order to group comics panels into scenes and extract visual semantic information to generate more robust text translations. In this way, the researchers propose to develop a fully automated approach to comics translation [11]. This is the only study of which we are aware that incorporates comics' visual features in generating translations.

2.2 Image Classification

A number of studies explore approaches to comics classification using convolutional neural networks (CNNs). Many of these studies use pre-existing labels to categorize manga pages among discrete genres, after which they train CNNs using this labeled data for image-based genre classification tasks [3,26,31]. User-labeled data for CNNs have also been used for attribution tasks. Essentially,

---

[1] https://github.com/ogkalu2/comic-translate.
[2] https://www.basiccat.org/imagetrans/.
[3] https://github.com/zyddnys/manga-image-translator.

artists are considered categories, to which CNNs attribute a given comics image according to visual features. Experiments for author classification show promise on page-level but not panel-level manga data. Author and series attribution have also been addressed in digital humanities research that explores computational methods for distinguishing artistic styles [16].

2.3 Text Segmentation

Text segmentation poses an acute problem for comics. One reason is the heavily stylized nature of comics typography, which can impede OCR models not familiar with the wide, seemingly limitless, array of text forms that appear in comics. Region detection is another issue, as OCR models often interleave transcriptions from separate speech bubbles.

An early unsupervised approach to manga text detection groups text according to the distance between detected text characters [15]. A later approach utilizes visual features (e.g. color, shape, topology, etc.) identify connected text blocks from which to produce independent transcriptions [24,25]. While this latter approach shows a marginally decreased performance in terms of F1-score (approx. 2%), it is tested on a larger dataset of Japanese *manga* and French *bandes dessinées*. Pixel-level text detection is another approach, albeit one with less success [4]. Research shows CNNs perform well for detecting speech balloons and other visual text markers in Bangla comics [7].

3 Method

We investigate methods based on image classification for identifying links between comics and their translations. As we shall see, translations of comics and even reprints in the same language, may change not only the text but also the visual layout.

Comics are a primarily visual medium. In light of this, text segmentation and translation experiments have asked how comics' visual components may improve text extraction. For instance, Hinami et al. use character detection and other image detection methods to improve text translations [11]. Text segmentation methods also leverage comics visual components (e.g., balloon boundaries, color, etc.) to improve text extraction. We examine the reverse: how do textual features affect image-based comics retrieval tasks? More specifically, we ask: to what extent do textual differences affect a model's performance in identifying otherwise identical comics?

Image classification applications for comics focus primarily on genre and artist attribution. Our paper breaks new ground by investigating the effect of text variants and extraction on image classification accuracy for American newspaper comic strips. In this way, we explore comics translation and classification through the lens of information retrieval. Our research thereby examines the interplay of text and image in comics, an issue which has been a primary concern of comics scholars [9,14,17,22].

3.1 Datasets

We compile a dataset of 19,526 images of American newspaper comic strips. These images are divided among five series-based classes: *Calvin & Hobbes*, *Luann*, *Nancy*, *Peanuts*, and *Garfield*. The set contains colorized and binary strips in landscape and portrait orientations according to how the strip was initially published. The set contains daily and Sunday strips. In each class, half of the images are English-language comics with the other half being their Spanish-language translations. We compile metadata to readily identify comics image files as linguistic pairs.

We generate a parallel dataset to examine how text extraction affects comics translation retrieval. Using the Tesseract OCR engine[4], we automatically identify and remove text with whitespace from the entire corpus of comics images.

Additionally, we create a third and final dataset to examine how more strictly visual features—specifically, color—affect comics translation retrieval. We use the Pillow library[5] to create a binarized version of the original comics dataset.

Test Data. We use a set of 500 comics images for evaluating our models. This set consists of 100 images from each of the *Calvin & Hobbes*, *Luann*, *Nancy*, *Peanuts*, and *Garfield* classes (500 images total). It contains both daily strips and Sunday strips. All 500 test images are unique English-language comics.

We use corresponding versions of the same 500 strips for evaluating models and embeddings across each dataset. This means that, whichever 500 comics we use for evaluating the original comics embeddings, we use those same comics' text-stripped versions for evaluating the redacted embeddings, and their binarized versions for evaluating the binarized embeddings.

3.2 Models and Embeddings

We use four pre-trained neural networks: ResNet50 [10], MobileNet-v2 [27], EfficientNet-B0 [29], and Vision Transformer (ViT) [5]. The former three are all CNNs that have been widely tested in image classification research, although, to our knowledge, ResNet is the only one of these to have been used in comics image classification research [3,31]. We include ViT for two reasons: 1) a number of recent experiments in computer vision compare vision transformers and CNNs with mixed results [2,23,32]; and 2) to our knowledge, no research investigates ViT's applications with regard to comics classification and retrieval.

We generate twelve sets of image embeddings. Specifically, we create three sets of embeddings for each of the pre-trained image classification models: one from the dataset of original comics, one from the dataset redacted, and one from the binarized dataset.

---

[4] https://github.com/tesseract-ocr/tesseract.
[5] https://pypi.org/project/pillow/.

3.3 Evaluation

Our primary concern is how well each model identifies a given comic's corresponding translation compared to other comics within and outside the same series and language. In other words, we ask: if provided a comic strip from a given linguistic domain, how often will the model return the strip's corresponding language variant rather than an unrelated comic?

To answer this question, we evaluate model retrieval performance using our test set of 500 comics. For each image in these two test sets, we compile the ten-closest image-vectors in a given model's embeddings. Upon compiling the ten-closest vectors for a given English-language test comic, we then determine for how many comics in both of our test sets the model identified the corresponding Spanish-variant as the closest vector.

We evaluate for two different comics retrieval scenarios. First, we evaluate for situations in which each comic's language is known beforehand. In this approach, when compiling the ten-closest image-vectors to a provided English-language comic, we consider only those comics from a different language. That is, since all of the comics in our test set are English-language comics, we only consider Spanish-language comics when retrieving the closest image-vectors. We also evaluate for situations in which the language of each comic is unknown beforehand. If we provide a given English-language comic, how well can the model survey all comics regardless of language and identify its corresponding Spanish-language variant? In this way, we evaluate model performance for different information retrieval scenarios.

Note that we never test for class-wise restrictions. That is, in both unrestricted and language-restricted search tasks, we evaluate model's performance in identifying a corresponding comic strip out of given every comic across all available classes.

We calculate retrieval accuracy for each test comic by using cosine similarity scores to determine recall at rank one and at rank ten.

Cosine Similarity. We use cosine similarity for comparing two image-vectors in a given model's embeddings. For two embedding vectors $\mathbf{x}$ and $\mathbf{y}$ of images of comic strips, cosine similarity is computed as:

$$\cos(\mathbf{x}, \mathbf{y}) = \frac{\sum_{i=1}^{n} \mathbf{x}_i \mathbf{y}_i}{\sqrt{\sum_{i=1}^{n} \mathbf{x}_i^2} \sqrt{\sum_{i=1}^{n} \mathbf{y}_i^2}} \quad (1)$$

Cosine similarity is the cosine of the angle (i.e., distance) between two points in a vector space, which we use to signify the model's judged degree of similarity between two comics strips. The higher two strips' cosine value, the more visually similar the model determines the strips to be.

Recall. The standard recall formula used in classification tasks is:

$$Recall = \frac{TP}{TP + FN} \quad (2)$$

TP signifies the total number of true positives in a model output, and FN is the total number of false negatives. In our task, a true positive is the corresponding Spanish-language comic for a given English-language comic. Everything else is a true negative.

We are interested in whether a model identifies corresponding translations as visually similar. We therefore calculate recall at one and recall at ten for each image in our text set.

Recall at one (R@1) is a modification of the standard recall metric for information retrieval that asks: how often is the first returned item a true positive? Using cosine similarity scores to determine the Spanish-language comic closest to a test comic, we calculate how often the closest Spanish-language comic is a true positive, i.e. the true corresponding translation of a given English-language comic. We also calculate recall at ten (R@10). This measures how often a true positive is among the ten highest ranked items in terms of similarity.

4 Results

Tables 1 and 2 show class-wise recall at 1 and 10 for a search task where the language of each strip is known. When given a query strip in English, therefore, the retrieval system will only consider candidate strips in Spanish. Tables 3 and 4 show class-wise recall at 1 and 10 for an search task not restricted by language. All tables show recall on three different versions of the comics: the original image, comics with redacted text, and binarized images. All tables display model recall for each comics series and average recall across all five series.

Table 1. Class-wise R@1 for language-restricted search task

	Calvin & Hobbes	Luann	Nancy	Peanuts	Garfield	Total
ResNet50						
original comics	1	.99	.12	.73	.98	.76
redacted	.99	.94	.16	.65	.97	.74
binarized	.97	.96	.82	.50	.94	.84
MobileNet-v2						
original comics	1	1	.94	.92	.98	.97
redacted	1	1	.88	.93	.97	.96
binarized	1	1	.99	.85	.97	.96
EfficientNet-b0						
original comics	1	1	.98	.87	.97	.96
redacted	1	1	.99	.78	.97	.95
binarized	1	1	.99	.88	.97	.97
ViT						
original comics	1	1	.11	.43	.97	.70
redacted	1	1	.11	.53	.97	.72
binarized	1	1	.98	.17	.97	.82

Table 2. Class-wise R@10 for language-restricted search task

	Calvin & Hobbes	Luann	Nancy	Peanuts	Garfield	Total
ResNet50						
original comics	1	1	.19	.90	.98	.81
redacted	1	.97	.16	.84	.98	.79
binarized	1	.98	.86	.75	.96	.91
MobileNet-v2						
original comics	1	1	.97	.98	.98	.99
redacted	1	1	.96	.97	.98	.98
binarized	1	1	.99	.92	.97	.98
EfficientNet-b0						
original comics	1	1	.99	.91	.97	.97
redacted	1	1	.99	.89	.98	.97
binarized	1	1	.99	.93	.97	.98
ViT						
original comics	1	1	.12	.61	.97	.74
redacted	1	1	.11	.71	.98	.76
binarized	1	1	.98	.36	.98	.86

MobileNet and EfficientNet markedly outperform ResNet and ViT on all three input conditions. Each model consistently performs well with the *Calvin & Hobbes*, *Luann*, and *Garfield* classes. By comparison, models evidence more variable performance with respect to both the *Nancy* and *Peanuts* classes.

Comparing Tables 1 and 2 with Tables 3 and 4 shows that all models perform better in the language-restricted search task rather than the unrestricted search task. If the model's search is limited to Spanish-language comics in the former, it has a more limited set of data with which to compare a given English-language comic. While R@10 is by definition higher than R@1, the gap is not very large in any of the comics series or models evaluated.

5 Discussion

The four tables reveal notable differences between model architectures in comics translations retrieval tasks. MobileNet and EfficientNet markedly outperform ResNet and ViT. One potential explanation for this may be their respective design purposes. Both MobileNet and EfficientNet are designed to work with limited data and computing constraints, which may have a regularizing effect on their representations of input images.

As previously mentioned, all four tables reveal class-wise differences in model performance. For instance, all models across all three input conditions reveal high accuracy in identifying translation pairs for *Calvin & Hobbes*, *Luann*, and *Garfield*.

Table 3. Class-wise R@1 for unrestricted search task

	Calvin & Hobbes	Luann	Nancy	Peanuts	Garfield	Total
ResNet50						
original comics	.98	.99	.10	.54	.98	.72
redacted	.98	.95	.10	.46	.97	.69
binarized	.97	.96	.80	.21	.94	.78
MobileNet-v2						
original comics	1	1	.82	.78	.97	.91
redacted	1	1	.60	.67	.97	.85
binarized	1	1	.99	.57	.97	.91
EfficientNet-b0						
original comics	1	1	.96	.76	.97	.94
redacted	1	1	.96	.64	.97	.91
binarized	1	1	.99	.67	.97	.93
ViT						
original comics	1	1	.11	.15	.97	.65
redacted	1	1	.12	.20	.97	.66
binarized	1	1	.97	.01	.94	.78

Table 4. Class-wise R@10 for unrestricted search task

	Calvin & Hobbes	Luann	Nancy	Peanuts	Garfield	Total
ResNet50						
original comics	1	.99	.11	.80	.98	.78
redacted	1	.95	.10	.67	.98	.74
binarized	1	.98	.83	.52	.94	.85
MobileNet-v2						
original comics	1	1	.95	.89	.98	.96
redacted	1	1	.88	.88	.98	.95
binarized	1	1	.99	.83	.97	.96
EfficientNet-b0						
original comics	1	1	.99	.85	.97	.96
redacted	1	1	.99	.80	.97	.95
binarized	1	1	.99	.78	.97	.95
ViT						
original comics	1	1	.11	.87	.97	.79
redacted	1	1	.11	.44	.97	.70
binarized	1	1	.98	.04	.97	.80

Our modifications to comics' visual features (i.e., text redaction and binarization) have a variable effect on model performance. ResNet and ViT model performance improves on *Nancy* comics once we binarize the dataset. This is undoubtedly due to that fact that, unlike other series, language variants of *Nancy* comics are colored differently. That is, English-language daily *Nancy* strips are colorized while their Spanish-language counterparts are black-and-white. ResNet, MobileNet, and ViT recall for *Peanuts* comics decreases with binarized comics however.

Text removal does not significantly affect model performance for any class. Nevertheless, model performance marginally decreases on the redacted dataset. We believe this slight decrease in performance may be attributed to the introduction of new visual features between language variants introduced by the text removal process. The Tesseract OCR engine does not identify all of the text in each comic, and so a select few words are left behind at times in our redacted dataset. But which lines and portions of text remain differ between languages. The Tesseract OCR engine does not fully strip corresponding regions and lines of text from different language variants. Thus, while Tesseract removes large portions of text, it nevertheless leaves different remnants of text. In this way, our text removal process essentially introduces new visual differences between linguistic variants of a single comic strip. We suspect these increased visual differences account for the marginal decrease in model performance between the original comics and redacted datasets. With access to more accurate text-detection models for comics, however, it might be possible to achieve more meaningful results with text redaction.

Differences in panel layouts between English and Spanish variants inconsistently affect model performance. Figure 1 shows the panel layouts for a given *Peanuts* Sunday strip in both its English and Spanish-language versions. The panels are numbered according to their corresponding content. The content of corresponding panels are the same; the only difference between these copies (aside from their language) is the panel arrangement The English variant is in portrait while the Spanish variant is in landscape. The landscape arrangement of panels also includes an additional panel (panel 2) not included in our English copy (revealing how some of our comics have been modified from their original syndication appearance). All of our models, however, correctly identify these two *Peanuts* strips as a translation pair.

Panel re-arrangement seems to negatively affect other cases, however. Across all four tables, model recall never goes about .99 for the *Nancy* class and .98 for the *Garfield* class. Indeed, every model consistently misidentifies the same one *Nancy* strip and two *Garfield* strips. Both are Sunday strips in which panel arrangements differ between the English and Spanish-language variants, and an additional title-panel is added to the top of the Spanish-language version.

In fact, panel layout provides some explanation for why the *Calvin & Hobbes* and *Luann* classes achieve perfect recall scores. Unlike many other American newspaper comics strips, *Luann* Sunday strips are a single row of panels. Thus, as with every comics' daily strips, *Luann* Sunday strip panels are not rearranged.

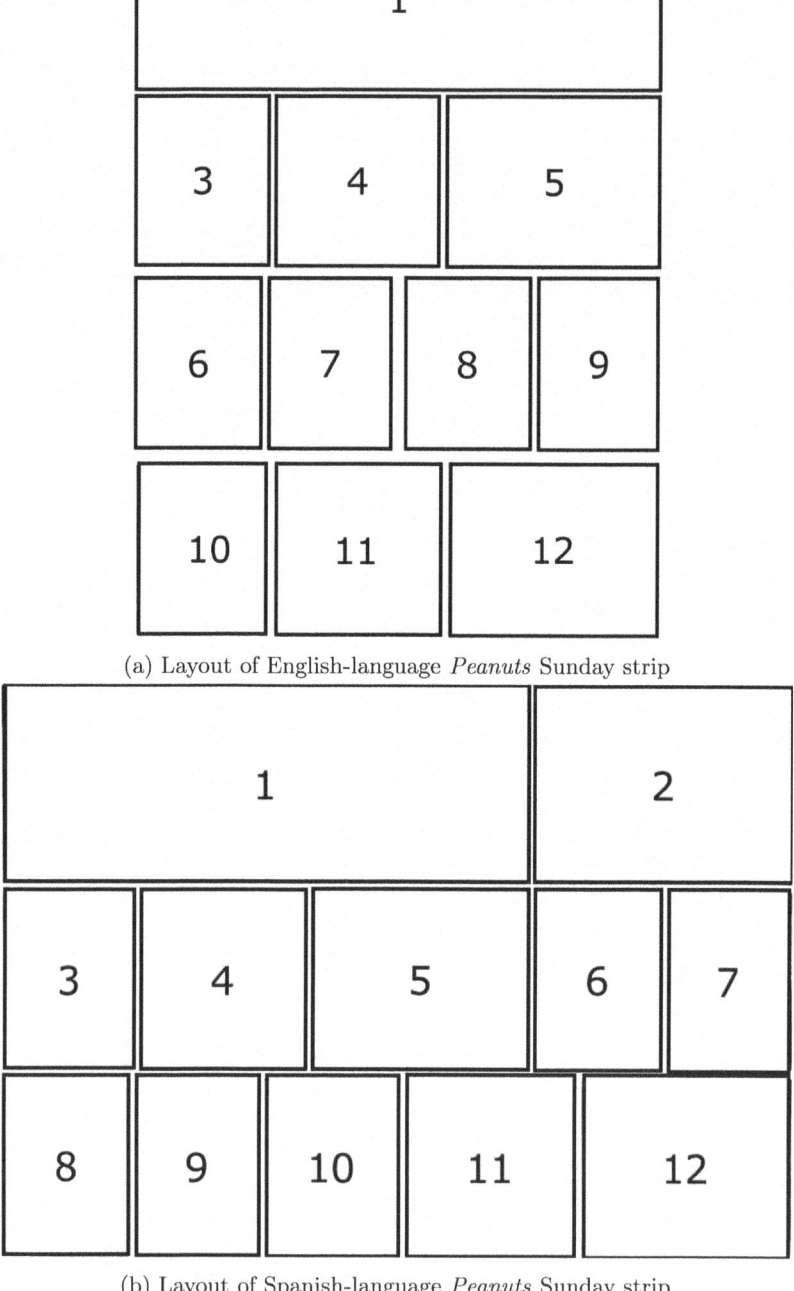

Fig. 1. Comparison of panel layouts for a pair of English and Spanish-language *Peanuts* Sunday strips

Calvin & Hobbes Sunday strips are likewise unmodified, albeit for different reasons. Bill Watterson, the creator of *Calvin & Hobbes*, explicitly designed many Sunday strips so that the panels cannot be rearranged by avoiding a traditional grid format [30]. Moreover, as revealed through correspondence with Andrews McMeel while attempting to acquire rights for republishing comics in this paper, Watterson has stipulated that republications of *Calvin & Hobbes* cannot alter their original created appearance. Unlike other newspaper comics strips, then, *Luann* and *Calvin & Hobbes* are less open to panel rearrangement, perhaps accounting for universally high model performance with these classes.

6 Conclusion

In this paper, we explore an image classification method for identifying translation pairs of American newspaper comic strips. We use four popular image classification models to generate embeddings of comics images. We use series/author as image classes. We use recall at one and ten to evaluate for different search tasks. Our experiments show that pre-trained image classification models can successfully identify translations of comics images, with some variable performance across classes.

Visual differences between linguistic variants seems to the central factor affecting model performance. Thus, binarizing comics images appears to improve model performance overall, while removing the comics text with a standard OCR engine hinders model performance. Panel rearrangements between translations remains a key factor negatively affecting models.

7 Future Work

One area for future work is developing a method for identifying comics translations despite rearranged panel layouts. One potential method for addressing panel rearrangements is to search for translation pairs at the panel level, or to re-rank the top k candidates at the strip level by using panel-level alignments. Panel segmentation is an ongoing and key research area in machine learning applications for comics [6,12,21,28]. Given the architectural differences between linguistic variants of comics, a comics retrieval pipeline could segment panels within a given strip and match one strip's panels with those of another. The projected similarity between strips—and so the probability that they are translations of one another—would then be a function of the degree of similarity between one strip's individual panels and those of another strip. One could also consider the interaction between text translation and visual content. Do some translators move text among panels to balance the layout? Or do they modify each panel's appearance to accommodate a panel-by-panel translation? What can be left unspecified in the text translation because of visual context? These are only a few avenues for further research in comics translation opened up by the methods presented here.

References

1. Augereau, O., Iwata, M., Kise, K.: A survey of comics research in computer science. J. Imaging **4**(7), 87 (2018)
2. Bai, Y., Mei, J., Yuille, A.L., Xie, C.: Are transformers more robust than CNNs? Adv. Neural Inf. Process. Syst. **34**, 26831–26843 (2021)
3. Daiku, Y., Iwata, M., Augereau, O., Kise, K.: Comics story representation system based on genre. In: 2018 13th IAPR International Workshop on Document Analysis Systems (DAS), pp. 257–262 (2018)
4. Del Gobbo, J., Herrera, R.M.: Unconstrained text detection in manga: a new dataset and baseline. In: Proceedings of the 16th European Conference on Computer Vision (ECCV) Workshops, pp. 629–646 (2020)
5. Dosovitskiy, A., et al.: An image is worth 16x16 words: transformers for image recognition at scale (2021)
6. Dutta, A., Biswas, S.: CNN based extraction of panels/characters from Bengali comic book page images. In: 2019 International Conference on Document Analysis and Recognition Workshops (ICDARW), vol. 1, pp. 38–43 (2019)
7. Dutta, A., Biswas, S., Kumar Das, A.: Cnn-based segmentation of speech balloons and narrative text boxes from comic book page images. Int. J. Doc. Anal. Recogn. **24**, 49–62 (2021)
8. Eisner, W.: Comics and Sequential Art. Poorhouse Press (1985)
9. Groensteen, T.: The System of Comics. University Press of Mississippi (2009)
10. He, K., Zhang, X., Ren, S., Sun, J.: Deep residual learning for image recognition. In: 2016 IEEE Conference on Computer Vision and Pattern Recognition (CVPR), pp. 770–778 (2016)
11. Hinami, R., Ishiwatari, S., Yasuda, K., Matsui, Y.: Towards fully automated manga translation. In: Proceedings of the Thirty-Fifth AAAI Conference on Artificial Intelligence, vol. 35, pp. 12998–13008 (2021)
12. Ho, A.K.N., Burie, J.C., Ogier, J.M.: Panel and speech balloon extraction from comic books. In: 2012 10th IAPR International Workshop on Document Analysis Systems, pp. 424–428 (2012)
13. Kato, J., Iwata, M., Kise, K.: Manga vocabulometer, a new support system for extensive reading with Japanese manga translated into English. In: Proceedings for the 25th International Conference on Pattern Recognition, Workshops and Challenges, pp. 223–235 (2021)
14. Lambeens, T., Pint, K.: The interaction of image and text in modern comics. In: Texts, Transmissions, Receptions: Modern Approaches to Narratives, pp. 240–257. Brill (2015)
15. Li, L., Wang, Y., Tang, Z., Lu, X., Gao, L.: Unsupervised speech text localization in comic images. In: 2013 12th International Conference on Document Analysis and Recognition, pp. 1190–1194 (2013)
16. Manovich, L.: How to compare one million images? In: Berry, D. (ed.) Understanding Digital Humanities, pp. 249–278. Palgrave Macmillan (2012)
17. McCloud, S.: Understanding Comics: The Invisible Art. Kitchen Sink Press (1993)
18. Miodrag, H.: Comics and Language: Reimagining Critical Discourse on the Form. University Press of Mississippi (2013)
19. Mitchell, W., Spiegelman, A.: Public conversation: what the comics? In: Critical Inquiry, pp. 20–35 (2012)
20. Nguyen, N.V., Rigaud, C., Burie, J.C.: Digital comics image indexing based on deep learning. J. Imaging **4**(7), 89 (2018)

21. Nguyen Nhu, V., Rigaud, C., Burie, J.C.: What do we expect from comic panel extraction? In: 2019 International Conference on Document Analysis and Recognition Workshops (ICDARW), vol. 1, pp. 44–49 (2019)
22. Postema, B.: Narrative Structure in Comics: Making Sense of Fragments. RIT Press (2013)
23. Raghu, M., Unterthiner, T., Kornblith, S., Zhang, C., Dosovitskiy, A.: Adv. Neural Inf. Process. Syst. **34**, 12116–12128 (2021)
24. Rigaud, C., Burie, J.C., Ogier, J.M.: Segmentation-free speech text recognition for comic books. In: 2017 14th IAPR International Conference on Document Analysis and Recognition (ICDAR), vol. 03, pp. 29–34 (2017)
25. Rigaud, C., Nguyen, N.V., Burie, J.C.: Text block segmentation in comic speech bubbles. In: Proceedings for the 25th International Conference on Pattern Recognition, Workshops and Challenges, pp. 250–261 (2021)
26. Rishu, Kukreja, V., Sharma, V.: Automated classification of comics into genres using CNN-SVM model: a study on visual storytelling. In: Second International Conference on Augmented Intelligence and Sustainable Systems (ICAISS 2023), pp. 122–127 (2023)
27. Sandler, M., Howard, A., Zhu, M., Zhmoginov, A., Chen, L.C.: Mobilenetv2: inverted residuals and linear bottlenecks. In: Proceedings of the IEEE Conference on Computer Vision and Pattern Recognition (CVPR), pp. 4510–4520 (2018)
28. Stommel, M., Merhej, L., Müller, M.: Segmentation-free detection of comic panels. In: Proceedings of the International Conference on Computer Vision and Graphics (ICCVG), pp. 633–640 (2012)
29. Tan, M., Le, Q.: EfficientNet: rethinking model scaling for convolutional neural networks. In: Proceedings of the 36th International Conference on Machine Learning. Proceedings of Machine Learning Research, vol. 97, pp. 6105–6114 (2019)
30. Watterson, B.: Calvin and Hobbes: Sunday Pages 1985–1995. Andrews McMeel (2001)
31. Xu, C., et al.: Panel-page-aware comic genre understanding. IEEE Trans. Image Process. **32**, 2636–2648 (2023)
32. Zhang, C., et al.: Delving deep into the generalization of vision transformers under distribution shifts. In: Proceedings of the IEEE/CVF Conference on Computer Vision and Pattern Recognition (CVPR), pp. 7277–7286 (2022)

Investigating Neural Networks and Transformer Models for Enhanced Comic Decoding

Eleanna Kouletou[1,2](✉) , Vassilis Papavassiliou[2] , and Vassilis Katsouros[2]

[1] National Kapodistrian University of Athens, Athens, Greece
[2] Athena Research Center, Marousi, Greece
eleni.kouletou@athenarc.gr

Abstract. Comic books, merging art with narrative, continue to captivate readers, cinema producers, and collectors, maintaining their allure as a cherished form of visual storytelling across decades. Comic image segmentation is a pivotal aspect in the digital transformation of comics. Leveraging heuristic approaches, neural network-based model (YOLO), and innovative transformer-based architectures (GroundingDINO, SAM), our research aims to autonomously segment comic pages into fundamental components: panels, comic characters, and text areas. To this end, we further trained YOLOv5 and YOLOv8 models to identify these components, while transformer-based models employed prompts to retrieve them. By comparing their outputs across three well-known datasets (eBDtheque, DCM772, Manga109) and using different metrics (Precision, Recall, Average Precision), we conclude that pre-trained self-supervised transformer models can competently outperform state of the art approaches, which often require further fine-tuning to achieve comparable results.

Keywords: Comics · Object Detection · Object Segmentation · Panel Detection · Character Detection · Text Area Detection · Neural Networks · Transformers

1 Introduction

Comic books have been a popular form of visual storytelling, captivating audiences with their unique blend of art and narrative. Comics had an extremely high penetration among readers in the previous decades and are still popular to either readers, cinema producers, or collectors. Many studies [5,10,30] have shown that comics help young children develop critical thinking, underscoring their significance for subsequent developmental stages. With the advent of digital platforms and the increasing digitization of media, the study and analysis of comic images have gained newfound significance. Comic images present a compelling challenge in image processing and computer vision due to their intricate artistic styles, diverse layouts, and the fusion of textual and visual content. Comics have a unique layout format that presents additional difficulties in specifying a query and finding results at the page level.

Comic image segmentation, which divides a comic page into meaningful regions, is key to unlocking a deeper understanding of a comic narrative's visual and textual elements. Accurate segmentation enables extracting individual panels, comic characters, faces, speech balloons, captions, and links between characters and balloons, facilitating various applications. Some examples of systems on comic books are comic translation [8], indexing [19], adaptive display on various devices (smartphones, tablet, laptops, etc.) [21], improved resolution [11] and even the automatic comic book generation [29].

This paper aims to investigate the boundaries of comic image segmentation using computer vision principles and the latest and promising transformer-based models. The central goal is to compare and test state-of-the-art (i.e., YOLO, [20,22]) and new well-presented models (i.e., GroundingDINO, SAM) for autonomously identifying and separating distinct elements within comic pages. Specifically, the proposed pipeline identifies the panels, the characters, and the locations of dialogues and narratives. In addition, a heuristic approach will be presented and tested for efficiency for the text area detection.

The paper organizes its investigation into applying advanced machine-learning techniques for decoding comics. It begins with thoroughly examining the 'Related Work' on panel, character, text area detection, and the available datasets, where it contextualizes its research within the existing scholarly landscape, addressing how datasets have been previously utilized and the methodologies employed for detecting various comic elements. Following this foundation, the 'System Overview' section describes the proposed system architecture, highlighting the integration of neural networks and transformer models to innovate comic decoding. The 'Experiments' section rigorously evaluates the system's performance through various tests and scenarios, offering quantitative and qualitative analyses to substantiate the research claims. Finally, the paper concludes with 'Conclusions and Future Work', summarizing the key findings and the implications of this study for the field of comic decoding and proposing directions for future research to build upon the groundbreaking work presented.

2 Related Work

Comic images are a non-ordinary part of document images. They contain small images as scenes of a story, called panels; inside them are the comic characters that "talk" to each other, think something, or even do an action, and there is a description. So, digitizing a comic book should solve many comic-related tasks. Many previous studies focused on specific tasks to propose an enhanced solution. However, [19] presents an end-to-end comic indexing tool based on deep learning. Its workflow involves panel detection, character/face detection, balloon localization and association with characters, and text recognition. The annotations produced are stored in an XML file following the Comic Book Markup Language.

2.1 Panel Detection

The most essential task is panel detection, the root task for all the following tasks. In previous works, panels were identified by traditional image processing techniques like connected components analysis and line detection. However, some comics do not have a frame around each panel and these algorithms fall in fault. Therefore, [19] proposed an object detection model with convolutional neural networks You Only Look Once (YOLO) version 2 for panel detection and trained it to identify the panels. To fine-tune the model, two approaches were followed based on the pre-trained weights from Pascal-VOC [3]. The first is Anchor Boxes Learning using k-means clustering (k = 5) to find five representatives of bounding box shapes, and the second one is Representation Learning, which is a type of transfer learning. In Representation Learning, the backneck (Darknet) is trained on a different classification task and gets the weights to YOLO. Their research showed that traditional methods [23] have a slight advantage, on average, compared to YOLO. However, the more complex options, i.e., panels without a frame around them, can be detected correctly only with deep learning. Another work is the Comic-MTL [20,21], which describes a model for multiple tasks on the same time (panel, face, character, narrative boxes, balloon & balloon association to its speaker). Multitask learning used in Comic-MTL reduces the computation time to analyze comic book images.

2.2 Character Detection

The next task is character detection, aimed at identifying the complete figures of comics' protagonists. For this task, [19] also, proposed YOLOv2. It was further trained with the DCM772 train set by adopting representative learning and anchor box learning strategies using page or panel images as input. Both traditional image processing and machine learning methods face difficulties in selecting features and generating heuristic rules for generalized characters because characters in each comic will vary significantly across comics (e.g., persons, animals, objects, etc.). Furthermore, a recent study [17] explored the enhancement derived from integrating two datasets (eBDtheque and Manga109) and examined the varying outcomes based on the resolution of page or panel inputs. This research implemented data augmentation techniques to equalize the sample representation between eBDtheque and Manga109. The findings indicate that models trained with a combined dataset exhibited superior performance. Additionally, it was concluded that training with panel-level data only (i.e., not including the whole comic page in training sets) could be more efficient and increase processing time but without a corresponding improvement in accuracy.

2.3 Text Area Detection

Another critical task is text area detection. Some research was focused on speech balloon segmentation, others on text line detection, and others on text body detection. Research-related work for all the above tasks helps to create the text area detection model. One previous work for balloon segmentation [24] used traditional techniques starting with adaptive threshold selection (for binarization), followed by balloon candidate selection (selecting white connected components, as candidate balloons, if they enclose black connected components, i.e. letters), and finally with balloon candidate analysis (selecting thresholds for removing "false alarms"). Another research obtains deep learning models for balloon segmentation. The model in [2] combines the VGG-16 CNN model in a U-Net architecture to predict a pixel-wise segmentation. Moreover, for text area detection the [22] suggests with high Average Precision, a proposed forked model addresses the assignment problem by creating multiple copies of the anchor set, each dedicated to a specific task, allowing for accurate detection of highly overlapped objects. Using SSD300 as the base, the model leverages a multi-scale feature extractor and detection layer to output object locations and class probabilities, ensuring precise object detection. Hence, the SSD300-fork model contains a shared feature extractor for panel, character, face, text area detection and different detection layer for each one task.

2.4 Datasets

To execute the tasks mentioned above, it is essential to utilize labeled data for the neural network training step and evaluation. The following sections provide descriptions of the most renowned datasets that are accessible. Each dataset exhibits a label format and aligns with a subset of comic-related tasks.

eBDtheque. [4] is the most compact dataset that contains labels for all comic tasks. The main drawback is that the images are insufficient for training deep-learning models. It consists of 100 images, each paired with a svg file. Annotations in this dataset cover four classes: Panel, Balloon (with tailDirection), Character, and Line (with textType and text inside). The dataset contains 100 images, 850 panels, 1550 characters, 1092 balloons, and 4691 text lines. In our work, we used Version 3 - July 2019[1].

Digital Comic Museum 772 (DCM772). [19] comprises 772 annotated images sourced from 27 comics available in the Digital Comic Museum. Each image is accompanied by a text file containing annotations in the format "class id and bounding box coordination". The dataset includes annotations for Panel, Character, and Face classes (4470 panels, 8385 characters, 5438 faces), which are publicly available[2]. The images could be downloaded from the Digital Comic Museum[3] site.

---

[1] http://ebdtheque.univ-lr.fr/download/v3/.
[2] https://gitlab.univ-lr.fr/crigau02/dcm-dataset/-/tree/master.
[3] https://digitalcomicmuseum.com/.

Manga109. [1,18] encompasses annotations for 109 different manga volumes. Each book has an XML file containing annotations for four classes: Frame (panel), Face, Body, and Text. Annotations for all classes are represented as bounding boxes with their coordinates. This dataset offers a comprehensive collection of manga book annotations, facilitating research and applications in computer vision and comics analysis. In our work, we used the 2021 released version[4], which contains 10130 annotated images, 103850 panels, 157234 characters, and 147887 text areas.

3 System Overview

This section outlines the proposed methodology for the analysis of comic images. To begin with, the pipeline employed in the paper was created having as baseline the workflow in [19]. Our approach is organized to encompass five fundamental components: panel detection, character detection, character re-identification, text area detection, and text recognition. These components collectively constitute the underlying framework for our strategy to analyze comic images, ensuring a comprehensive and efficient process. Notably, these tasks are interdependent, with panel and character detection of paramount importance. Each panel represents a distinct scene, typically containing one or more characters and speech balloons with text. Figure 1 illustrates the proposed pipeline's workflow. Initially, both panels and characters are identified. Subsequently, text areas of each panel are detected. Following this, text recognition via OCR is performed. Simultaneously, character detection results for each comic book are used to conduct clustering to identify the same characters across different panels.

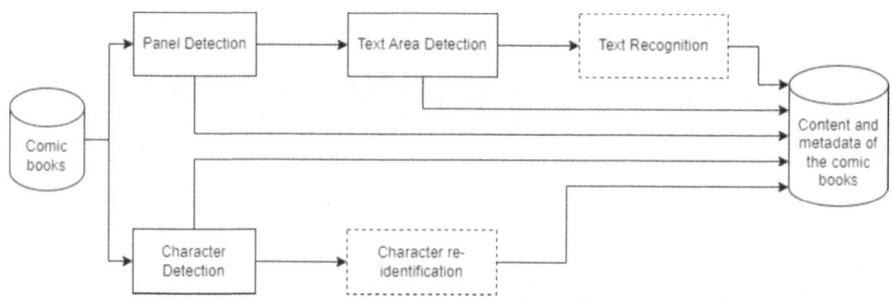

Fig. 1. System pipeline

In this paper, the tasks examined are panel, character, and text area detection using different models and comparing them with previous works. Character re-identification and text recognition can be the future work to finalize the comic

---

[4] http://www.manga109.org/.

analysis pipeline and get coherent content for each comic book. Having identified the critical components of a comic through this pipeline, additional components could be employed to bring comics into the digital era and improve the user-reader's experience. For instance, a machine translation subsystem could translate the identified and recognized text into another language and replace it within the designated area it should cover (i.e., into a "balloon or text caption"). Additionally, a text-to-speech component, separating the comics' text information into dialogues and narratives, and character identification could transform comics from an illustrated text into an audiovisual material/product and thus improve the user-reader's experience.

3.1 YOLO-Based Approach

For the first approach, given the fact that the paper [19] used the YOLOv2 architecture to solve panel/character and face detection, we experimented with the fifth and eighth versions of YOLO [28]. The training dataset contains a combination of the three open datasets mentioned in Sect. 2.4, which have the bounding boxes of both panels and characters to be used as ground truth labels with the purpose of handling multiple classes and not each one detected with a different model.

3.2 Transformer-Based Approach

The second approach focuses on two recently released state-of-the-art zero-shot Transformer-based models for object detection. Our research examined the pre-trained models Segment-Anything-Model (SAM) [7] and GroundingDINO [16], which have open-source code and pre-trained weights by the Meta AI team. SAM can separate an image into masks and use prompts to focus on a specific area. In our case, this model cannot be used stand-alone because the pre-trained weights of the model with the text prompt as input have yet to be publicly available. GroundingDINO is closer to our problem. It gets an input, text prompt, and image and gives the bounding boxes of elements like the text prompt. Finding segmented masks of each detected component is valid in the character and text detection case. For that reason, we performed the SAM, giving as input the image and the GroundingDINO bounding boxes as prompt. We would like to mention that the results of segmented masks highly depend on GroundingDINO results. If GroundingDINO does not find the bounding box, the segmented mask cannot be detected. However, it was noticed that sometimes, even though a box was found from the GroundingDINO, SAM could not find any mask. For that reason, we investigated the results of the GroundingDINO & SAM as a different model.

3.3 Panel Detection

Panel detection can be examined as an object detection process. Two types of model architectures were used to develop the detection model, described in the above Sects. 3.1, 3.2. First of all, a visual inspection was performed in the three datasets (Fig. 2).

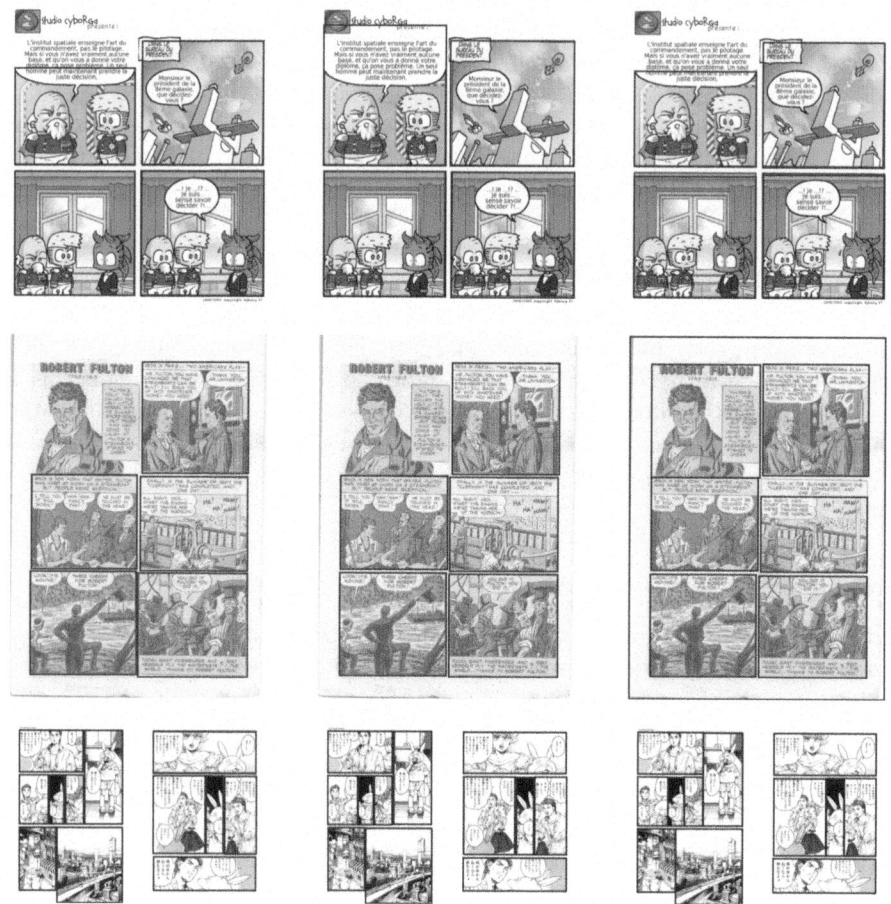

Fig. 2. Panel Detection on eBDtheque (1st row), DCM772 (2nd row) and Manga109 (3rd row). The first column contains the dataset labels, the second the results of YOLOv8 and the third the results of GroundingDINO.

The results show that both the trained model YOLO and the pre-trained model GroundingDINO can detect panels highly accurately. For example, in the second row of Fig. 2 the YOLO understands, as the ground truth indicates, that the first scene is not a panel, but the GroundingDINO has a different opinion. So, it is obvious that the ground truth is questionable for those cases, and we cannot criticize any of them as wrong. What is more, the GroundingDINO considers the whole page as a panel in the scanned images of the DCM772 (Fig. 2 second row). This prompted us to investigate its results (as GroundingDINO-post), removing the panels that are fully involved with the other panels. Furthermore, both models must perform well in Manga, which has more unstructured panels.

3.4 Character Detection

The model architecture and processing we used are described in Sects. 3.1, 3.2. Examples for each dataset are depicted in Fig. 3.

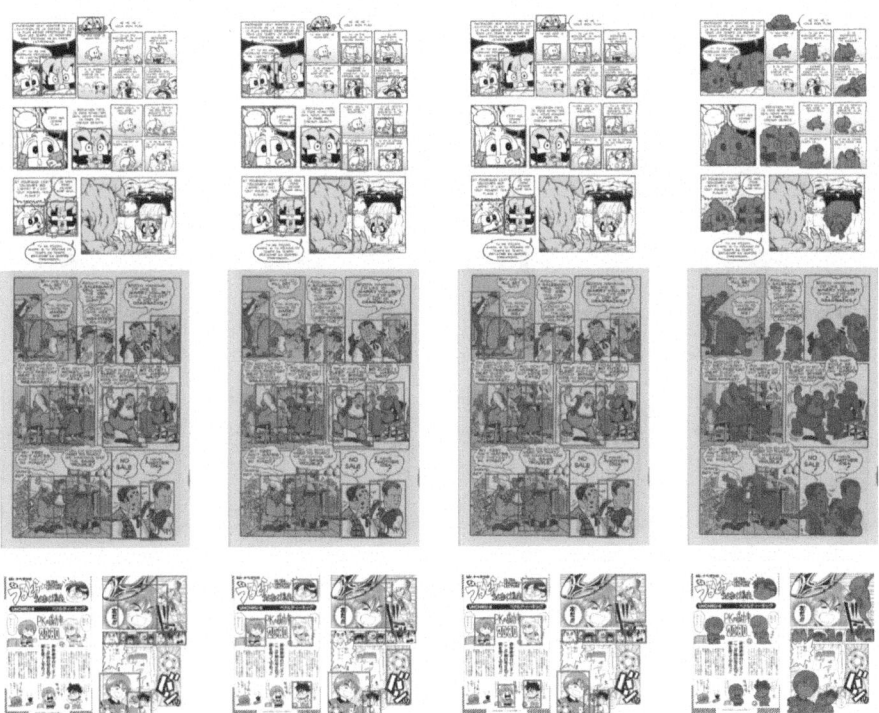

Fig. 3. Character Detection on eBDtheque (1st row), DCM772 (2nd row) and Manga109 (3rd row). The first column contains the dataset labels, the second the results of YOLOv8, the third the results of GroundingDINO and the fourth the results of GroundingDINO & SAM.

The main characters of each image are detected by every model. Only some small characters in Fig. 3 in the first row were not detected and one character in the third row and third and fourth column. Additionally, the models found extra characters that are not labeled in the original dataset (Fig. 3 first and third row). It is worth mentioning that there are some missing objects in the ground truth datasets, which should be taken under consideration in the discussion about low recall values.

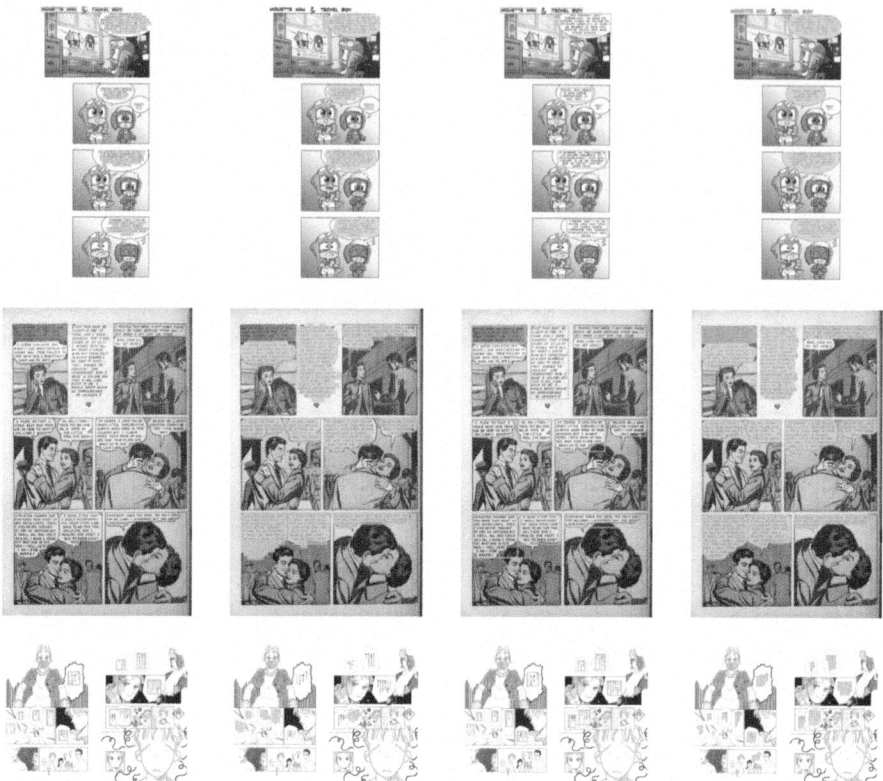

Fig. 4. Text Area Detection on eBDtheque (1st row), DCM772 (2nd row), Manga109 (3rd row). The first column contains the dataset labels of the text lines/area (except for the DCM772, which does not have text labels), the second the results of our heuristic approach, the third the results of GroundingDINO, and the fourth the results of GroundingDINO & SAM.

3.5 Text Area Detection

The primary challenge encountered pertains to the detection of text areas within images. In our endeavors, a significant limitation was the need for labeled images, attributed to the inadequacy of one of the three datasets (DCM772), which lacks annotations for text areas. Conversely, the eBDtheque dataset includes annotations for text lines and speech balloons, whereas the Manga109 dataset provides labels for text area regions differing from text lines or balloons. The challenge of training exclusively with the eBDtheque dataset is compounded by its limited volume of images, and training with Manga109 is problematic due to its specificity to the Japanese language, rendering generalization across different languages challenging. Consequently, we explored the use of GroundingDINO and GroundingDINO & SAM, and introduced a heuristic text detection method inspired by the methodology presented in [24]. Hence, we did not produce any further training for text area detection. The heuristic approach that we design, contains only classical computer vision algorithms like filters and edge detection algorithms (i.e. Sobel) without any machine learning model. We decided to do that experiment in order to compare and indicate how much a heuristic method can compete with transformers.

The method is applied on each detected panel separately. First, the panel is binarized to create a distinction between potential text areas (foreground) and the rest of the content (background). Connected components analysis is employed by using kernel density estimation to map out the distribution of component areas. This analysis facilitates the identification of patterns indicative of text (lobes in the pdf graph), with the search for local maxima and minima within the distribution aiding in pinpointing probable text regions. The algorithm applies to a multi-scale image pyramid, adjusting the scale of the image and repeating the binarization and analysis until greater than two text regions are detected. Once a sufficient quantity of potential text areas is detected, a specialized filter is applied for refinement. This filter aims to enhance the precision of the identified text regions. It operates by assessing the distinct connected components. These components are evaluated within a kernel. The size of this kernel is not fixed; it dynamically adjusts. The adjustment is based on the median area of the connected components.

Some examples are shown in Fig. 4. We can observe that transformers detect the text of all languages (i.e., French, Japanese, and English). The errors of the transformers concern a small text in Fig. 4 first row, and a small false positive area in Fig. 4 second row. As far as the heuristic approach is concerned, the Japanese language struggled to be identified. Moreover, the words above the first panel in Fig. 4 first row cannot be predicted because the heuristic pipeline depends on panel detection and can identify only the text inside the panel. Furthermore, the narrative box in Fig. 4 third row on the panel above left does not match the specifications we perform in the heuristic approach (i.e., the letters have high contrast with the background), so the algorithm fails. Finally, some false positive masks are also presented.

4 Experiments

The main concept of the experiments is to evaluate the convolutional neural network YOLO retrained in a comic-specific combined dataset, the zero-shot pretrained transformer-based models, and a heuristic image processing approach in text area detection, compared with previous work. The hardware used for the experiments is a Linux-based machine with the processor of 2 Intel Xeon CPU @ 2.20 GHz, memory 13 GB RAM and one GPU NVIDIA Tesla P100 with 16 GB GPU memory.

For the neural networks, we trained a YOLOv5-large and YOLOv8-large model using a combined dataset of the three available datasets (eBDtheque, DCM772, Manga109) separating the same test set as the [19,22] mentioned for the DCM772 and Manga109, respectively. For the eBDtheque, 10 of 100 images were randomly selected for the test set. The training set was randomly separated into train and validation sets with a ratio of 90%/10% in each dataset. After transforming the ground truth labels to the YOLO input format, the model trained for 250 epochs, batch size 32, and resized the images with a maximum size of 256 pixels. The same model was introduced to identify panels and characters with different class identifiers.

The next model we tested on panel, character, and text area detection is the GroundingDINO. This model was used without further training on comic-oriented images. As input, the model gets the text prompt 'panel', 'character' and 'text' respectively and returns the bounding boxes that were similar to that prompt. We experimented using SwinB and SwinT weights, box, and text thresholds. After experiments, the thresholds selected are box threshold 0.35 and text threshold 0.25, the backbone of SwinB[5] and keep the results with confidence greater than 35%. The GroundingDINO was trained using data from O365 [27], VG [9], RefCOCO [6], COCO [15], OpenImage [12], Cap4M [14] and ODinW-35 [13].

The combined GroundingDINO and SAM model was also used to find the segmented characters and text areas. The pipeline of this model is to get the bounding boxes of the GroundingDINO predictions and passed to SAM as a prompt to find the proper mask. The parameters selected for SAM are the Visual Transformer (ViT) huge weights[6] and multimask output argument equal False to return only the best result for each bounding box.

To evaluate the performance of the models, the Precision and Recall metrics were calculated to compare them with previous work. We also used the Average Precision metric as PASCAL-VOC [3] mentioned. In order to provide comparable results, we adopted Intersection over Union (IoU) for calculating these metrics and set the success threshold to 50%. We performed a small change in this threshold for GroundingDINO & SAM, and the heuristic approach, due to the fact that the masks are included in the bounding boxes, without covering all their areas. Experimentally, we decided to reduce the threshold to 30%.

4.1 Panel Detection

The models we evaluate for panel detection in the three datasets are YOLOv5, YOLOv8, GroundingDINO and GroundingDINO combined with a post-processing step to reduce this error by removing the proposed panels that fully involved the others. The results are presented in Tables 1, 2, 3, 4 comparing with previous work. We used this structure in the Tables in order to compare as fare as we can the models with the related work in the same datasets and tasks.

In Tables 1 and 4 the results concern evaluation on the same test set as the previous work mentioned. In DCM772 (Table 1), YOLO models have the highest Precision and Recall, and GroundingDINO has remarkable results, too. Only

---
[5] https://github.com/IDEA-Research/GroundingDINO/releases/download/v0.1.0-alpha2/groundingdino_swinb_cogcoor.pth.
[6] https://dl.fbaipublicfiles.com/segment_anything/sam_vit_h_4b8939.pth.

Table 1. Panel and Character detection Precision/Recall for YOLO, GroundingDINO and GroundingDINO & SAM for DCM772 same test set based on [20].

Model	Panel Precision	Panel Recall	Character Precision	Character Recall
Rigaud [25]	86.78	74.84	–	–
Rigaud [26]	85.22	74.41	–	–
Nguyen-YOLOv2 [19]	84.75	86.62	–	–
Faster R-CNN [20]	92.10	93.21	78.93	65.25
Comic MTL - optimized [20]	96.84	97.76	76.21	67.56
YOLOv5	**98.25**	98.9	61.56	68.09
YOLOv8	97.41	**99.12**	**82.99**	63.58
GroundingDINO	90.97	92.97	79.77	84.04
GroundingDINO-post	93.17	92.97	–	–
GroundingDINO & SAM	–	–	77.8	**87.09**

Table 2. Panel, Character & Text Area detection Precision/Recall GroundingDINO with previous work for eBDtheque. The results are for the whole dataset without training on eBDtheque.

Model	Panel Precision	Panel Recall	Character Precision	Character Recall	Text Area Precision	Text Area Recall
Nguyen-YOLOv2 [19]	83.44	58.96	–	–	–	–
Rigaud [25]	86.55	81.24	40.5	21.6	–	–
SSD300-fork [22]	73.30	76.40	58.0	42.2	–	–
Heuristic approach	–	–	–	–	33.64	61.07
GroundingDINO	92.23	**83.76**	83.52	67.59	**93.92**	**82.77**
GroundingDINO-post	**93.52**	83.18	–	–	–	–
GroundingDINO & SAM	–	–	**86.26**	**72.47**	74.61	80.47

Table 3. Panel and Character detection Precision/Recall for eBDtheque test set for YOLO, GroundingDINO, GroundingDINO & SAM compared [20,21] used 5-fold cross-validation.

Model	Panel Precision	Panel Recall	Character Precision	Character Recall
Comic MTL [21]	73.19	76.95	71.79	62.17
Faster R-CNN [20]	91.52	90.77	71.23	61.56
Comic MTL-optimized [20]	92.11	90.91	71.79	62.17
YOLOv5	95.08	89.23	63.91	**92.39**
YOLOv8	**98.39**	**93.85**	**82.18**	90.22

Table 4. Panel, Character and Text Area detection AP@50 YOLO and GroundingDINO with previous work based on [22] for Manga109 same test set

Model	Panel	Character	Text Area
Faster R-CNN	96.1	63.9	23.8
SSD300	**97.1**	79.1	82.0
YOLOv2	90.2	46.9	64.6
SSD300-fork	96.9	**79.6**	**84.1**
YOLOv5	80.5	58.5	–
YOLOv8	83.5	61.9	–
GroundingDINO	85.2	77.6	45.9
GroundingDINO & SAM	–	78.7	59.8

Comic MTL outperformed GroundingDINO, but it should noticed that Comic MTL - optimized, Faster R-CNN, and YOLOv2 are trained using the DCM772 training dataset. Regarding the results on Manga109 (Table 4), it is obvious that the dedicated models (i.e. SSD300) outperform the others. Nevertheless, GroundingDINO has slightly better results than YOLO models, aside from the fact that it has never seen Manga images. YOLO models show high precision and recall at specific thresholds, AP considers the model's performance across a range of thresholds. Fluctuations in precision and recall across these thresholds can result in a lower AP, even if there are points of high precision and recall. On the other hand, the GroundingDINO does not perform fluctuations in precision and recall, so the mAP is better.

Table 2 shows the metrics about the 100 images of eBDtheque, and the comparison is with models that are not trained using this dataset. We can clearly identify that GroundingDINO has better results, and the exclusion of the identified panels that involved others, contributes on reducing false positives (precision increased by 1.2%) while removing only a few true positive predictions (recall reduced by 0.6%).

Table 3 shows the results of our models in the test set compared to the 5-fold results of the other models. Obviously, YOLO versions attain higher scores but it should be noted that their evaluation is based on only 10 comic images (10% of the dataset). However, because the other models performed cross-validation and our results do not, our results are not highly reliable.

4.2 Character Detection

For the character detection, we evaluate YOLOv5, YOLOv8, GroundingDINO, and GroundingDINO & SAM in the three datasets. Tables 1, 2, 3, 4 present the results. The GroundingDINO & SAM outperforms both in precision and recall the other models in the eBDtheque dataset, has a higher F1 score in DCM772, and is closely (-0.9%) to the best-performing model (SSD300-fork). Once more, we mention that SSD300-fork is trained specifically for Manga images, while

GroundingDINO has never seen Manga comics. The training dataset of GroundingDINO is mentioned in the Sect. 4. The SAM model helps GroundingDINO to accelerate its performance reducing the false positives and false negative results. Our YOLO models also have a good position after GroundingDINO, except for the Manga109. However, as we mentioned in the panel detection, Table 3 is not highly reliable for our results because our test set is small and the previous work performed cross-validation.

4.3 Text Area Detection

The last component investigated is text area detection. We evaluate the results of the heuristic approach compared with GroundingDINO and GroundingDINO & SAM on the eBDtheque dataset. The eBDtheque has labels about the text lines and the speech balloons. On the other hand, our models locate the text region either as a bounding box (GroundingDINO), or a segmented mask of (heuristic approach and GroundingDINO & SAM). To compare the results with the ground truth, we applied morphological closing on the textlines in order to merge them in case they are relatively close (i.e. less than 10 pixels). The first approach compares the labels and results using the IoU threshold of more or equal to 50% as in the previous tasks (See Table 2). The second one uses pixel-wise metric segmentation accuracy, i.e., the ratio of true positives to the sum of true positives, false positives, and false negatives (See Table 5). For that comparison, we observed that GroundingDINO and GroundingDINO & SAM managed to detect a good amount of text area. In contrast, the heuristic approach underperformed, finding out around many false positive components (low precision in Table 2) and many false positive and false negative pixels in Table 5. So, it may detect text but not the accurate text area mask close to the labels.

Furthermore, the GroundingDINO and GroundingDINO & SAM evaluated in Manga109 using AP getting around 60% compared with SSD300-fork that achieved 84.1%. Text areas in the Japanese language have different letters and orientations than the Latin languages. It is important to remember that SSD300-fork is trained on Manga dataset, whereas the GroundingDINO and SAM are not, as their authors mentioned.

Table 5. Text area detection on eBDtheque dataset pixel-level metrics.

Model	Segmentation accuracy	Precision	Recall
Heuristic	31.78	47.66	48.82
GroundingDINO	**66.49**	**93.70**	69.60
GroundingDINO & SAM	60.30	70.84	**80.21**

5 Conclusions and Future Work

Based on our experiments, it becomes clear that zero-shot transformers GroundingDINO and SAM accomplished impressive results without any fine-tuning to comic-specific datasets. The most challenging dataset for them is the Manga images. This may be explained by the distinctiveness of these comics in terms of complex layout and the Japanese language. However, we consider that the performance would be improved after feeding transformers with comic images.

Thus, future work could be fine-tuning GroundingDINO and SAM in comic images. A good approach to creating a larger training dataset is to collect more comic images, use the GroundingDINO as a semi-supervised annotator, and then train them in a large amount of comic data. These models seem to have immense potential, but the process requires many computational resources and costs.

References

1. Aizawa, K., et al.: Building a manga dataset "manga109" with annotations for multimedia applications. IEEE Multimedia **27**(2), 8–18 (2020). https://doi.org/10.1109/mmul.2020.2987895
2. Dubray, D., Laubrock, J.: Deep CNN-based speech balloon detection and segmentation for comic books. In: 2019 International Conference on Document Analysis and Recognition (ICDAR), pp. 1237–1243. IEEE (2019)
3. Everingham, M., Van Gool, L., Williams, C.K., Winn, J., Zisserman, A.: The pascal visual object classes (VOC) challenge. Int. J. Comput. Vision **88**, 303–338 (2010)
4. Guérin, C., et al.: ebdtheque: a representative database of comics. In: Proceedings of the 12th International Conference on Document Analysis and Recognition (ICDAR), pp. 1145–1149 (2013)
5. Kang, H.S.: Comic book project as a tool for teaching multimodal argument and fostering critical thinking skills: implications for the 12 writing classroom. In: The College English Association Forum (2017)
6. Kazemzadeh, S., Ordonez, V., Matten, M., Berg, T.: Referitgame: referring to objects in photographs of natural scenes. In: Proceedings of the 2014 Conference on Empirical Methods in Natural Language Processing (EMNLP), pp. 787–798 (2014)
7. Kirillov, A., et al.: Segment anything. arXiv preprint arXiv:2304.02643 (2023)
8. Ko, U.-R., Cho, H.-G.: SickZil-machine: a deep learning based script text isolation system for comics translation. In: Bai, X., Karatzas, D., Lopresti, D. (eds.) DAS 2020. LNCS, vol. 12116, pp. 413–425. Springer, Cham (2020). https://doi.org/10.1007/978-3-030-57058-3_29
9. Krishna, R., et al.: Visual genome: connecting language and vision using crowdsourced dense image annotations. Int. J. Comput. Vision **123**, 32–73 (2017)
10. Krusemark, R.: Comic books in the American college classroom: a study of student critical thinking. J. Graphic Novels Comics **8**(1), 59–78 (2017)
11. Kumar, A., Srivastava, S., Chattopadhyay, P.: Machine and deep-learning techniques for image super-resolution. In: Machine Learning Algorithms for Signal and Image Processing, pp. 89–113 (2022)
12. Kuznetsova, A., et al.: The open images dataset v4: unified image classification, object detection, and visual relationship detection at scale. In: IJCV (2020)

13. Li, C., et al.: Elevater: a benchmark and toolkit for evaluating language-augmented visual models. Adv. Neural. Inf. Process. Syst. **35**, 9287–9301 (2022)
14. Li, L.H., et al.: Grounded language-image pre-training. In: CVPR (2022)
15. Lin, T.-Y., et al.: Microsoft COCO: common objects in context. In: Fleet, D., Pajdla, T., Schiele, B., Tuytelaars, T. (eds.) ECCV 2014. LNCS, vol. 8693, pp. 740–755. Springer, Cham (2014). https://doi.org/10.1007/978-3-319-10602-1_48
16. Liu, S., et al.: Grounding dino: marrying dino with grounded pre-training for open-set object detection. arXiv preprint arXiv:2303.05499 (2023)
17. Lucas, J., Gallego, A.J., Calvo-Zaragoza, J., Martinez-Sevilla, J.C.: Automatic detection of comic characters: an analysis of model robustness across domains. In: Coustaty, Mickael, Fornés, Alicia (eds.) ICDAR 2023, Part I, pp. 151–162. Springer, Cham (2023). https://doi.org/10.1007/978-3-031-41498-5_11
18. Matsui, Y., et al.: Sketch-based manga retrieval using manga109 dataset. Multimedia Tools Appl. **76**(20), 21811–21838 (2017). https://doi.org/10.1007/s11042-016-4020-z
19. Nguyen, N.V., Rigaud, C., Burie, J.C.: Digital comics image indexing based on deep learning. J. Imaging **4**(7), 89 (2018)
20. Nguyen, N.V., Rigaud, C., Burie, J.C.: Comic MTL: optimized multi-task learning for comic book image analysis. Int. J. Doc. Anal. Recognit. **22**, 265–284 (2019)
21. Nguyen, N.-V., Rigaud, C., Burie, J.-C.: Multi-task model for comic book image analysis. In: Kompatsiaris, I., Huet, B., Mezaris, V., Gurrin, C., Cheng, W.-H., Vrochidis, S. (eds.) MMM 2019. LNCS, vol. 11296, pp. 637–649. Springer, Cham (2019). https://doi.org/10.1007/978-3-030-05716-9_57
22. Ogawa, T., Otsubo, A., Narita, R., Matsui, Y., Yamasaki, T., Aizawa, K.: Object detection for comics using manga109 annotations (2018)
23. Rigaud, Christophe: Segmentation and indexation of complex objects in comic book. ELCVIA Electron. Lett. Comput. Vision Image Anal. **14**(3), 59–60 (2016). https://doi.org/10.5565/rev/elcvia.833
24. Rigaud, C., Burie, J.-C., Ogier, J.-M.: Text-independent speech balloon segmentation for comics and manga. In: Lamiroy, B., Dueire Lins, R. (eds.) GREC 2015. LNCS, vol. 9657, pp. 133–147. Springer, Cham (2017). https://doi.org/10.1007/978-3-319-52159-6_10
25. Rigaud, C., Guérin, C., Karatzas, D., Burie, J.C., Ogier, J.M.: Knowledge-driven understanding of images in comic books. Int. J. Doc. Anal. Recognit. **18**, 199–221 (2015)
26. Rigaud, C., Tsopze, N., Burie, J.-C., Ogier, J.-M.: Robust frame and text extraction from comic books. In: Kwon, Y.-B., Ogier, J.-M. (eds.) GREC 2011. LNCS, vol. 7423, pp. 129–138. Springer, Heidelberg (2013). https://doi.org/10.1007/978-3-642-36824-0_13
27. Shao, S., et al.: Objects365: a large-scale, high-quality dataset for object detection. In: Proceedings of the IEEE/CVF International Conference on Computer Vision, pp. 8430–8439 (2019)
28. Ultralytics. YOLOv5: a state-of-the-art real-time object detection system (2021). https://docs.ultralytics.com
29. Yang, X., et al.: Automatic comic generation with stylistic multi-page layouts and emotion-driven text balloon generation. ACM Trans. Multimedia Comput. Commun. Appl. **17**(2), 1–19 (2021)
30. Yonanda, D.A., Yuliati, Y., Saputra, D.S.: Development of problem-based comic book as learning media for improving primary school students' critical thinking ability. In: Elementary School Forum (Mimbar Sekolah Dasar), vol. 6, pp. 341–348. ERIC (2019)

Comics Datasets Framework: Mix of Comics Datasets for Detection Benchmarking

Emanuele Vivoli[1,2](✉), Irene Campaioli[2], Mariateresa Nardoni[2], Niccolò Biondi[2], Marco Bertini[2], and Dimosthenis Karatzas[1]

[1] Computer Vision Center, UAB, Barcelona, Spain
[2] MICC, University of Florence, Florence, Italy
evivoli@cvc.uab.cat

Abstract. Comics, as a medium, uniquely combine text and images in styles often distinct from real-world visuals. For the past three decades, computational research on comics has evolved from basic object detection to more sophisticated tasks. However, the field faces persistent challenges such as small datasets, inconsistent annotations, inaccessible model weights, and results that cannot be directly compared due to varying train/test splits and metrics. To address these issues, we aim to standardize annotations across datasets, introduce a variety of comic styles into the datasets, and establish benchmark results with clear, replicable settings. Our proposed Comics Datasets Framework standardizes dataset annotations into a common format and addresses the overrepresentation of manga by introducing Comics100, a curated collection of 100 books from the Digital Comics Museum, annotated for detection in our uniform format. We have benchmarked a variety of detection architectures using the Comics Datasets Framework. All related code, model weights, and detailed evaluation processes are available at https://github.com/emanuelevivoli/cdf, ensuring transparency and facilitating replication. This initiative is a significant advancement towards improving object detection in comics, laying the groundwork for more complex computational tasks dependent on precise object recognition.

Keywords: Comics · Manga · Detection · Benchmarks · Unification · annotations

1 Introduction

Comics represent a unique form of media that integrates text with graphical elements. This medium has gained widespread popularity as a form of cultural expression globally, from "American Comics" in the US, to "Bandes Dessinées" in France and Belgium and "Manga" in Japan. Despite their apparent simplicity and accessibility, even for children, the intricate layouts of comic pages

---

E. Vivoli, I. Campaioli and M. Nardoni—Annotation team.

© The Author(s), under exclusive license to Springer Nature Switzerland AG 2024
H. Mouchère and A. Zhu (Eds.): ICDAR 2024 Workshops, LNCS 14935, pp. 154–167, 2024.
https://doi.org/10.1007/978-3-031-70645-5_11

present substantial computational challenges. The classic components such as panels, balloons, characters, text, and onomatopoeia are highly influenced by the author's creativity and artistic style, making the task of comic image analysis a non-trivial endeavor.

In recent works [15, 25], authors approached Comics research to more complex tasks starting from *detection* to *comics dialog generation*. However, many of the more difficult tasks are based on the simplest one, i.e. generating the dialog for a comics [15] requires the model to recognize the elements, link the text to the characters, identify multiple instances of the same character and sort textboxes within panels and the panels themself. One mistake in this process will propagate to the next stages of the pipeline making an erroneous final dialog generation.

Despite these "classical" computer vision tasks being well-explored in other domains, they are far from being solved in comics. This challenge can be attributed primarily to two factors: (i) datasets' annotations and size, and (ii) model availability. If we look at comics field, the available datasets are relatively small in size and available annotations do not always agree. In fact, the biggest annotated dataset is Manga109 [8] which provide panels, characters, text, face, onomatopoeias and dialog for the 10.6k images available. Another very famous dataset is DCM [18], which instead has 772 pages, fully annotated and with comics style. Lastly, eBDtheque [9] provide similar annotations for just 100 pages of mixed styles (Japanese, American and French comics). The second issue, regarding model availability, mainly impact the reproducibility of experiments. Due to this second factor, little progress is possible and new works cannot compare with previous state-of-the-art models due to non-standard train/test splits, and the unavailability of models.

In this work we aim to solve these two issues: we collected four of the main available datasets, and provided unified annotations of common objects, fixing the missing ones. As the majority of them are Manga-styled images, we annotated 100 American comic books resulting in more than 5k images. Finally, we benchmark available models for detection, and fine-tuned three CNNs model classes (namely Faster R-CNN, SSD and YOLO). We provide the research community with the code to manage the dataset's images[1], the corrected annotations and to evaluate against CoMix, the collection of more than 29k comic book pages of multiple styles and languages.

2 Related Works

Comics Datasets. The field of comics analysis includes various datasets aimed at processing and studying comic media. However, many of these datasets are either no longer available or require specific permissions from research groups to use. This kind of restriction is common with copyrighted materials, which are generally available only for research purposes. If redistribution is prohibited, it is usual to offer annotations along with a script that facilitates the downloading and organization of the dataset, as exemplified in [25]. The datasets vary greatly

---

[1] for copyright issues we cannot share the datasets' images.

in quality and size. For instance, the dataset *Manga109* [8] is a well-annotated and modest-size manga dataset (109 books, 10.6k images of double-sided pages), with recent works expanding it to include annotations for onomatopoeias [2] and dialogue [15]. In contrast, the *COMICS* dataset by Nguyen et al. [17] is larger (3.9k books, 198k pages) but has limited, automatically generated labels and has seen only minimal updates in recent years [1,29]. Other highly-curated but very small datasets are *eBDtheque* [9], a collection of 100 pages over 20 books, designed for detection and segmentation, and *DCM* [18], collection of 772 pages from Digital Comic Museum[2] containing panels, characters and faces bounding boxes. Lastly, a recent dataset of English Manga-style comics has been proposed in [25] named PopManga, containing almost 2k test split with detection annotations and text-character and character-character links for dialog transcription. The datasets discussed primarily consist of comics in English, French, and Japanese, and notably lack recent publications (post-2010)[3] due to copyright restrictions. The creation of private, small-scale datasets and the challenges in replicating studies are significant issues in the comic domain. In our work, we tackle these issues by focusing on enhancing the accessibility of datasets and facilitating the sharing of models.

Detection Models in Comics. Object detection is a prevalent task in comic analysis. It was first introduced in [9] (2013), but recently appears to be one of the main task in [5] (2022) and [25] (2024). In Table 1, we present the highest performance metrics by task and dataset, citing the model type and the original works. However, comparisons of state-of-the-art (SOTA) performances are often not straightforward: access to code is frequently unavailable, and the only models for which weights are accessible are the 2022 DASS [28] and 2024 Magi [25], leaving many model architecture un-replicable. Moreover, other missing information such as training/testing splits and metric calculation methods, makes any comparison unfair. As highlighted in [22], significant differences exist among standard detection algorithms and their metric calculations. Keeping this in mind, we have selected the model families for our detection tasks (Faster R-CNN, SSD and YOLO) and benchmarked them uniformly. The evaluation script is part of our contribution to the *CDF* repository.

3 Comics Datasets Framework

In this section, we describe the structure of the *CDF* repository (Sect. 3.1), the unified annotations format (Sect. 3.2), the supported datasets (Sect. 3.3), and the evaluation procedure (Sect. 3.4).

3.1 Structure

The two crucial stages of our pipeline include (i) adapting datasets from any format to our standardized structure, and (ii) transforming this unified structure into widely recognized formats such as CVAT, COCO, and others. The

---

[2] www.digitalcomicmuseum.com.
[3] Only PopManga contains a few chapters of 2020 volumes.

Table 1. Overview of existing Benchmarks. *Only the highest values are reported.* Abbreviations represent [tasks]: Detection (det), Segmentation (seg), graph matching (gm), subgraph spotting (subg-s), Text-cloze (T-c), Visual-cloze (T-c), Character-coherence (C-c), classification (cls), recognition (rec), link-prediction (l-p). These tasks can be applied to different objects: Character [C], balloon [B], face [F], panel [P], text [T], emotion [E], and onomatopoeia [O]. Metrics are: mean Average Precision (mAP), Precision (P), Recall (R), and Accuracy (Acc).

DATASETS INFO			BENCHMARKS			
name	format	task	model type	work	metric	perf (%)
Fahad18 [13]	-	det [C]	-	[13]	mAP	41,7
Ho42 [7]	-	gm	graph	[7]	P	91,5
					R	71,5
		det [C]	graph	[7]	Acc	71,4
eBDtheque [9]	SVG/XML	seg [B]	custom- CNN	[6]	P	93,5
					R	96,2
					F1	94,8
					P	75,2
		det [F]	Faster R-CNN	[23]	R	49,8
					F1	60
sun70 [27]	-	det [C]	SIFT	[27]	P	97,8
					R	47
SSGCI [14]	XML	subg-s	graph	[14]	P	75,4
					R	9,8
					ScoreP	82,18
					ScoreR	80,71
COMICS [12]	TXT	T-c [easy]	ComicVT5	[29]	Acc	79,1
		T-c [hard]	ComicVT5	[29]	Acc	71,3
		V-c [easy]	CNN + LSTM	[12]	Acc	85,7
		V-c [hard]	CNN + LSTM	[12]	Acc	63,2
		C-c	CNN + LSTM	[12]	Acc	70,9
Comics3w [10]	-	det [P]	custom- Faster R-CNN	[11]	P	99,24
					R	99,16
					F1	99,2
JC2463 [23]	-	det [F]	Faster R-CNN	[23]	P	95
					R	93,2
					F1	94,1
AEC912 [23]	-	det [F]	Faster R-CNN	[23]	P	82,4
					R	73,1
					F1	77,5
GNC [4]	CSV	det [B]	U-Net (VGG-16)	[3]	P	95,58
					R	94,04
					F1	94,48
DCM772 [18]	TXT	det [T]	custom- CNN	[6]	P	92,7
					R	96,9
					F1	94,7
		seg [B]	custom- CNN	[6]	P	93,56
					R	95,49
					F1	94,51
	XML	det [P]	SSD300	[21]	Acc	97,1
		det [T]	custom- SSD300	[21]	Acc	84,1
		det [F]	custom- SSD300	[21]	Acc	76,2
		det [C]	custom- SSD300	[21]	Acc	79,6
Sequencity4k [19]	-	seg [B]	U-Net (VGG-16)	[19]	P	91,01
					R	91,23
					F1	91.12 (+ 5.44)
EmoRecCom [20]	CSV	cls [E]	CNN + BERT	[20]	AUC	68,49
BCBld (Bangla) [5]	TXT/XML	seg [B]	custom- CNN	[6]	P	97,05
					R	98,81
					F1	97,92
		det [T]	custom- CNN	[6]	P	95,63
					R	98,52
					F1	97,05
COO [2]	XML	det [O]	MTSv3	[2]	P	69,8
					R	65,9
					H	67,8
		rec [T]	CNN + BiLSTM	[2]	Acc	81
		l-p [O]	custom- M4C	[2]	P	77,2
					R	68,7
					H	72,7

ability to support multiple conversions is essential for specific applications: the CVAT standard is utilized by the corresponding annotation tool, which is instrumental for handling and updating existing annotations; the COCO standard is widely adopted for computer vision annotations and predictions, extensively used for evaluation with cocotools[4]. As shown in Fig. 1, our annotation conversion pipeline necessitates a distinct adapter for each dataset structure received, and a converter for each target format supported.

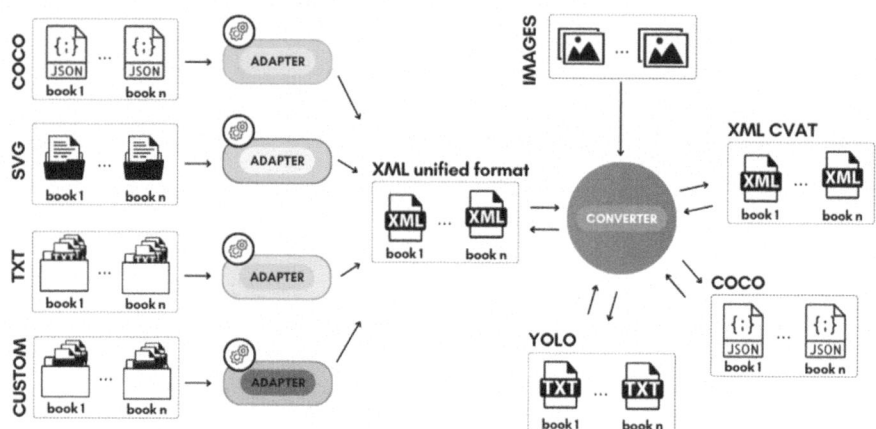

Fig. 1. Unification pipeline schema: given a dataset in an origin format, through a specialized adapter we obtain the XML unified format. This can be converted to CVAT, COCO, or any format required.

3.2 UCA Format

In an effort to standardize the analysis of comic book datasets, we introduce the *Unified Comics Annotation* (UCA) format. UCA is an XML-based format, inspired by Manga109 [8], that systematically categorizes various elements commonly found in comics. The primary goal of this format is to facilitate detailed and structured annotations that support a broad spectrum of comic analysis tasks. The core structure of UCA is hierarchical, beginning with the root element <book>, which encompasses all subsequent data. This encapsulates metadata such as the book title, and sub-elements that detail <characters>, <stories>, and <pages>. Each page within a book is represented by a <page> element, within the <pages> tag, which includes dimensions and other metadata. It serves as a container for finer annotations like panels, text, characters, and speech balloons. These are described using polygonal coordinates to define their spatial positioning on the page. For instance, <panel> and <text> elements use a four-point polygon to delineate their boundaries, while <balloon>

---

[4] https://cocodataset.org/detection-eval.

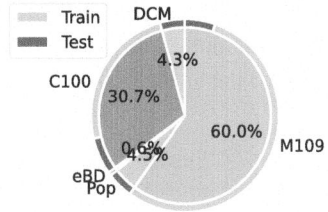

Table 2. Details of images per datasets.

Split	DCM	C100	eBD	M109	Pop
Train	–	4336	–	9675	–
Test	762	1086	100	927	789
tot	762	5422	100	10602	789

Fig. 2. Dataset composition.

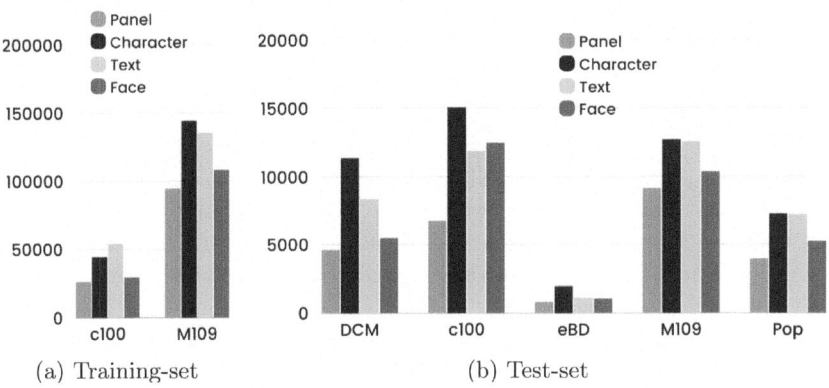

Fig. 3. Number of annotation types per dataset.

elements can utilize polygons with multiple points for irregular shapes. Moreover, UCA accommodates annotations for character reidentification and speaker identification within the comic pages, fostering more complex tasks. Elements like <characters> belonging to the <book> contain the character ID and names which are then used in the character annotation within the pages. Similarly, <link_sbsc> provides detailed annotations of the textual interactions between characters, enabling dialog tasks. The UCA format's comprehensive approach allows researchers to meticulously annotate and analyze the complex interplay of visual and textual elements in comics. By standardizing comic book annotations, UCA aims to improve the accessibility and comparability of comic research, providing a robust foundation for advancing the field.

3.3 Datasets and Annotation Quality

As presented in our collection of reported benchmarks in Table 1, various datasets have been proposed so far. However, many of them are not available and, for others, the data and annotation quality are not sufficient to be considered in this study. Therefore, we selected 4 datasets.

eBDtheque (eBD). The eBDtheque[5] [9] is a collection of 100 pages over 20 books, mainly in French comics with also some English and Japanese percentage. Rigaud et al. [24] provide annotations by domain experts for 850 panels, 1092 balloons, 1550 characters, and 4691 text lines.

DCM772 (DCM). The DCM772[6] [18] contains 772 page images from 27 golden-age comic books from Digital Comic Museum. Annotations contain panels, characters, and face bounding boxes.

Manga109 (M109). The Manga109[7] [8] dataset contains 109 manga volumes from 93 different authors. This dataset has been further extended by COO [2] which added onomatopoeias polygons annotations and links among truncated onomatopoeias. Another set of annotations comes from Manga-Dialog [15] which added a text-character link for (almost) every speaking text box.

PopManga (Pop). The PopManga[8] [25] dataset contains English manga titles from the most popular mangas. The dataset contains two test splits: test-seen with 1100 pages and test-unseen with 980 pages, which correspond to pages of books that the Magi model has seen and has not seen during training, respectively. In this study, we only consider the unseen split, namely "pop" across the images and tables reported.

A notable challenge in our study is the variability in annotation quality across diverse datasets. For instance, Manga109 [8] is equipped with high-quality annotations for object detection categories such as panels, characters, text, faces, and onomatopoeia, and it provides additional metadata like character names and text-character links that are supported by our data format. In contrast, the COMICS [12] dataset offers only pseudo-labels generated by a YOLO model fine-tuned on a limited set of 500 examples for panel and text detection, resulting in annotations that lack the precision required for dependable usage. The more recently developed PopManga dataset [25] includes comprehensive annotations for panel, character, and text detection, and it establishes links between text and characters. Given the complexity of tasks such as transcribing dialogue from comic pages, it is essential to assess models against a test data pool that is not only diverse and substantial but also of high quality. To address these needs, we have harmonized the annotations across datasets, specifically targeting object detection classes like panels, characters, text, and faces. Figure 2 details the statistics of the proposed *Comics Datasets* unification, which is composed, in the test split, of an equal mix of American-style and Manga-style comics, with a minor inclusion of French comics. Moreover, the dataset's multilingual nature is reflected in Manga109's inclusion of Japanese comics, while PopManga, DCM and Comics100 feature English comics, and eBDtheque, though smaller in scale, includes comics in English, Japanese, and French. In Fig. 3 are provided comprehensive annotation counts per dataset and per split.

---

[5] http://ebdtheque.univ-lr.fr/registration.
[6] https://git.univ-lr.fr/crigau02/dcm_dataset.
[7] http://www.manga109.org/index_en.php.
[8] https://github.com/ragavsachdeva/Magi.

3.4 Evaluation

The main goal of this work is to be able to properly evaluate models on common settings. For this, we provide an evaluation system within our framework which relies on cocotools, a set of common evaluation functions used across various projects among which *"torchmetrics"*[9], *"ultralytics"*[10] and more. By using the CoMix framework converter, one can obtain a JSON COCO-format file from XML, YOLO, or any other supported format. With the COCO-format predictions, we can evaluate per-class and global detection metrics such as precision at different IoU thresholds $\{.50, .50 - .95\}$, and recall with gradually increasing the number of objects detected $\{1, 10, 100\}$, considering single classes AP, AR, or over multiple classes mAP, mAR. Generally, everything starts with the two common metrics: precision and recall. Then, average precision (per class) and mean average precision (over classes) are calculated. Finally, these metrics are calculated at varying thresholds of IoU.

Intersection over Union (IoU) is a metric used to evaluate the accuracy of a single prediction, and measures the overlap between the predicted bounding box and the ground truth bounding box:

$$IoU = \frac{\text{area of overlap}}{\text{area of union}}$$

The IoU threshold is crucial in determining whether a detection is considered a true positive. It plays a significant role in the evaluation of both the spatial accuracy of the bounding box and the correctness of the object classification.

Precision measures the accuracy of the positive predictions made by the model, defined as the proportion of predicted positive detections that were both correctly classified and met a minimum IoU threshold with the ground truth:

$$Precision = \frac{TP}{TP + FP}$$

where TP represents true positives (correct class and sufficient IoU overlap) and FP represents false positives (incorrect class or insufficient IoU overlap).

Recall measures the model's ability to detect all relevant instances in the dataset, considering correct classification and IoU:

$$Recall = \frac{TP}{TP + FN}$$

where FN represents false negatives (missed detections or incorrect classifications).

Average Precision (AP) quantifies the trade-off between precision and recall at various decision thresholds. AP is calculated as the area under the precision-recall curve (PR curve). The PR curve is derived by adjusting the

---

[9] https://github.com/Lightning-AI/torchmetrics.
[10] https://github.com/ultralytics/ultralytics.

decision threshold on the detection scores provided by the model, considering each class separately:

$$AP = \sum_n (R_n - R_{n-1})P_n$$

where P_n and R_n are the precision and recall at the nth threshold.

The **Mean Average Precision (mAP)** is the mean of the AP scores calculated for all classes:

$$mAP = \frac{1}{N}\sum_{i=1}^{N} AP_i$$

where N is the number of classes. The mAP can be calculated at different IoU thresholds to provide insights into the model's performance across various levels of localization strictness.

In some benchmarks, such as those used in competitions like COCO[11] (Common Objects in Context), the mAP is calculated at multiple IoU thresholds (e.g., 0.5 to 0.95 in steps of 0.05). This involves computing the precision and recall at each of these IoU levels to get a more nuanced understanding of the model's performance across different criteria of spatial accuracy.

4 Benchmarking

The selected models for benchmarking comprehend convolution- and transformer-based models finetuned on comics data or in a zero-shot setting. Among the zero-shot, GroundingDino [16] is our main choice. This model is trained for the task of open-set object detection, an extended version of object detection with natural language object classes.

We used GroundingDino with different class names to detect the four classes of objects: panels, characters, text, and faces. For each of these objects, we have merged results obtained with similar object classes that could help the model to resemble comics elements, such as: (panels) "comics panels", "manga panels", "frames", "windows"; (characters) "characters", "comics characters", "person", "girl", "woman", "man", "animal"; (text) "text box", "text", "handwriting"; and finally (face) "face", "character face", "animal face", "head", "face with nose and mouth", "person's face". All the other models, instead, have been specifically trained on comics data.

DASS is a convolution-based model (YOLOX) trained in a self-supervised approach distilling a teacher network with the OHEM loss. The model comes with three weight versions: dcm, manga, and mix. The three different model instances have been fine-tuned on the corresponding datasets, which should correspond to higher scores in the same distribution of comic style.

Additionally, we have selected three model classes that have been used multiple times in previous works: Faster R-CNN, SSD, and YOLO. For each of these convolution-based models, we have trained on "comics-style", "manga-style", and "mixed comics-styles", following [28]. The "comics-style" training

---
[11] https://cocodataset.org/#detection-eval

set is obtained by the Comics100 train split (almost 4.5k pages), which we will release in the *CDF* repository as part of this work. The "manga-style" train set is the one provided by Manga109, which corresponds to 9k images. When a model has been trained on the "mixed comics-style" set, we refer to the fusion of Comics100 and Manga109 train splits. This can show us the impact of different training data distributions on performances in different comic styles. Our Faster R-CNN initialized with the available weights in "torchvision" with ResNet-50 backbone. We have modified the last layer, box predictor, to support our number of classes (4) instead of the default coco classes. The model has been fine-tune for 50 epochs with SGD with a *learning rate* $5e^{-3}$, *momentum* and *weight_decay* of 0.9 and $5e^{-4}$, respectively. We have used a *StepLR* scheduler inter-epochs and a *CosineAnnealingLR* scheduler at every epoch. We have rescaled the images to (1024×1024) and applied *RandomHorizontalFlip* with probability 0.5. The SSD model, instead, has been trained using the "mmdetection" framework[12], using the default configurations for ssd300, and scaling input images to (1024×1024), as for Faster R-CNN. We have fine-tuned SSD on the three datasets, all with the same configuration. Finally, the YOLOv8 model has been fine-tuned from the pre-trained weights available in "ultralytics", following the default yolov8x configuration. All these experiments were carried out with a batch size of 32, in a single A40 GPU.

The last model we have tested on our *Comics Datasets* collection is the recent Magi [25], a transformer-based model with DeTr backbone and two MLP heads for character re-identification and text-character linking, inspired from [26]. Authors pre-trained Magi on GroundingDINO annotated Mangadex[13] noisy dataset and fine-tuned on an *un-available* manually annotated popmanga dev-set.

The detection results are reported in Tables 7, where the used metric is the mean Average Precision (mAP) averaged across the four classes considered. However, some model (DASS and Magi) only considers two and three classes, respectively. Thus, we provide the mAP score averaging only detectable classes in Table 8. In the Appendix, we reported mAP detailed for panels (table 3), characters (table 4), faces (table 5) and text 6. Other metrics are reported in the repository, across which the mAP averaged across the 0.5%-0.95% IoU, Average Recall at 1 (AR@1) and 100 (AR@100), which are common metrics for object detection tasks.

In the tables, we emphasize that GroundingDino is used in a zero-shot setting, not specifically trained for the detection of our classes; the Faster R-CNN, SSD, and YOLOv8 correspond to our fine-tuned models; while DASS and Magi are the available models specifically tailored for comics and/or manga. An additional note is that DASS presents in its training a subset of the DCM dataset split. This is because we consider DCM (the whole dataset) as a test-set split. The choice of considering the whole DCM for testing in our *Comics Datasets* collection is due to its small size and availability of well-curated annotations.

---

[12] https://github.com/open-mmlab/mmdetection.
[13] https://mangadex.org/.

Table 3. Panel detection.

	DCM	c100	eBD	M109	Pop	avg
G.Dino	63,4	62,5	56,9	61,8	73,7	64,7
R-CNN	86,3	**88,9**	65,4	64,9	77,6	**79,5**
SSD	12,1	9,1	28,4	34,6	4,6	15,6
YOLO	81,4	75,0	**67,0**	76,8	64,5	74,3
DASS	–	–	–	–	–	–
Magi	**89,0**	73,9	62,1	65,3	**92,8**	78,5

Table 4. Characters detection.

	DCM	c100	eBD	M109	Pop	avg
G.Dino	57,7	62,1	40,1	25,3	46,0	48,2
R-CNN	50,6	61,0	34,7	4,7	50,9	42,2
SSD	52,4	54,1	39,5	55,8	32,6	49,3
YOLO	45,6	55,4	30,1	9,4	42,0	38,6
DASS	**75,1**	76,0	**60,9**	**84,4**	70,5	**76,3**
Magi	71,8	**76,7**	56,6	50,4	**79,7**	69,3

Table 5. Face detection.

	DCM	c100	eBD	M109	Pop	avg
G.Dino	66,5	58,9	37,3	38,1	62,0	55,4
R-CNN	43,0	38,9	20,7	8,7	43,0	32,7
SSD	60,1	60,0	30,9	76,4	75,4	66,5
YOLO	43,1	48,8	20,6	16,2	42,1	37,5
DASS	**78,8**	**62,7**	**61,1**	**87,8**	**78,0**	**75,3**
Magi	–	–	–	–	–	–

Table 6. Text detection.

	DCM	c100	eBD	M109	Pop	avg
G.Dino	20,7	23,0	17,8	9,9	27,6	20,1
R-CNN	64,2	**83,1**	41,9	14,4	48,5	54,0
SSD	58,5	70,2	38,5	70,8	31,7	59,1
YOLO	68,3	73,0	38,7	42,2	12,7	50,9
DASS	–	–	–	–	–	–
Magi	**84,0**	77,9	**73,6**	49,2	**93,4**	**75,2**

In these tables, we present several significant values for mean average precision. Notably, the most revealing insights arise from the analysis of failures. The SSD model demonstrates the ability to detect characters, faces, and text with around 60% average precision (AP.5), yet it significantly underperforms in identifying panels across various datasets. This issue does not stem from an excess of bounding boxes, as indicated by the low recall rates (recall and other metrics are reported in the repository). It is more likely related to the specific characteristics of the datasets employed for pertaining. For instance, the comic datasets (referred to in the tables as c100) are generated through automatic annotation from multiple sources. Panels are identified using a Yolov8 model fine-tuned on 500 pages solely annotated for panels as detailed by [12]. Character and face detections are enhanced by the DASS-dcm, which is finetuned on DCM books. Textboxes are identified using text extracted by Amazon Textract as reported in [29], where lines within the same balloon are combined. Therefore, we have already established various upper-bounds for different classes.

Table 7. Averaged mAP across all classes.

	DCM	c100	eBD	M109	Pop	avg
G.Dino	34,7	36,9	25,3	27,0	36,8	33,7
R-CNN	40,7	**58,2**	27,1	18,5	44,2	41,1
SSD	30,5	33,3	22,9	**47,5**	17,2	32,6
YOLO	39,7	50,8	26,1	28,9	29,8	38,1
DASS	25,7	19,0	20,3	34,4	17,6	23,9
Magi	**40,8**	57,1	**32,1**	33,0	**66,5**	**49,2**

Table 8. Averaged mAP on detected classes.

	DCM	c100	eBD	M109	Pop	avg
G.Dino	52,1	51,1	38,0	33,8	52,0	46,9
R-CNN	61,0	66,0	40,7	23,1	59,0	52,3
SSD	45,8	44,5	34,3	59,4	23,0	43,7
YOLO	59,6	63,0	39,1	36,1	54,1	53,2
DASS	77,0	67,6	61,0	**86,1**	73,3	**75,1**
Magi	**81,6**	**75,6**	**64,1**	54,9	**88,0**	74,0

Additional unexpected failures include Faster R-CNN's performance on Japanese Manga109 images and Yolov8 (m109 version) in detecting characters, text, and panels outside its domain. With Faster R-CNN, the model performs moderately for panels but is less effective with other classes, detecting only sporadic texts and faces. Conversely, Yolov8, specifically tuned for Manga109, shows average performance on manga-style comics within its domain but falls short on other comic styles.

5 Conclusion

In conclusion, our paper introduces the *Comics Datasets Framework*, designed to facilitate the management of datasets with restricted access through conversion scripts that transform custom annotations and folder structures into a Unified Comics Annotation. This framework not only corrects and extends annotations for existing datasets but also introduces new annotations for comic styles that are less represented compared to manga styles. Additionally, a primary contribution of our work is establishing comparable and replicable experiments through common setting baselines and making models available. We have constructed the largest test set of various comic styles, namely *Comics Datasets*, by enhancing and integrating existing annotations, and we provide a standardized evaluation setting to enable fair comparison of both existing and new methods. We demonstrate how to effectively benchmark both existing and new comic detection models, providing access to the model weights and code to ensure reproducibility. We believe this initiative is crucial for clarifying the comics research landscape and enabling consistent reproducibility and benchmarking.

References

1. Agrawal, H., Mishra, A., Gupta, M., Mausam: multimodal persona based generation of comic dialogs. In: Rogers, A., Boyd-Graber, J., Okazaki, N. (eds.) Proceedings of the 61st Annual Meeting of the Association for Computational Linguistics (Volume 1: Long Papers), pp. 14150–14164. Association for Computational Linguistics. https://doi.org/10.18653/v1/2023.acl-long.791, https://aclanthology.org/2023.acl-long.791
2. Baek, J., Matsui, Y., Aizawa, K.: COO/ Comic Onomatopoeia Dataset for Recognizing Arbitrary or Truncated Texts. arXiv. https://doi.org/10.48550/arXiv.2207.04675, http://arxiv.org/abs/2207.04675
3. Dubray, D., Laubrock, J.: Deep CNN-based Speech Balloon Detection and Segmentation for Comic Books. arXiv, http://arxiv.org/abs/1902.08137
4. Dunst, A., Hartel, R., Laubrock, J.: The graphic narrative corpus (GNC): design, annotation, and analysis for the digital humanities. In: 2017 14th IAPR International Conference on Document Analysis and Recognition (ICDAR). vol. 03, pp. 15–20. https://doi.org/10.1109/ICDAR.2017.286
5. Dutta, A., Biswas, S., Das, A.K.: BCBId: First Bangla comic dataset and its applications **25**(4), 265–279. https://doi.org/10.1007/s10032-022-00412-9

6. Dutta, A., Biswas, S., Das, A.K.: CNN-based segmentation of speech balloons and narrative text boxes from comic book page images **24**(1-2), 49–62. https://doi.org/10.1007/s10032-021-00366-4, https://link.springer.com/10.1007/s10032-021-00366-4
7. family=Ho, given=HN, g.i., Rigaud, C., Burie, J.C., Ogier, J.M.: Redundant structure detection in attributed adjacency graphs for character detection in comics books. https://www.semanticscholar.org/paper/Redundant-structure-detection-in-attributed-graphs-Verview/172855949a669fd6d5f8049f1a8ec6f7614cbccd
8. Fujimoto, A., Ogawa, T., Yamamoto, K., Matsui, Y., Yamasaki, T., Aizawa, K.: Manga109 dataset and creation of metadata. In: Proceedings of the 1st International Workshop on coMics ANalysis, Processing and Understanding (Manpu), pp. 1–5. https://doi.org/10.1145/3011549.3011551, https://dl.acm.org/doi/10.1145/3011549.3011551
9. Guérin, C., et al.: eBDtheque: a representative database of comics. In: 2013 12th International Conference on Document Analysis and Recognition, pp. 1145–1149. https://doi.org/10.1109/ICDAR.2013.232, https://www.scopus.com/inward/record.uri?eid=2-s2.0-84889595169&doi=10.1109%2fICDAR.2013.232&partnerID=40&md5=38719ad5d5966b1a143ef2020aa8ee73
10. He, Z., Zhou, Y., Wang, Y., Tang, Z.: SReN: shape regression network for comic storyboard extraction. In: Proceedings of the AAAI Conference on Artificial Intelligence. vol. 31. https://doi.org/10.1609/aaai.v31i1.11074, https://ojs.aaai.org/index.php/AAAI/article/view/11074
11. He, Z., Zhou, Y., Wang, Y., Wang, S., Lu, X., Tang, Z., Cai, L.: An end-to-end quadrilateral regression network for comic panel extraction. In: Proceedings of the 26th ACM International Conference on Multimedia, pp. 887–895. MM '18, Association for Computing Machinery. https://doi.org/10.1145/3240508.3240555
12. Iyyer, M., et al.: The amazing mysteries of the gutter: drawing inferences between panels in comic book narratives. In: Proceedings - IEEE Conference on Computer Vision and Pattern Recognition, CVPR. vol. 2017-January, pp. 6478–6487. arXiv. https://doi.org/10.1109/CVPR.2017.686, https://www.scopus.com/inward/record.uri?eid=2-s2.0-85044502769&doi=10.1109%2fCVPR.2017.686&partnerID=40&md5=53fdaad2abea6dc7342a1e0639ddb132
13. Khan, F.S., Anwer, R.M., van de Weijer, J., Bagdanov, A.D., Vanrell, M., Lopez, A.M.: Color attributes for object detection. In: 2012 IEEE Conference on Computer Vision and Pattern Recognition, pp. 3306–3313 (2012). https://doi.org/10.1109/CVPR.2012.6248068
14. Le, T.N., et al.: Subgraph spotting in graph representations of comic book images **112**, 118–124. https://doi.org/10.1016/j.patrec.2018.06.017, https://www.sciencedirect.com/science/article/pii/S0167865518302629
15. Li, Y., Aizawa, K., Matsui, Y.: Manga109Dialog A Large-scale Dialogue Dataset for Comics Speaker Detection. arXiv. https://doi.org/10.48550/arXiv.2306.17469, http://arxiv.org/abs/2306.17469
16. Liu, S., et al.: Grounding DINO: Marrying DINO with Grounded Pre-Training for Open-Set Object Detection, http://arxiv.org/abs/2303.05499
17. Nguyen, N.V., Rigaud, C., Burie, J.C.: Comic characters detection using deep learning. In: 2017 14th IAPR International Conference on Document Analysis and Recognition (ICDAR). vol. 3, pp. 41–46. IEEE Computer Society. https://doi.org/10.1109/ICDAR.2017.290, http://ieeexplore.ieee.org/document/8270235/
18. Nguyen, N.V., Rigaud, C., Burie, J.C.: Digital Comics Image Indexing Based on Deep Learning **4**(7), 89. https://doi.org/10.3390/jimaging4070089, https://www.mdpi.com/2313-433X/4/7/89

19. Nguyen, N.V., Rigaud, C., Revel, A., Burie, J.C.: A learning approach with incomplete pixel-level labels for deep neural networks **130**, 111–125. https://doi.org/10.1016/j.neunet.2020.06.025, https://linkinghub.elsevier.com/retrieve/pii/S0893608020302409
20. Nguyen, N.-V., Vu, X.-S., Rigaud, C., Jiang, L., Burie, J.-C.: ICDAR 2021 competition on multimodal emotion recognition on comics scenes. In: Lladós, J., Lopresti, D., Uchida, S. (eds.) ICDAR 2021. LNCS, vol. 12824, pp. 767–782. Springer, Cham (2021). https://doi.org/10.1007/978-3-030-86337-1_51
21. Ogawa, T., Otsubo, A., Narita, R., Matsui, Y., Yamasaki, T., Aizawa, K.: Object Detection for Comics using Manga109 Annotations. arXiv.https://doi.org/10.48550/arXiv.1803.08670, http://arxiv.org/abs/1803.08670
22. Padilla, R., Passos, W.L., Dias, T.L.B., Netto, S.L., family=Silva, given=Eduardo A. B., p.u.: A Comparative Analysis of Object Detection Metrics with a Companion Open-Source Toolkit **10**(3), 279. https://doi.org/10.3390/electronics10030279, https://www.mdpi.com/2079-9292/10/3/279
23. Qin, X., Zhou, Y., He, Z., Wang, Y., Tang, Z.: A Faster R-CNN Based Method for Comic Characters Face Detection. In: 2017 14th IAPR International Conference on Document Analysis and Recognition (ICDAR). vol. 01, pp. 1074–1080. https://doi.org/10.1109/ICDAR.2017.178
24. Rigaud, C.: Segmentation and indexation of complex objects in comic book **14**(3). https://doi.org/10.5565/rev/elcvia.833, http://elcvia.cvc.uab.es/article/view/v14-n3-rigaud
25. Sachdeva, R., Zisserman, A.: The Manga Whisperer: Automatically Generating Transcriptions for Comics, http://arxiv.org/abs/2401.10224
26. Shit, S., et al.: Relationformer: A Unified Framework for Image-to-Graph Generation, http://arxiv.org/abs/2203.10202
27. Sun, W., Burie, J.C., Ogier, J.M., Kise, K.: Specific comic character detection using local feature matching. In: 2013 12th International Conference on Document Analysis and Recognition. pp. 275–279. https://doi.org/10.1109/ICDAR.2013.62, http://ieeexplore.ieee.org/document/6628627/
28. Topal, B.B., Yuret, D., Sezgin, T.M.: DASS-Detector: Domain-Adaptive Self-Supervised Pre-Training for Face & Body Detection in Drawings, http://arxiv.org/abs/2211.10641
29. Vivoli, E., Baeza, J.L., Llobet, E.V., Karatzas, D.: Multimodal Transformer for Comics Text-Cloze, http://arxiv.org/abs/2403.03719

A Comprehensive Gold Standard and Benchmark for Comics Text Detection and Recognition

Gürkan Soykan[1,2]( ), Deniz Yuret[1,2], and Tevfik Metin Sezgin[1,2]

[1] Computer Engineering Department, Koç University, Istanbul, Turkey
{gsoykan20,dyuret,mtsezgin}@ku.edu.tr
[2] KUIS AI Center, Istanbul, Turkey
https://ai.ku.edu.tr/

Abstract. This study focuses on improving the optical character recognition (OCR) data for panels in COMICS [18], the largest dataset containing text and images from comic books. To do this, we developed a pipeline for OCR processing and labeling of comic books and created the first text detection and recognition datasets for Western comics, called "COMICS Text+: Detection" and "COMICS Text+: Recognition". We evaluated the performance of fine-tuned state-of-the-art text detection and recognition models on these datasets and found significant improvement in word accuracy and normalized edit distance compared to the text in COMICS. We also created a new dataset called "COMICS Text+", which contains the extracted text from the textboxes in COMICS. Using the improved text data of COMICS Text+ in the comics processing model from *COMICS* resulted in state-of-the-art performance on cloze-style tasks without changing the model architecture. The *COMICS Text+* can be a valuable resource for researchers working on tasks including text detection, recognition, and high-level processing of comics, such as narrative understanding, character relations, and story generation. All data, models, and instructions can be accessed online (https://github.com/gsoykan/comics_text_plus).

Keywords: Optical Character Recognition (OCR) · Text Detection · Text Recognition · Comic Processing

1 Introduction

Comics are a multimodal structure and medium that use writing and images in a spectrum, from just images to purely writing, to convey a story or an idea. Comics are at the intersection of education [17], sociocultural studies [4], linguistics and cognitive sciences [2,23]. Understanding comics can help advance progress in these fields. Additionally, methods developed for processing and understanding comics can be applied to other forms of *Visual Language* [5], including but not limited to cave paintings, mangas, graphic novels.

© The Author(s), under exclusive license to Springer Nature Switzerland AG 2024
H. Mouchère and A. Zhu (Eds.): ICDAR 2024 Workshops, LNCS 14935, pp. 168–197, 2024.
https://doi.org/10.1007/978-3-031-70645-5_12

The unique structure of comics can be analyzed at the low-level and high-level. At the low-level, natural language processing techniques can extract text from dialogues, narratives, and onomatopoeias. Computer vision methods can detect and segment motion lines, characters, and the location of objects and components [14,31,32,34]. At the high-level, the relationships between characters, storytelling, narrative understanding, inter-panel and intra-panel events must be studied. While the high-level features of natural images and scenes are widely studied [28], the same cannot be said for comics. This is due to the great variety of comics across time, location, and different styles and the lack of annotated data (see Table 1 for an overview of datasets). The limited availability of datasets in this domain can also be attributed to copyright and access rights challenges.

The low amount of annotated data creates a bottleneck for studies on comics using deep learning methods. Although there have been works on panel and speech bubble detection and segmentation, such as in [14,31,32,34], and onomatopoeia detection and recognition, such as in [3], the only extensive dataset with OCR-extracted comics texts is called COMICS [18]. However, when the text quality of COMICS is measured over a sample, it can be seen as unreliable (see Subsect. 3.1 and Table 6). There is a limited amount of study on the OCR of dialogue and narrative texts in comics to produce high-quality text data. Hartel and Dunst [14], presented such a pipeline, but the data and models could not be shared due to copyright issues.

This study presents a pipeline for OCR of comics, allowing us to extract text from speech bubbles in panels. This pipeline can also be used for general OCR purposes. We use COMICS [18] to curate a reliable and comprehensive dataset of American golden age comics. These comics do not show variety in style, allowing our work to be directly applied to previous studies in the literature. We inspect the reported and unreported problems in COMICS and develop separate solutions for text detection and recognition using the MMOCR toolbox [22]. To improve the performance of text detection and recognition, we follow a loop of annotating panels and speech bubbles, training models and evaluating them using ground truth data. Once we have achieved stable performance with the trained models, we select the best end-to-end text detection and recognition model pair to process COMICS and extract OCR texts. Finally, we apply post-processing techniques to correct systematic errors in the OCR data and create the *COMICS TEXT+* dataset. Our contributions can be summarized as follows:

- We release a substantially improved version of COMICS OCR results, called *COMICS Text+*, which significantly outperforms its predecessor in terms of text quality. The word accuracy, including symbols, has been increased from 13% to 40%. It is the best quality and most comprehensive publicly available dataset, with over two million transcriptions of text boxes.
- We release text detection and text recognition datasets created from Golden Age comics called respectively, *COMICS Text+: Detection* and *COMICS Text+: Recognition* along with ground truth (GT) to validate end-to-end OCR pipelines. *COMICS Text+: Detection* contains more than 20,000

annotations from 1112 images. Whereas, *COMICS Text+: Recognition* contains more than 17,000 annotations from 1006 images.
- As a benchmark, we train and evaluate the performance of 14 text detection and 10 text recognition models on COMICS Text+ datasets.
- We present text detection and recognition models, based on FCENet [46] and MASTER [30], respectively, that have been trained on the *COMICS Text+: Detection & Recognition* datasets. There are no other publicly available models and datasets for OCR of Western comics, so we hope this work will serve as a baseline for future studies and lower the barrier to entry for those interested in this domain.
- We improved an annotation tool forked from LabelMe [39] by integrating text detection and recognition models to aid and speed up the annotation process. Our tool converts annotations into datasets for training OCR models. We hope this tool will facilitate the creation of high-quality datasets for OCR research of comics and advance the SOTA in this field.
- The comics processing backbone for cloze-style tasks proposed in the [18] is reproduced, and the reproduced model is trained using both COMICS and *COMICS TEXT+* datasets. The test results of the two models are compared, and it is shown that the model trained with *COMICS TEXT+* outperforms the other model and achieves state-of-the-art results on most cloze-style tasks.

2 Related Works

2.1 OCR for Comics

In a preliminary study, [35], the challenges of speech text recognition for comics are outlined, and several approaches for addressing these challenges are implemented. The study concludes that typewritten-like text is better recognized by generic OCR systems, while specifically trained OCR models can perform better for skewed, uppercase, and cursive fonts. However, our study shows that specifically trained models can outperform generic OCR systems, demonstrating the progress that has been made in the field since the previous study.

In [10], pixel-level text annotations were curated for unconstrained text detection in the manga, and special evaluation metrics were developed to measure performance. In a more recent study, the Comic Onomatopoeia Dataset (COO) [3] presents a dataset of onomatopoeia (textual representations of sounds, states, or objects) for text detection, text recognition, and link prediction tasks. COO provides onomatopoeia in Japanese, pushing the limits of current irregular text detection and text recognition models. While our study focuses on providing an extensive dataset for comics, COO focuses more on the performance of models for their specific tasks. However, combining our research on both datasets can provide a more comprehensive understanding of the text modality of comics.

In [14], Hartel and Dunst constructed an OCR pipeline for comics that begins with segmenting comics page components, such as panels and speech bubbles, and then extracts text from these components. They also analyzed the textual properties of their OCR results in terms of genre affiliation and page length.

COMICS TEXT+   171

Table 1. Overview of datasets in the domain of the comic. *REQ* means available upon request.

Dataset	Release Dates	Type	Count	Annotation Types	Available
GNC	1975 - 2018	Comic Book	240	panel, balloon, caption, text, diegetic text, character, object	NO
COMICS	1938 - 1956	Comic Book	3948	panel, balloon, caption, text	YES
Manga109	1970 - 2010	Manga	109	panel, balloon, caption, text, onomatopoeia, character, face, body	REQ
DCM772	1938 - 1956	Comic Book	27	panel, balloon, caption, character, face, body, type of character	YES
COO	1970 - 2010	Manga	109	onomatopoeia, text	REQ
eBDtheque	1905 - 2012	Comic Book Page	100	panel, balloon, text	REQ
EmoRecCom	1938 - 1956	Panel	8258	multiple emotion labels for panels	YES

Our work differs from theirs because we use state-of-the-art deep learning-based text detection and text recognition models for OCR rather than the open-source software Calamari [42]. Additionally, we make our models, training datasets, and OCR results for more than 2 million textboxes publicly available, whereas their data cannot be shared due to copyright restrictions.

2.2 Datasets

Although there are not many datasets available for the comic domain due to copyright issues, a few datasets have been made publicly available. In Table 1, you can see a complete list of datasets focused on comic books and manga within our knowledge. The Graphic Narrative Corpus (GNC) [7] is a unique dataset to serve both computational sciences and social sciences. The dataset provides a variety of metadata, such as genre, author, and geographic origins, along with annotated data to the extent of eye gaze data. However, unfortunately, the dataset can not be accessed publicly. COMICS [18] is the dataset with the most comic books and is the dataset we use for our study. They also propose downstream tasks for the domain of the comic. These are text cloze, visual cloze, and character coherence. The Manga109 [1] dataset contains 109 volumes of manga by 93 different authors. These volumes were released from the 1970s to the 2010s and fall into 12 different genres, such as fantasy, humor, and sports. DCM772 [31] is one of the few publicly available datasets because of how it was collected. The comic books in the dataset come from a free public domain collection called the Digital Comic Museum[1]. This dataset annotates different character types, such as animal-like or human-like. Comic Onomatopoeia (COO) [3] is a dataset that extends the Manga109 dataset with onomatopoeia annotations. Because COO focuses on onomatopoeia, it provides a challenging text detection and recognition dataset with various shapes, styles,

---

[1] http://digitalcomicmuseum.com.

and extreme curves. They also introduce a novel task of linking truncated texts. The eBDtheque dataset [12] is a pioneering dataset for comic books, including American comics, mangas, and Franco-Belgian comic book albums. It provides panels, balloons, text lines, and pages. The Multimodal Emotion Recognition on Comics scenes (EmoRecCom) [33] dataset is a small subset of COMICS with labels for eight different emotions for each scene. Our COMICS Text+ dataset may be able to improve the text quality of this dataset, as its panels are sourced from COMICS.

3 Methodology

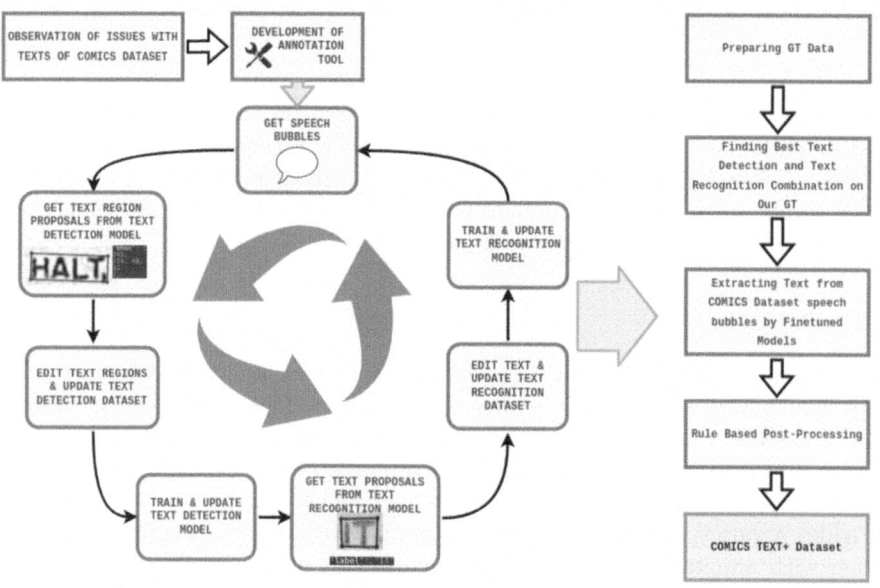

Fig. 1. The complete process and pipeline of our approach and dataset.

Our approach to improving the text data in COMICS [18] consists of three main stages, as shown by the green arrows in Fig. 1. In the first stage, we identified the shortcomings and issues with the text data in COMICS and developed an annotation tool by forking the LabelMe tool [39]. Our motivation for doing this is that the previous works that proposed COMICS did not provide any metrics [18,19].

In the second stage, we used the panels and speech bubbles cropped out of the pages of comics in COMICS to initiate the machine learning lifecycle. During this stage, we annotated the data and trained out-of-the-box models successively until the model performance was stable. At the end of this stage, we created text detection and text recognition datasets using more than 1000 speech bubbles (see Table 2).

In the third stage, we used our text detection and text recognition datasets to train text detection and text recognition models provided in the MMOCR framework [22]. We then benchmarked the performance of these models on the dataset. We selected the top-performing model combination for detection and recognition and used it to measure the peak performance on a carefully selected set of speech bubbles, referred to as GT. Finally, we used the selected model combination to run over the entire set of speech bubbles in COMICS and post-processed the resulting data to create *COMICS Text+*.

3.1 Textual Flaws of COMICS

It is stated in [18,19] that the OCR data of COMICS has detection and recognition flaws which were attempted to be mitigated through post-processing. Those flaws are as follows: detection of short words, detection of punctuation marks, and recognition of the first letter of the word. In addition to those cases, we realized more issues with the dataset. Such as: skipping lines, total misrecognition of words, and difficulty recognizing certain fonts.

The quantitative evaluation results of COMICS OCR compared to GT are shown in Table 6, where it can be seen that the character recall of COMICS is lower than our final results, supporting our qualitative analysis presented in Appendix A.

3.2 Creation of New Comics Text Detection & Recognition Datasets

To improve the text data in COMICS, we developed an annotation tool based on the LabelMe tool [39]. In addition to the features of the original tool, we added two new modes, "Detect Text" and "Detect and Recognize Text", which use inference from text detection and text recognition models to speed up the annotation process. As we continued to fine-tune our models with more data, this functionality helped to reduce the time required for annotation. For detailed information, see Appendix B.

This tool is essential to our approach, as it allows us to create new text detection and recognition datasets and continuously train text detection and recognition models. The cyclic process of using the tool is as follows:

1. Randomly crop speech bubbles or panels, selecting at least a couple of images from each comic book in COMICS.
2. Use a pretrained model to generate text region proposals for the selected images and annotate them for text detection.
3. Edit the proposed text regions as needed.
4. Train a text detection model (we used the DBNet++ model [27]) and update it for use in the text recognition pipeline.

5. Use the updated text detection model along with the pretrained text recognition model to generate text region and text proposals for the selected images and annotate them for text recognition.
6. Edit the proposed text regions and texts as needed.
7. Train a text recognition model (we used the NRTR model [36]) and update it for use in the text recognition pipeline.
8. Repeat the process for additional images to continue improving the performance of the models.

After each training cycle, we evaluate our models' performance using the corresponding test sets and GT. This allows us to monitor the accuracy of our annotations and make any necessary adjustments to our annotation policy. By regularly assessing the performance of our models, we can ensure that our annotation process is effective and that our models are learning from the data as intended.

COMICS Text+: Detection and Recognition Datasets. The *COMICS Text+: Detection* and *COMICS Text+: Recognition* datasets were created using our annotation tool and pipeline. The text detection dataset consists of annotated text regions in speech bubbles or panels. Because of the nature of the domain, these text regions are mostly individual words. However, defining the boundaries of text regions can be challenging in this domain. Our annotation policy was refined through several iterations. Initially, we annotated text regions with large margins and included punctuation marks within a text region that contained a word. This led to models trained on this data producing erroneous predictions of text regions that were often intersecting. Our final policy for text detection annotations involves using a low margin for words and excluding punctuation marks if they are not close enough to the words (i.e., their distance from the words must be no more than twice the letter spacing). Another factor that complicates annotations is comic books' varying spacing between words and hyphens. We follow the same strategy with hyphens as we do with punctuation marks.

For the text recognition dataset, we annotate text regions with their corresponding text. In this case, the focus is on recognizing the text within the text regions, so we carefully select regions with clear text. The challenges of text detection do not apply in this context. Another difference between the two datasets is that we sometimes skip some text regions within an image because of their irregular text shapes or poor quality (e.g., if the text is obscured or worn). Since we cannot skip these text regions in the text detection dataset, the text detection and text recognition datasets may have different images and annotations. The final statistics of our text detection and text recognition datasets are shown in Table 2.

COMICS TEXT+   175

Table 2. Statistics describing splits of dataset sizes constituting *COMICS Text+: Detection & Recognition* datasets. The dataset is used to train and evaluate text detection and recognition models. In our experiments, the training set size can be different, but the test split is always used as it is. *Images* stands for the number of images whereas *Annot.* denotes the number of annotations.

	Training		Validation		Test	
	Images	Annot.	Images	Annot.	Images	Annot.
Text Detection	1012	18494	50	790	50	986
Text Recognition	906	15635	50	759	50	714

3.3 Creating Ground Truth

To evaluate the combined performance of fine-tuned text detection and text recognition models, GT is needed. Since no such data was available for Golden Age comics, we created and organized our own GT. This data consists of tuples of speech bubbles or narrative boxes and their corresponding text. Unlike text detection and recognition datasets, the GT does not include labels for individual text regions; instead, it covers all the text in an image resulting in pairs of images and full-transcriptions. We carefully selected the speech bubbles for the GT to ensure high-quality annotations. This GT allows us to thoroughly evaluate the performance of our models on COMICS Text+.

Our policy for selecting speech bubbles for the GT is to ensure that all the text in the image is clearly visible and readable. The selected images come from different comic books and have various artistic styles. In some cases, there may be residual characters on the margins of the speech bubbles, but this is acceptable as long as the characters are not fully visible. This allows us to see how the text detector performs in the presence of residual text and ensures that it does not incorrectly label these as text regions. However, we do not include examples with half-words or full characters in the GT data because this would introduce irregular text and speech bubble detection problems that would skew the evaluation of our models' performance.

To create our GT data, we randomly cropped multiple images from different comic series using the coordinates of the text boxes provided in COMICS. We then filtered this data to refine our GT. We did not include images that had one or more of the following features: partial words or texts, parts of advertisement pages no text, multiple panels, irregular or specialized fonts.

By applying these filters, we were able to create a GT dataset with high-quality annotations that accurately reflect the text in the images. This allows us to accurately evaluate the performance of our text detection and recognition models. For visual samples, see Appendix C.

3.4 Benchmarking the Detection and Recognition Models

We trained and benchmarked fourteen text detection, and ten text recognition models using the MMOCR toolkit [22] on our *COMICS Text+: Detection & Recognition* datasets. Benchmarking results are shown in Table 3 and Table 4. After finding the top-performing models on the test split, we evaluated the combination of the top three models on the GT data to determine which combination of models gave us the best results. This allows us to identify the most effective models for detecting and recognizing text in Golden Age comic images. For complete benchmarking results, see Appendix D.

Finding Best Text Detection and Text Recognition Models for OCR. Our text detection and text recognition benchmarks showed that we have close-performing models after training on our text detection and text recognition datasets. Even though their results are close to each other in metrics, we realized their failure points differ. Because of this reason, we decided to measure the OCR performance of the top three of their combinations. Even though we chose DBNET++ in our annotation-training-evaluation cycle due to qualitative evaluation, after this process, we saw that the FCENET text detection backbone works best by a high margin for the OCR metrics in the domain of the comics (see Table 5). From our results, we can clearly say that for the current OCR of comics, the deciding factor is the text detection model used. The OCR performance does not vary by high values when we keep the same text detection model and change the text recognition model. However, the OCR results change dramatically when keeping the text recognition model the same and changing the text detection model. In addition, Appendix E can be seen for a detailed analysis of the effect of dataset size.

Table 3. The top-5 text detection model results trained on our dataset sorted by H-Mean. Models are as follows: **DBPP_r50**: DBNet++ [27] with ResNet-50 [16] backbone pretrained on ICDAR2015. **FCE_CTW_DCNv2**: FCENet [46] with ResNet-50 and deformable convolutional networks (DCN) [6] backbone pretrained on CTW-1500. **MaskRCNN_IC17** : Mask R-CNN [15] pretrained on ICDAR2017 [11]. **PS_IC15**: Progressive Scale Expansion Network (PSENet) [40] pretrained on ICDAR2015. **MaskRCNN_CTW**: Mask R-CNN pretrained on CTW-1500.

Model	Recall	Precision	H-Mean
DBPP_r50	93.3	97.1	95.2
FCE_CTW_DCNv2	92.9	96.7	94.8
MaskRCNN_IC17	92.5	96.8	94.6
PS_IC15	93.1	96.0	94.5
MaskRCNN_CTW	93.8	94.4	94.1

Table 4. Top-5 text recognition model results trained on our dataset, sorted by 1-N.E.D.. Model descriptions: **MASTER** [30]: Pretrained on Synth90k, SynthText, and SynthAdd [25]. **NRTR_1/8–1/4** and **NRTR_1/16–1/8**: NRTR [36] variants with different backbone feature heights, pretrained on Synth90k and SynthText. **RobustScanner** [43] and **SAR** [25]: Pretrained on multiple ICDAR datasets and synthetic datasets. Metrics are *Recall*: Character Recall, *Precision*: Character Precision, *Accuracy*: Word Accuracy, *W.Acc.Ig.*: Word Accuracy Ignoring Symbols, *NED*: 1-Normalized Edit Distance.

Model	Recall	Precision	Accuracy	W.Acc.Ig.	NED
MASTER	99.6	99.5	95.9	98.3	99.2
NRTR_1/8–1/4	99.5	99.4	95.8	98.0	99.2
NRTR_1/16–1/8	99.4	99.2	95.1	97.4	99.2
RobustScanner	99.2	99.2	94.3	97.1	98.6
SAR	99.3	99.3	94.5	97.1	98.3

Table 5. End-to-end, from panel to text, the results of the top three text detection and text recognition models trained on our dataset are measured against the ground truth data. The detection models are listed in the leftmost column, and the recognition models are listed in the top row. The metrics used are as follows: *W.Acc.Ig.*: Word Accuracy Ignoring Symbols, *NED*: Normalized Edit Distance (1-N.E.D.).

	NRTR 1/16–1/8		NRTR 1/8–1/4		MASTER	
	W.Acc.Ig	NED	W.Acc.Ig	NED	W.Acc.Ig	NED
MaskRCNN IC17	7.6	75.6	7.4	75.7	7.6	76.3
FCE CTW_DCNv2	71.4	97.6	70.4	97.7	**72.2**	**97.8**
DBPP_r50	51.6	94.8	59.3	96.2	60.5	96.31

3.5 Creating *COMICS Text+* Dataset

After determining and training the best models for OCR, we applied them to the images cropped from panels using the coordinates of the textboxes provided in COMICS [18]. However, the number of final textboxes with texts in our COMICS Text+ is different from COMICS for several reasons. Firstly, we filtered out advertisement panels and their textboxes, as well as textboxes that did not have matched texts, from our dataset. Secondly, some of our inferences resulted in erroneous or empty texts, which were also removed in order to improve the overall quality of the dataset. After this step, we were left with 2,208,006 textboxes (86.7% of COMICS).

As a final preprocessing step, we filtered out textboxes that overlapped. In COMICS, some textboxes had highly intersecting textboxes, which can lower the quality of the dataset. We removed the smaller textbox in cases where they had more than 0.8 Intersection over Union (IoU) within the same panel. This step reduced the total number of textboxes to 2,188,853, with 1,702,140 dialogue and 486,713 narratives. In comparison, COMICS has 2,545,728 textboxes. We believe that this preprocessing and filtering improves the overall quality of the dataset.

Postprocessing on Raw OCR Output. We improved the process in [19] for rule-based OCR post-processing by adding extra steps. Our approach targets mistakes in recognizing the first letter of a word and errors in single characters, which can be caused by OCR systems or textbox detection errors. We tokenized the OCR output, created a dictionary of frequently occurring words, and used that dictionary to replace misspelled words. Our algorithm also checks for errors in the starting and ending letters of words, and uses NLTK's[2] default corpus and the Brown Corpus [9] to check for errors in context. This is because the Golden Age of comics and the publication date of the Brown Corpus are closer in time than other corpora, which date from the 1950s and 1961. We also detect errors in single characters by checking for punctuation marks and valid characters. Our dataset includes both the raw OCR output and the post-processed version, and 110,838 and 131,833 textboxes were affected by the first and second parts of the post-processing respectively. For more detailed information and the algorithm, see Appendix F.

4 Results and Discussion

4.1 Finalized Dataset Results

Table 6 compares the OCR results of COMICS [18] and our method, COMICS Text+. Our method outperforms COMICS on all evaluation metrics, particularly character recall and word accuracy. We attribute these results to using labeled data specific to the comic book domain and fine-tuning state-of-the-art text detection and recognition models. The initial performance of these models is poor on the comic book data due to the unique characteristics of comic book text, such as speech bubbles and varied fonts. This demonstrates the need for fine-tuning using domain-specific data to achieve accurate OCR results in the comic book domain. Although labeling data is time-consuming even a limited amount of data (around 400 images) can improve the performance of SOTA models significantly. This study highlights the importance of both qualitative and quantitative analysis in evaluating OCR datasets and the need for domain-specific data and fine-tuning models to achieve accurate results in the comic book domain. For detailed information and qualitative results, see Appendix G.

Error Analysis. The most common errors in COMICS TEXT+ dataset include:

- Difficulty in detecting and recognizing symbols, particularly commas and dots.
- Text detection errors due to uncommon textures and background colors.

---

[2] https://www.nltk.org/.

COMICS TEXT+ 179

Table 6. Final comparison of our dataset *COMICS Text+* (CT+) and COMICS [18] OCR results, measured on ground truth data. Our dataset outperforms COMICS OCR in all metrics. Our dataset increases word accuracy by approximately 210% and word accuracy by ignoring symbols by approximately 40% on the speech bubbles of COMICS. Metrics are as follows: *W.Acc.Ig.*: Word Accuracy Ignoring Symbols, *NED*: Normalized Edit Distance (1-N.E.D.).

	Recall	Precision	Accuracy	W.Acc.Ig.	NED
COMICS	92.9	97.2	13.0	51.6	93.0
CT+	**97.5**	**98.0**	**40.5**	**72.3**	**97.7**

– Poorly detected bounding boxes lead to OCR issues. Speech bubble and narrative box detection or segmentation should be solved before the OCR step. Repercussions of this issue can lead to totally omitted or partially detected words, word groups, or multiple partial textboxes.

To address the issues with bounding boxes, improved textbox detection or segmentation methods may be necessary. However, addressing the other problems will likely require more extensive text detection and recognition datasets for fine-tuning.

4.2 Results with Cloze Style Tasks

We have reproduced the models and experimental setup described in [18]. For most of the experiment cases, our reproduction results are similar to those reported by a low margin, see Appendix H for exact reproduction results. We think the overall difference in results can stem from the dataset generation since the dataset itself is not shared. Although a script for generating the dataset is shared, it does not include explicit instructions on setting the hyperparameters. This may contribute to the discrepancies between our reproduction results and those reported in the original study.

Iyyer et al. [18] propose three tasks (text cloze, visual cloze, and character coherence) to evaluate a model's ability to understand the closure of a comic book narrative based on information from the previous three panels. A model that performs well on these tasks is able to make inferences between panel transitions and estimate the features of the following panel. This can be seen as a measure of a model's ability to understand the overall narrative structure of a comic book.

The general setup for the tasks involves using the features of context panels $p_{i-3}, p_{i-2}, p_{i-1}$ as input for a model that predicts the features of panel p_i from a single modality. This setup can be enhanced by using an arbitrary number of context panels, but that is a topic for future study. The final panel's features are

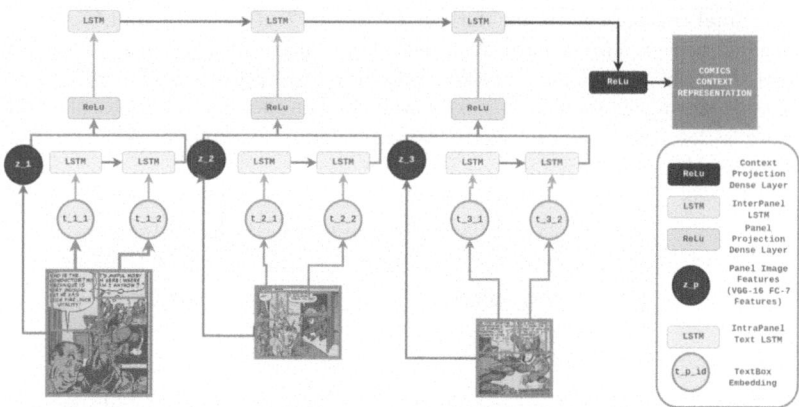

Fig. 2. Context representation extraction from the model's architecture using both image and text modality. It is called **image-text** architecture. Single modality architectures are named as **image-only** and **text-only**. Those models keep the same structure, disregarding the processing of other modalities.

predicted using a cloze-style framework. For all experiments, three candidates are presented, and the models use the extracted context information to score the correct candidate higher than the others. Figure 2 illustrates the context extraction process. Each task has two variations based on difficulty. In the *easy* case, the wrong candidates are selected from comic books that are different from the correct candidates. In contrast, the candidates of the *hard* case come from the same comic book.

Using our implementation, the experiments' text data is replaced with *COMICS TEXT+*. The test results of this training are shared in Table 7. Compared to our reproduction results using COMICS OCR data, an increase in accuracy can be observed for all text cloze task cases. We achieved state-of-the-art (SOTA) results using the same models compared to Iyyer et al.'s [18]. However, there was no or only a slight improvement in the visual cloze tasks. This indicates that when working with this architecture, the text modality has limited effectiveness on visual cloze.

While the improvement in the text-cloze task is notable, it is not groundbreaking, even if it is SOTA. This finding suggests that the current model architecture and experimental setup have reached their maximum potential. In that, it is necessary to develop new model architectures and experimental setups better suited to the domain to make further progress in the multimodal processing of comics. We hope to see such approaches soon.

Table 7. Our cloze-style task results with replacing text dataset with *COMICS TEXT+* dataset but keeping the same model architecture and training procedure. The results are represented using symbols: a star indicates SOTA performance, a single upward arrow indicates a slight improvement and double upward arrows indicate a significant improvement compared to our replication results. The top row illustrates the tasks and their levels of difficulty.

	Text Cloze		Visual Cloze		Char.
	Easy	Hard	Easy	Hard	
Text-Only	64.5 ⇈ ⋆	55.0 ⇈ ⋆	51.6 ↑	45.5 ↑	67.0
Image-Only	52.6 ⇈ ⋆	50.3 ⇈ ⋆	–	–	–
Image+Text	64.1 ⇈	61.4 ⇈ ⋆	83.5 ⋆	59.0 ⋆	68.0

5 Conclusion

We introduce the *COMICS TEXT+*, an advanced version of the OCR data derived from the panels of COMICS [18]. Unlike the original dataset, we utilized an open-source end-to-end OCR pipeline for generating text detection and text recognition data, known as *COMICS Text+: Detection* and *COMICS Text+: Recognition*. These datasets were then used to fine-tune state-of-the-art deep learning-based models for text detection and recognition, allowing text extraction from raw panels. Finally, we applied rule-based post-processing to the OCR results to further augment the data. The resulting dataset contains more than 2 million transcriptions of textboxes and has 1 - NED of 97.7, making it the most extensive and accurate dataset on this scale.

Overall, there is significant potential for further research in the computational studies of comics. By addressing the challenges of text detection, comics component processing, and unified OCR, we can enable the development of more effective computational methods for understanding the content of comics and develop high-level processing methods for tasks such as narrative understanding and story generation.

Acknowledgments. This project is supported by Koç University & İş Bank AI Center (KUIS AI). We would like to thank KUIS AI for their support.

Disclosure of Interests. The authors have no competing interests to declare that are relevant to the content of this article.

A Textual Flaws of COMICS

Examples of systematic errors within COMICS [18] can be seen in Table 8.

Table 8. Results of qualitative error analysis on COMICS [18] OCR data.

Speech Bubble	COMICS OCR	Ground Truth
	/ 7 ./ anoa < s 70 get you ready for 7 % e - vro // / yo	- - this is just to prove it ! and also to get you ready for the homicide squad , vironi !
	ok ,	click !
	is slaughter ,	this ghost army is slaughtering us !
	my oww wae , mamma , comantted	" my own wife , hannah , committed suicide ! i never understood that . . . "
	a sameersalaltano	in a flash , quick ! a somersault and
	that black id ride like of on to plummer out mule	i ' d like to ride plummer out of town on that black mule

B Annotation Tool

B.1 Detection Mode

In detection mode, a text detection model is used to infer the polygons of the text regions in an image. The user can then edit, add, or remove these polygons to create finely detailed annotations (see Fig. 3). This is the approach we took when

Fig. 3. Text detection mode of our annotation tool. Text regions, mostly words, can be estimated by a backbone text detection model and edited.

creating the *COMICS Text+: Detection*. By allowing for manual adjustment of the detected polygons, we were able to create a dataset with high-quality annotations that are well-suited for training text detection models.

B.2 Recognition Mode

Fig. 4. Text recognition mode of our annotation tool. Recognition results can be estimated by a backbone text recognition model and edited.

In recognition mode, a pipeline consisting of a text detection model and a text recognition model is used to infer the polygons of the text regions in an image and the corresponding text within those regions. The user can then edit, add, or remove these annotations to create finely detailed ones (see Fig. 4). This is the approach we took when creating the *COMICS Text+: Recognition*. Using a combination of text detection and recognition models, we were able to generate initial annotations that were then refined by manual editing, resulting in a dataset with high-quality annotations suitable for training text recognition models.

B.3 Conversion to Dataset Format

Once all the annotations are complete, we can export those in dataset formats with the conversions scripts in our pipeline, shared with this anon-repository. For the detection dataset, we use the *IcdarDataset* [21][3] format and for the

---
[3] https://mmocr.readthedocs.io/en/latest/tutorials/dataset_types.html#icdardataset.

recognition dataset, we use the $OCRDataset^4$ format. Using these standardized formats, we can easily integrate our datasets into existing pipelines for training and evaluating text detection and recognition models.

C Creating Ground Truth

Figure 5 shows some examples of our selected GT data and Fig. 6 shows some examples that were not included in GT.

Fig. 5. Five examples from selected speech bubbles for ground truth data.

Fig. 6. Five examples of speech bubbles that are not eligible for ground truth data.

D Benchmarking the Detection and Recognition Models

D.1 Detection Model Benchmark

We benchmarked fourteen recently published, high-performing pre-trained text detection models on our *COMICS Text+: Detection* dataset. The results are shown in Table 9. The models we used are as follows:

- DB_r18: DBNet [26] with ResNet-18 [16] backbone pretrained on ICDAR2015 [21] dataset
- DB_r50: DBNet with ResNet-50 backbone pretrained on ICDAR2015 dataset
- DBPP_r50: DBNet++ [27] with ResNet-50 [16] backbone pretrained on ICDAR2015 dataset

---
[4] https://mmocr.readthedocs.io/en/latest/tutorials/dataset_types.html#ocrdataset

- DRRG: Deep Relational Reasoning Graph Network [45] pretrained on CTW-1500 dataset [44]
- FCE_IC15: FCENet [46] with ResNet-50 backbone pretrained on ICDAR2015 dataset
- FCE_CTW_DCNv2: FCENet [46] with ResNet-50 and deformable convolutional networks (DCN) [6] backbone pretrained on CTW-1500 dataset
- MaskRCNN_IC17 : Mask R-CNN [15] pretrained on ICDAR2017 [11] dataset
- MaskRCNN_IC15: Mask R-CNN pretrained on ICDAR2015
- MaskRCNN_CTW: Mask R-CNN pretrained on CTW-1500 dataset
- PANet_CTW: Pixel Aggregation Network (PAN) [41] pretrained on CTW-1500 dataset
- PANet_IC15: PAN pretrained on ICDAR2015
- PS_IC15: Progressive Scale Expansion Network (PSENet) [40] pretrained on ICDAR2015
- PS_CTW: PSENet pretrained on CTW-1500 dataset
- TextSnake: TextSnake [29] pretrained on CTW-1500 dataset

Training Details. During training, we used a single NVIDIA V100 GPU. The batch sizes for each training were varied to maximize the use of the GPU. We used the Adam optimizer with a learning rate of 0.0001 and applied gradient clipping with a max norm value of 0.5. We also used a step learning rate scheduler at the third and fourth epochs, reducing the learning rate by 0.1. The models were trained for six epochs, and the best-performing model in each case was selected for evaluation on the test set based on the H-Mean. All experiments were configured using the MMOCR [22] training scripts.

D.2 Recognition Model Benchmark

We benchmark recently published and performant fourteen pre-trained text recognition models on our *COMICS Text+: Recognition* dataset. The results are shown in Table 10. The models we used are as follows:

- ABINet: ABINet [8] pretrained on Synth90k [20] and SynthText [13]
- CRNN: Convolutional Recurrent Neural Network (CRNN) [37] pretrained on Synth90k
- MASTER: MASTER [30] pretrained on Synth90k, SynthText, and SynthAdd [25]
- NRTR_1/8--1/4: NRTR [36] with the height of feature from the backbone is 1/16 of the input image, where 1/8 for width and pretrained on Synth90k and SynthText
- NRTR_1/16--1/8: NRTR with the height of feature from the backbone is 1/8 of the input image, where 1/4 for width and pretrained on Synth90k and SynthText

Table 9. Full list of text detection model results trained on our dataset sorted by H-Mean.

Model	Recall	Precision	H-Mean
DBPP_r50	93.3	97.1	95.2
FCE_CTW_DCNv2	92.9	96.7	94.8
MaskRCNN_IC17	92.5	96.8	94.6
PS_IC15	93.1	96.0	94.5
MaskRCNN_CTW	93.8	94.4	94.1
MaskRCNN_IC15	92.6	94.4	93.5
DB_r50	91.1	94.5	92.8
PS_CTW	91.6	93.9	92.7
DB_r18	89.7	95.5	92.5
PANet_IC15	90.3	93.9	92.0
TextSnake	85.9	94.8	90.1
FCE_IC15	90.2	89.3	89.8
DRRG	85.1	94.7	89.6
PANet_CTW	84.2	92.6	88.2

- `RobustScanner`: RobustScanner [43] pretrained on multiple ICDAR datasets, Synth90k, SynthText, SynthAdd and more[5]
- `SAR`: Show-Attend-and-Read (SAR) [25] pretrained on multiple ICDAR datasets, Synth90k, SynthText, SynthAdd and more[6]
- `SATRN`: Self-Attention Text Recognition Network (SATRN) [24] pretrained on Synth90k and SynthText
- `SATRN_sm`: SATRN with a reduced number of encoder and decoder layers pretrained on Synth90k and SynthText
- `CRNN-TPS`: CRNN-STN [38] pretrained on Synth90k

Training Details. During training, we used a single NVIDIA V100 GPU. The batch sizes for each training were varied to maximize the use of the GPU. We used the Adam optimizer with a learning rate of 0.0001, and applied gradient clipping with a max norm value of 0.5. We also used a step learning rate scheduler at the third and fourth epochs, reducing the learning rate by 0.1. The models were trained for six epochs, and the best-performing model in each case was selected for evaluation on the test set based on the 1-N.E.D. metric. All experiments were configured using the MMOCR [22] training scripts.

---

[5] https://mmocr.readthedocs.io/en/latest/textrecog_models.html#robustscanner.
[6] https://mmocr.readthedocs.io/en/latest/textrecog_models.html#sar.

Table 10. Full list of text recognition model results trained on our dataset sorted by 1-N.E.D..

Model	Recall	Precision	Accuracy	Ignored	NED
MASTER	99.6	99.5	95.9	98.3	99.2
NRTR_1/8–1/4	99.5	99.4	95.8	98.0	99.2
NRTR_1/16–1/8	99.4	99.2	95.1	97.4	99.2
RobustScanner	99.2	99.2	94.3	97.1	98.6
SAR	99.3	99.3	94.5	97.1	98.3
SATRN	99.1	99.2	94.3	96.5	98.1
SATRN_sm	98.7	98.8	93.0	95.0	97.7
ABINet	99.5	78.7	72.4	72.9	84.5
CRNN-TPS	98.6	78.5	70.8	71.5	84.1
CRNN	98.4	78.4	69.7	70.6	84.0

E OCR Performances by Dataset Size

We investigated the performance of our most successful text detection and recognition models based on OCR results of GT data, as the number of training data varies. This allows us to determine the minimum amount of data needed to achieve optimal model performance. Since annotating and labeling data can be labor-intensive and time-consuming, we aimed to provide a baseline for data count in different comic domains and inform future studies on this topic. We focused on understanding the relationship between the number of training data and model performance in the context of text detection and recognition in comics.

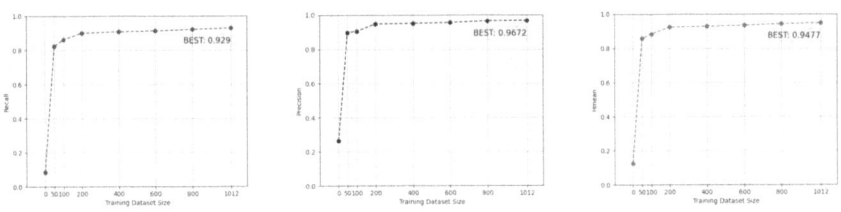

Fig. 7. Recall, precision, and H-Mean graphs of the most performant detection backbone on the test dataset, FCE_CTW_DCNv2, for varying training dataset size.

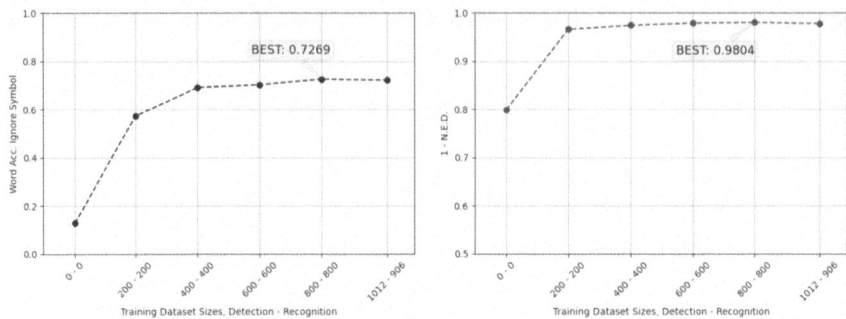

Fig. 8. Word accuracy ignoring symbols, and 1 - N.E.D. graphs of the most performant recognition backbone on the test dataset, MASTER, for varying training dataset size.

For text detection we show the results in Fig. 7 for precision, recall, and H-Mean. The interesting finding here is that since we already use a pretrained model, they adapt our domain with as low as fifty training samples and within a few epochs. Usually, it takes hundreds of epochs for the text detection model to learn from random initialization. After the training sample count is increased to two hundred, results do not change much, although as we increase the data, the model becomes more and more performant.

The same trend is observed in our text recognition model, as shown in Fig. 8, where we measured word accuracy by ignoring symbols and using one minus normalized edit distance (1-N.E.D.). However, the impact of domain adaptation on model performance is not as significant in this case, with around two percent improvement in performance. In contrast, we see a much greater increase in performance for text detection, with an improvement of more than eight times.

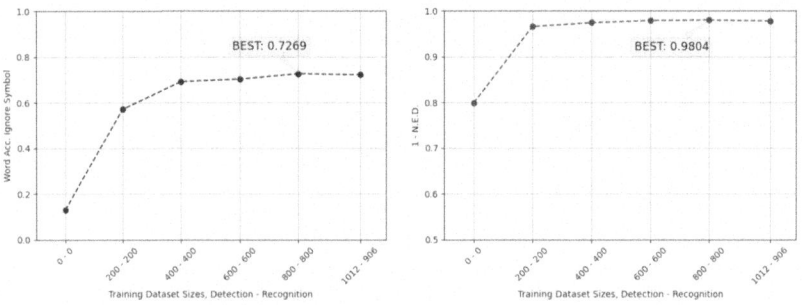

Fig. 9. Word accuracy ignoring symbols, and 1 - N.E.D. graphs of the most performant detection and recognition model type, FCE_CTW_DCNv2 - MASTER, on the ground truth data that are trained on varying training dataset size combinations.

In our final step of performance measurement, we varied the dataset size and measured the performance of a combination of text detection and text recognition models (see Fig. 9). We used the 1-N.E.D. metric to evaluate the model's performance and found that near-top performance was achieved with two hundred training sample sizes. However, when measuring word accuracy while ignoring symbols, we found that this level of performance was reached with four hundred training samples. This demonstrates the sensitivity of word accuracy to the model's performance and suggests that four hundred training samples for both text detection and text recognition can be a good baseline for comics.

F Creating *COMICS Text+* Dataset

F.1 Prefiltering of Some Textboxes

Table 11. Some pairs of overlapping text boxes in the third comic series of COMICS [18]. The one with a smaller area within the pairs is filtered, and the other is kept. As expected, most of their characters are shared.

Page No	Panel No	Textbox No	Text	x1	y1	x2	y2
5	2	0	i quickly drag the yorn body into ...	21	17	326	165
5	2	1	i quickly drag the yorn body into ...	24	21	326	165
20	0	1	lovers in all the dark corners ...	1	0	410	132
20	0	0	lovers in all the dark corners ...	0	0	410	121
30	2	2	not know the fish had been fed	0	529	384	587
30	2	3	not know ! the fish had been fed	0	540	384	587

Table 12. Statistics of filtered textboxes from the COMICS dataset [18], which are excluded from COMICS Text+ due to being empty or erroneous. These mostly include non-textbox images and misdetections on advertisement pages, which inflate the average character and word counts.

# Boxes	Avg. Chars.	Avg. Words	Median Chars.	Median Words
9,508	6.43	2.12	3	1

Some of our inferences resulted in erroneous or empty texts, which were also removed in order to improve the overall quality of the dataset. The statistics of the textboxes filtered for these reasons can be seen in Table 12, and some examples can be seen in Fig. 10.

As a concluding preparatory step for the creation of *COMICS TEXT+*, we implemented a filter to eliminate textboxes that exhibited overlap. In the context of COMICS [18], some textboxes were found to have significant levels of intersection with other textboxes, which could potentially compromise the quality of

Fig. 10. Some examples of textboxes are filtered from our COMICS Text+ dataset.

the dataset. To address this issue, we eliminated the smaller of the two textboxes in instances where they exhibited more than 0.8 Intersection over Union (IoU) within the same panel, as illustrated in Table 11.

F.2 Postprocessing on Raw OCR Output

We improved the process described in [19] for the rule-based approach by adding several additional steps. In Iyyer's work, two types of mistakes were targeted: mistakes in recognizing the first letter of a word (e.g., "eleportation" instead of "teleportation"), and errors in single characters. While these types of errors can be introduced by the OCR system, we believe that some are also the result of faulty textbox detection coordinates. As a result, we extended and modified their post-processing approach by considering errors that include not just the first letter but also letters from starting or ending.

Algorithm 1. Our algorithm for creating a dictionary for replacing tokens in our OCR output with PyEnchant suggestions.

---
procedure CREATETOKENREPLACEMENTDICT
  $corpus \leftarrow set(nltk\_corpus + brown\_corpus)$
  for each $token \in tokens$ **do**
    $con\_one \leftarrow token.length \geq 4$
    $con\_two \leftarrow token \notin corpus$
    $con\_three \leftarrow$ CHECKPYENCHANT(token)
    if con_one **and** con_two **and** con_three **then**
      $suggestions \leftarrow$ GETSUGGESTIONS(token)
      for each $s \in suggestions$ **do**
        if S.START(token) **or** S.END(token) **then**
          ADDTODICT(token, suggestion)
        end if
      end for
    end if
  end for
end procedure

To address the first type of error, we tokenized the OCR output and created a vocabulary of tokens sorted by the number of occurrences in descending order. We used the NLTK Punkt Tokenizer and Word Tokenizer[7] for tokenization. We

---
[7] https://www.nltk.org/.

then applied Algorithm 1 to tokens ranked between 10,001 and 100,000, assuming that misspelled words are less frequent. This allowed us to create a dictionary that could be used to replace those tokens in the output.

Our algorithm differs from Iyyer's [19] approach in several ways. Firstly, we check whether the tokens are part of a corpus, using NLTK's default corpus and the Brown Corpus [9] specifically. This is because the Golden Age of comics and the publication date of the Brown Corpus are closer in time than other corpora, which date from the 1950s and 1961. Secondly, instead of checking the token difference for the first letter of PyEnchant suggestions, we check whether the suggestion starts or ends with the token. This allows us to fix errors introduced by textbox detection errors.

Algorithm 2. Our algorithm for applying rule-based post-processing to OCR output to finalize our COMICS Text+ dataset.

    procedure APPLYPOSTPROCESSING
        valids ← {a, d, i, m, s, t, x}
        for each token ∈ all_tokens do
            if not PUNCBEFOREORAFTER(token) then
                token ← REPLACETOKEN(token)
            if token.length = 1 and token ∉ valids then
                REMOVE(token)
            end if
        end if
    end for
    end procedure

Erroneous single characters are detected by checking whether they have punctuation marks before or after them. If they do not have punctuation marks, we check whether they are in the set of valid single characters, which includes a, d, i, m, s, t, x. These characters frequently occur in the text and are likely to be part of a word. We check for punctuation marks because there may be abbreviations of special names in the text (e.g., *F.B.I.*, *U.S.A.*). We then apply Algorithm 2 to post-process all of our OCR output and prepare it for its final form.

Our dataset shares both the raw OCR output and the post-processed version. A total of 110,838 textboxes were affected by the first part of our post-processing, where some of their tokens were replaced by PyEnchant. A total of 131,833 textboxes were updated with the second part of the process, where some of their single-character tokens were deleted. Both of these post-processing steps affected 18,580 textboxes as their intersections. It is difficult to measure the effects of the rule-based post-proc

G Finalized Dataset Results

Our method outperforms COMICS [18] on all evaluation metrics, particularly character recall and word accuracy. This can be seen in the qualitative comparison in Table 14.

G.1 Error Analysis

Table 15 presents examples of the most common error types in the *COMICS TEXT+* dataset.

H Replication Results with Cloze Style Tasks

We have reproduced the models and experimental setup described in [18] and present our results in Table 13. The details of the tasks are as follows:

- The **Text Cloze** task involves predicting the text of the final panel p_i. To reduce the task's complexity and measure the models' ability to learn from inter-panel transitions, the final panels are selected from those with only a single text box. Architectures that make use of the visual modality, such as *image-text* and *image-only*, apply masking to the text box areas so that the model cannot "see" the text content and provides the artwork features along with the context to score the text box candidates.
- The **Visual Cloze** task involves predicting the artwork of the final panel p_i. Unlike the *Text Cloze* task, the models are not provided with the text features of the final panel.
- The **Character Coherence** task involves reordering two dialogue text boxes for the final panel.

Table 13. Comparison of our replication results Iyyer et al.'s [18] results based on accuracy. Human baseline is also shared for only hard tasks since the easy task is trivial for humans. *NC-Image-text* models use only the final panel's image without any information from preceding context panels.

MODEL	Text Cloze				Visual Cloze				Char. Coherence	
Task Difficulty	EASY		HARD		EASY		HARD			
	Original	Ours	Original	Ours	Original	Ours	Original	Ours	Original	Ours
Text-only	63.4	62.8	52.9	51.7	55.9	51.8	48.4	45.0	68.2	70.9
Image-only	51.7	50.7	49.4	45.6	85.7	79.5	63.2	56.1	70.9	70.7
Image-text	68.6	62.1	61.0	57.2	81.3	83.6	59.1	59.5	69.3	71.5
NC-Image-text	63.1	57.3	59.6	56.5	–		–		65.2	75.6
Human	–		84		–		88		87	

COMICS TEXT+ 193

Table 14. Qualitative comparison between samples of COMICS Text+ and COMICS OCR. The examples shown are selected by their difference in 1-N.E.D. metric from ground truth data.

Speech Bubble	COMICS OCR	COMICS Text+	G. T.
	we seou₢ht reinforce	- - - we brought re- inforce - ments ! a poose !	- - - we brought rein- force - ments ! a posse !
	ooor ! 000f ,	ooof !	ooof !
	l / gh / ugh	ugh !	ugh !
	why ,	why you . . . i	why you . . ! * '
	hon ' s business ?	hi , ferdie ! how ' s business ?	hi , ferdie ! how ' s business ?
	c / se that odd weapon against us	look , lie ' s going to use that odd weapon against us !	look , he ' s going to use that odd weapon against us !
	and there ' s	and there ' s your pay . . . you	and there ' s your pay . . . you fool !
	am crazy ea take him	im crazy , eh ? take him to the roof !	im crazy , eh ? take him to the roof !
	don ' t be a ...	don ' t be a . . ooooff	don ' t be a . . ooooff !
	aboard . gone	come aboard - he ' s gone !	come aboard - he ' s gone !
	did !	he did !	he did !
	erry ' s 74lew<unk>acr throwhag forma snowballs comer navamdy -	terry ' s talent for throwing snowballs comes in handy !	jerry ' s talent for throwing snowballs comes in handy !

Table 15. Results of qualitative error analysis on COMICS Text+ dataset. The examples shown are selected from the least performing speech bubbles based on the 1-N.E.D. metric from ground truth data. The COMICS [18] OCR results for the selected speech bubbles are also added for reference.

Speech Bubble	COMICS OCR	COMICS Text+	G. T.
	suddenly , ogg sees haga approaching	sudo ogg	suddenly , ogg sees haga approaching . . .
	pow	ma dow	pow
	there ' s the no , the	no the there ' s the	there ' s the no , the
	asluckx ste / s - int the room he i spelledby a terrific blow from behind	into felled terrific blow from	as lucky steps into the room he is felled by a terrific blow from behind
	weird jungle beast pould bring a big america bob morgan yelded 7 tema	weird jungle beast would bring a big yielded to temp . america . ! bob morgan	weird jungle beast would bring a big america ! bob morgan yielded to temp .
	itis susan	it ! susan !	it is susan !
	and there ' s	and there ' s your pay . . . you	and there ' s your pay . . . you fool !
	your brother still clings to his mad ideas , monsieur uarnac ! he says he ' s going to break out and return to his experiments .	your brother still clings monsieur to his mad ideas , jarnac ! he says he ' s going to break out and return to his experiments , !	your brother still clings to his mad ideas , monsieur jarnac ! he says he ' s going to break out and return to his experiments !
	williston parki	willist . park !	williston park !
	didnt enough ryants	i didn ' t enough suryant ' s	didn ' t enough bryant ' s

References

1. Aizawa, K., et al.: Building a manga dataset "manga109" with annotations for multimedia applications. IEEE Multimedia **27**(2), 8–18 (2020). https://doi.org/10.1109/mmul.2020.2987895

2. Augereau, O., Iwata, M., Kise, K.: An overview of comics research in computer science. In: 2017 14th IAPR International Conference on Document Analysis and Recognition (ICDAR). vol. 3, pp. 54–59. IEEE (2017)
3. Baek, J., Matsui, Y., Aizawa, K.: Coo: Comic onomatopoeia dataset for recognizing arbitrary or truncated texts (2022). arXiv preprint arXiv:2207.04675
4. Brienza, C.: Producing comics culture: a sociological approach to the study of comics. J. Graph. Novels Comics **1**(2), 105–119 (2010)
5. Cohn, N.: The Visual Language of Comics: Introduction to the Structure and Cognition of Sequential Images. A&C Black (2013)
6. Dai, J., et al.: Deformable convolutional networks. In: Proceedings of the IEEE International Conference on Computer Vision, pp. 764–773 (2017)
7. Dunst, A., Hartel, R., Laubrock, J.: The graphic narrative corpus (GNC): design, annotation, and analysis for the digital humanities. In: 2017 14th IAPR International Conference on Document Analysis and Recognition (ICDAR). vol. 3, pp. 15–20. IEEE (2017)
8. Fang, S., Xie, H., Wang, Y., Mao, Z., Zhang, Y.: Read like humans: Autonomous, bidirectional and iterative language modeling for scene text recognition (2021)
9. Francis, W.N., Kucera, H.: Brown corpus manual. Lett. Editor **5**(2), 7 (1979)
10. Del Gobbo, J., Matuk Herrera, R.: Unconstrained text detection in manga: a new dataset and baseline. In: Bartoli, A., Fusiello, A. (eds.) ECCV 2020. LNCS, vol. 12537, pp. 629–646. Springer, Cham (2020). https://doi.org/10.1007/978-3-030-67070-2_38
11. Gomez, R., et al.: ICDAR2017 robust reading challenge on COCO-text. In: 2017 14th IAPR International Conference on Document Analysis and Recognition (ICDAR). vol. 01, pp. 1435–1443 (2017). https://doi.org/10.1109/ICDAR.2017.234
12. Guérin, C., et al.: eBDtheque: a representative database of comics. In: Proceedings of the 12th International Conference on Document Analysis and Recognition (ICDAR), pp. 1145–1149 (2013)
13. Gupta, A., Vedaldi, A., Zisserman, A.: Synthetic data for text localisation in natural images. In: IEEE Conference on Computer Vision and Pattern Recognition (2016)
14. Hartel, R., Dunst, A.: An OCR pipeline and semantic text analysis for comics. In: Del Bimbo, A., et al. (eds.) ICPR 2021. LNCS, vol. 12666, pp. 213–222. Springer, Cham (2021). https://doi.org/10.1007/978-3-030-68780-9_19
15. He, K., Gkioxari, G., Dollár, P., Girshick, R.: Mask R-CNN. In: 2017 IEEE International Conference on Computer Vision (ICCV), pp. 2980–2988 (2017). https://doi.org/10.1109/ICCV.2017.322
16. He, K., Zhang, X., Ren, S., Sun, J.: Deep residual learning for image recognition. In: Proceedings of the IEEE Conference on Computer Vision and Pattern Recognition, pp. 770–778 (2016)
17. Herbst, P., Chazan, D., Chen, C.L., Chieu, V.M., Weiss, M.: Using comics-based representations of teaching, and technology, to bring practice to teacher education courses. ZDM **43**(1), 91–103 (2011)
18. Iyyer, M., et al.: The amazing mysteries of the gutter: Drawing inferences between panels in comic book narratives. In: Proceedings of the IEEE Conference on Computer Vision and Pattern Recognition (CVPR) (2017)
19. Iyyer, M.N.: Discourse-Level Language Understanding with Deep Learning. Ph.D. thesis (2017)
20. Jaderberg, M., Simonyan, K., Vedaldi, A., Zisserman, A.: Reading text in the wild with convolutional neural networks. Int. J. Comput. Vis. **116**(1), 1–20 (2016)

21. Karatzas, D., et al.: ICDAR 2015 competition on robust reading. In: 2015 13th International Conference on Document Analysis and Recognition (ICDAR), pp. 1156–1160. IEEE (2015)
22. Kuang, Z., et al.: MMOCR: A comprehensive toolbox for text detection, recognition and understanding (2021). arXiv preprint arXiv:2108.06543
23. Laubrock, J., Dunst, A.: Computational approaches to comics analysis. Top. Cogn. Sci. **12**(1), 274–310 (2020)
24. Lee, J., Park, S., Baek, J., Oh, S.J., Kim, S., Lee, H.: On recognizing texts of arbitrary shapes with 2D self-attention. In: Proceedings of the IEEE/CVF Conference on Computer Vision and Pattern Recognition Workshops, pp. 546–547 (2020)
25. Li, H., Wang, P., Shen, C., Zhang, G.: Show, attend and read: a simple and strong baseline for irregular text recognition. In: Proceedings of the AAAI Conference on Artificial Intelligence. vol. 33, pp. 8610–8617 (2019)
26. Liao, M., Wan, Z., Yao, C., Chen, K., Bai, X.: Real-time scene text detection with differentiable binarization. In: Proceedings of the AAAI Conference on Artificial Intelligence, pp. 11474–11481 (2020)
27. Liao, M., Zou, Z., Wan, Z., Yao, C., Bai, X.: Real-time scene text detection with differentiable binarization and adaptive scale fusion. IEEE Trans. Pattern Anal. Mach. Intell. **45**, 919–913 (2022)
28. Liao, W., Rosenhahn, B., Shuai, L., Ying Yang, M.: Natural language guided visual relationship detection. In: Proceedings of the IEEE/CVF Conference on Computer Vision and Pattern Recognition Workshops (2019)
29. Long, S., Ruan, J., Zhang, W., He, X., Wu, W., Yao, C.: TextSnake: A flexible representation for detecting text of arbitrary shapes, pp. 20–36 (2018)
30. Lu, N., et al.: MASTER: multi-aspect non-local network for scene text recognition. Pattern Recogn. **117**, 107980 (2021)
31. Nguyen, N.V., Rigaud, C., Burie, J.C.: Digital comics image indexing based on deep learning. J. Imaging **4**(7), 89 (2018). https://doi.org/10.3390/jimaging4070089, http://www.mdpi.com/2313-433X/4/7/89
32. Nguyen, N.V., Rigaud, C., Burie, J.C.: Comic MTL: optimized multi-task learning for comic book image analysis. Int. J. Doc. Anal. Recogn. (IJDAR) **22**(3), 265–284 (2019)
33. Nguyen, N.-V., Vu, X.-S., Rigaud, C., Jiang, L., Burie, J.-C.: ICDAR 2021 competition on multimodal emotion recognition on comics scenes. In: Lladós, J., Lopresti, D., Uchida, S. (eds.) ICDAR 2021. LNCS, vol. 12824, pp. 767–782. Springer, Cham (2021). https://doi.org/10.1007/978-3-030-86337-1_51
34. Rigaud, C., et al.: Speech balloon and speaker association for comics and manga understanding. In: 2015 13th International Conference on Document Analysis and Recognition (ICDAR), pp. 351–355. IEEE (2015)
35. Rigaud, C., Pal, S., Burie, J.C., Ogier, J.M.: Toward speech text recognition for comic books. In: Proceedings of the 1st International Workshop on coMics ANalysis, Processing and Understanding, pp. 1–6 (2016)
36. Sheng, F., Chen, Z., Xu, B.: NRTR: a no-recurrence sequence-to-sequence model for scene text recognition. In: 2019 International Conference on Document Analysis and Recognition (ICDAR), pp. 781–786. IEEE (2019)
37. Shi, B., Bai, X., Yao, C.: An end-to-end trainable neural network for image-based sequence recognition and its application to scene text recognition. IEEE Trans. Pattern Anal. Mach. Intell. **39**, 2298–2304 (2016)
38. Shi, B., Wang, X., Lyu, P., Yao, C., Bai, X.: Robust scene text recognition with automatic rectification (2016)

39. Wada, K.: labelme: Image polygonal annotation with python (2018). https://github.com/wkentaro/labelme
40. Wang, W., et al.: Shape robust text detection with progressive scale expansion network. In: Proceedings of the IEEE/CVF Conference on Computer Vision and Pattern Recognition, pp. 9336–9345 (2019)
41. Wang, W., et al.: Efficient and accurate arbitrary-shaped text detection with pixel aggregation network. In: ICCV, pp. 8439–8448 (2019)
42. Wick, C., Reul, C., Puppe, F.: Calamari - A high-performance Tensorflow-based deep learning package for optical character recognition. Digital Humanit. Q. **14**(1) (2020)
43. Yue, X., Kuang, Z., Lin, C., Sun, H., Zhang, W.: RobustScanner: dynamically enhancing positional clues for robust text recognition. In: European Conference on Computer Vision (2020)
44. Yuliang, L., Lianwen, J., Shuaitao, Z., Sheng, Z.: Detecting curve text in the wild: New dataset and new solution (2017). arXiv preprint arXiv:1712.02170
45. Zhang, S.X., et al.: Deep relational reasoning graph network for arbitrary shape text detection, pp. 9699–9708 (2020)
46. Zhu, Y., Chen, J., Liang, L., Kuang, Z., Jin, L., Zhang, W.: Fourier contour embedding for arbitrary-shaped text detection. In: CVPR (2021)

Toward Accessible Comics for Blind and Low Vision Readers

Christophe Rigaud[1,2](✉) , Jean-Christophe Burie[1] , and Samuel Petit[2]

[1] L3i Laboratory, SAIL joint Laboratory, 17042 La Rochelle CEDEX 1, France
{christophe.rigaud,jean-christophe.burie}@univ-lr.fr
[2] Comix AI (a subsidiary of De Marque Group), 5, place des Coureauleurs, 17000 La Rochelle, France
samuel.petit@demarque.com

Abstract. This work explores how to fine-tune large language models using prompt engineering techniques with contextual information for generating an accurate text description of the full story, ready to be forwarded to off-the-shelf speech synthesis tools. We propose to use existing computer vision and optical character recognition techniques to build a grounded context from the comic strip image content, such as panels, characters, text, reading order and the association of bubbles and characters. Then we infer character identification and generate comic book script with context-aware panel description including character's appearance, posture, mood, dialogues etc. We believe that such enriched content description can be easily used to produce audiobook and eBook with various voices for characters, captions and playing sound effects.

Keywords: comics understanding · large language model · prompt engineering · character identification · comic book script · accessible comics

1 Introduction

Visual arts play an important role in cultural life, providing access to social heritage and self-enrichment [3]. However, most works of art are inaccessible to the visually impaired, whether they are legacy blind, blind, with eye movement disorder or having cognitive eye disease [32]. People with such disease could largely benefit from "intelligent" computer vision tools in order to also get precise information about visual arts [40] such as comics [7], manga and webtoon [12]. Visual details and contexts which are essential to understand and feel the beauty of these artworks are often missing in current experimental tools.

Researchers have considered various ways to make visual art such as comic albums accessible to visually impaired people, including automatic image and/or text/audio descriptions [28] and tactile graphics. However, in the systematic review [26], Oh et al. conclude that image description was out of the scope of interest for most studies, suggesting that automatic retrievals of image-related information is one of the bottlenecks for making images accessible at scale.

© The Author(s), under exclusive license to Springer Nature Switzerland AG 2024
H. Mouchère and A. Zhu (Eds.): ICDAR 2024 Workshops, LNCS 14935, pp. 198–215, 2024.
https://doi.org/10.1007/978-3-031-70645-5_13

To address this challenge, one strategy is to create a comprehensive description of comic books, including layout details, transcriptions of text, and detailed textual descriptions of the graphical content of each panel [15]. This comprehensive description can then be utilized by text-to-speech tools for playback [45]. The current state of Text-to-Speech (TTS) models has reached an impressive level of sophistication, allowing them to generate audio that is nearly indistinguishable from human speech. They are characterized by their ability to produce natural-sounding synthetic voices with emotions, pauses, and realistic tone. In that scenario, the quality, details and relations of textual transcription and description are key issues for both textual and audio versions of the books. Recent advances using (multimodal) large language models (LLM) based on transformer architecture, sparked a research frenzy by highlighting the impressive capabilities of not only processing text, but also to understand and detail image content [30].

In this work, we investigate prompt engineering techniques such as chain-of-prompt to make the most of this technique and dramatically improves prediction quality without requiring additional data or annotation time. The generated description is intended to mirror the image content by providing an overall scene description with named characters, dialogues and interaction following the natural reading order [4]. We propose to use previous computer vision techniques and optical character recognition to extract information from the comic strip images, such as the panels, characters, text, reading order and the association of bubbles and characters. We then use this information as additional context for instruction-tuning large language models and generate a precise description of each panel. The main contributions of this work are as follows:

- Text type classification: sound, caption, dialogue
- Automatic character's names inference
- Contextual panel description
- Script generation including all textual and visual elements

2 Related Works

In this section, we first review comics accessibility and then comics image analysis. Finally, we give a focus to recent transformer-based method known as Large Language Model (LLM) applied to comics analysis and understanding.

2.1 Comics Accessibility

In 2017, Rayar *at al.* called on the document analysis community to address the issues of visually impaired people [31]. Then several experimental tools have been proposed such as an accessible comic reader for people with low vision [32] and a study on the user experience of accessible comic book reader for textual sound effects [15,27]. When the participants were asked to select the most important information they desired while reading a comic, the majority prioritized scene descriptions, followed by transcriptions and facial expressions of characters, etc. [35]. Both studies are compared in [8] (in Spanish). There is an ongoing collaboration at San Francisco State University about comics studies, a program for visual impairment, and the Accessible Comics Collective which explores ways of making comics accessible for blind and low-vision readers[1]. An approach on Webtoon proposed to utilize comments to improve non-visual Webtoon accessibility. It is based on a panel-anchored adaptive descriptions [12]. Synopsis generation has been explored to have quick and crisp understanding of a comic story [6]. We propose to consider panel-based description, sound effects, script generation and extend them with other text type and character-related information to produce a richer reading experience for Blind and Low-Vision (BLV) readers.

2.2 Comics Image Analysis

Recently, a decade systematic literature review has been proposed by [38] for comics image segmentation, classification and recognition methods. Earlier, other surveys presented an extended overview of computational approaches for comics analysis [14] and for comic research in computer science [1]. They highlighted numerous methods for most common element extraction like panels, speech balloons and text. Also, some methods have been proposed for very specific elements such as inferring unseen actions [13,44] or balloon tail detection [16,24,35] which are really helpful for accurate text-to-character association [17]. To our knowledge, automatic text type classification and script generation have not been addressed yet. We review character clustering and identification (naming) in the two following subsections.

Character Clustering. Character clustering (or re-identification) consists in recognizing characters consistently across all different panels in a comics or series of comics. It is subsequent to character detection for which several methods are proposed in the literature (see surveys introduced in the previous section). It presents significant challenges due to limited annotated data and complex variations in character appearances [41].

Initially, k-means based method [43] was proposed, but they require specifying the number of clusters (character instance groups), which is unknown in our case. To solve this problem, more advanced clustering method which automatically determine the number of clusters (Manga characters) have been proposed [25,46–48]. Ramaprasad et al. [30] used large pre-trained vision encoder Contrastive Language-Image Pre-Training (CLIP) [29] a multi-modal large language model that can perform zero-shot image classification based on natural language prompts. Sachdeva et al. compared their proposition also to CLIP image feature for representing manga characters [35]. An extended clustering can be achieved by combining visual features with spatial-temporal information, an approach chosen by [48] to achieve unsupervised person re-identification in Japanese manga. Their method relies on design rules such as characters tend to

---
[1] https://spinweaveandcut.com/blind-accessible-comics/.

appear on adjacent pages one after another, characters in the same frame tend to belong to different identities, some characters tend to appear in pairs, and so on. Also, characters' unique signature is highlighted, such as special accessories, hairstyles, costumes, etc.

Character Identification. Regarding character identification, Ramaprasad *et al.* [30] used CLIP to predict character's names from a given list of names. Pre-defined list of name is a strong limitation that we can not afford in our context because character names of little-known or new comics series can not be determined in advance. From our knowledge, there is only one very recently published method that extend character clustering to character naming and associate them with speech balloon [18]. This method requires characters names list as input and then associate these names to detected character based on nearby text region analysis using LLM. The ultimate task of matching texts and speakers is challenging in comics analysis, often necessitating an understanding of conversation history and context to disambiguate speakers [18,35].

2.3 Comics and Large Language Model

Very recently, Sachdeva *et al.* tackled the problem of diarisation i.e. generating a transcription of who said what and when, in a fully automatic way for visual impairment. The result of this work is similar to our proposition for the dialogue part but does not consider panel scene description. The work [18] proposes to identify and name characters based on a given name list as input and analyse text blocks with LLM. In our proposition, we extend LLM capacities to automatically build the name list from dialogues.

The work of Ramaprasad *et al.* is the most similar to our contribution by proposing to generate accessible text descriptions for comic strips including panel description using LLM prompt engineering [36] as well [30]. In this study, they use famous comics which allows LLM to use their previous knowledge for character name inference but also generate hallucination effects regarding extra details. MaRU method [39] incorporates a vision-text encoder that combines textual and visual information into a unified embedding space, enabling the retrieval of relevant scenes based on user scene description queries (even in another language). This method excels in end-to-end dialogue retrieval and exhibits promising results for scene retrieval which enhance the understanding of and improve accessibility of Manga. To extend this study, the authors propose as future work to recognise comic characters for enhancing attribution and understanding of visual content. We address the latter in Sect. 3.2.

In [11], a method to complement missing comic (manga) text content during the manga creation/translation process is proposed. This manga argumentation method mines event knowledge within the comics with large language models. Then, fine-grained visual prompts support manga complement.

3 Proposed Method

The aim of the proposed method is to automatically generate a structured comic book script-like description of each panel including scene, action and dialogues. This is intended to facilitate enriched text-to-speech (or text-to-braille) conversion for enhanced comic accessibility. The main challenges are to associate each dialogue to corresponding named characters and generate a detailed text description using natural language. Our contribution can be seen as one component within a broader pipeline, outlined as follows:

Fig. 1. Proposed contributions within an example of complete pipeline. Proposed blocks are represented with solid rectangles while complementary ones are dashed.

3.1 Script Generation

Considering that all the visual and textual elements have been extracted by pre-processing image analysis algorithms, we generate a script of the album gathering all these elements in a single structured text file according to the original layout of each page. We encode the script using Markdown markup language which is appropriated for both comics script[2] and LLM inference on structured documents [22]. Then, the script will be used as context in the next step for further inferences such as character naming (see Sect. 3.2).

Page Layout. Each comic book page is usually composed by several panels that should be read in a certain order e.g. left-to-right and down the "Z-path" for English comics books [5]. Assuming that the panel have been previously extracted and ordered during the content extraction pre-processing stage, we propose to list them in a single text file according to their reading order (see script sample at this end of this section).

Text Type Classification. The script is then completed with textual information from previous content extraction module and inserted into each `Panel` section of the script following the natural text reading order. The detection of speech balloon tail with algorithms from the literature allows us to identify most of the spoken text, and we propose a set of rules to also label the onomatopoeia (sound effects) from detected text. For comics genres that do not make use of tail, balloon contour shape combined with text analysis can be used instead. We qualify as "caption" all the remaining text (not classified as spoken or sound

---

[2] https://www.codymarkelz.com/posts/2022-01-20-rmarkdown-comic-script.html

effect). It is important in our study to qualify text as much as possible in order to improve textual description and accessibility (e.g. using different synthetic voices to render different type of text).

Sound Effect. We consider as sound effect (SFX) single text lines with an important height or slope. If an onomatopoeia is composed of several text lines, they will be considered as separated and different at this stage. We set the minimum text line height $minH$ - to be considered as sound effect - to $minH = 0.025 \times imagewidth$ (2.5%) and the minimum slope $minS$ to 0.1 (10%) compared to horizontal line, based on empirical experiments. Note that these rules may consider a shopfront or other drawing (big or sloped) as onomatopoeia.

Caption. In comics, caption or narration boxes are used for narration, transitional text (e.g. "Meanwhile..."), or off-panel dialogue. Captions usually have rectangular borders, but they can also be border-less or floating letters. Here, we simplify the definition and consider as caption all text blocks that are not classified as sound effect and that haven't been associated to a comic character (no associated tail). Assuming an error-free result from the previous tail detection algorithm, we expect this category to be composed by usual rectangular text region providing contextual narrative information throughout the story.

Dialogue. The dialogue sets are composed of all the remaining text, each dialogue being composed by one or more lines of text contained in a speech balloon associated with a comic character (with a tail pointing to it). Each line of dialogue is preceded by the identifier of the corresponding character i.e. $c0$, $c1$. Character's identifier computation comes from character clustering which is detailed below.

Character Clustering. We consider spatial-temporal relationships, as explored in [48], during the instruction-tuning phase. This involves using a generated script that contains ordered panels and text as context (see Sect. 3.2).

For character clustering, we compute image feature vectors on each character instance using a variant of Contrastive Language-Image Pre-training (CLIP) associated with Vision Transformer (ViT) like in [30,39]. This model is designed to perform zero-shot image classification. It has been trained on a large-scale general purpose image dataset and a text dataset. CLIP uses ViT to get visual features and a causal language model to get the text features. Both the textual and visual features are projected into a latent space of identical dimensions. In our case, we use only its image embedding part and reduce the dimensionality of its output feature vectors. This simplifies the embedding into a form that traditional clustering methods, as presented here[3], can handle more effectively. Then we cluster reduced image feature vectors to automatically group the characters by visual similarities, encoded in image vectors.

To reduce the size of image vectors, we use Uniform Manifold Approximation and Projection (UMAP) [2] like in [25] with a size of 5 (see parameter validation

---

[3] https://blog.bruun.dev/semantic-image-clustering-with-clip/.

in experiment Sect. 4.3). For the clustering, we employed Hierarchical Density-Based Spatial Clustering of Applications with Noise (HDBSCAN) [21] algorithm, which gives state-of-the-art results on comics according to [25,46]. The minimum cluster size is the only parameter of HDBSCAN, we set it to 15, assuming that the main characters appear many times in comic books. Groupings smaller than this size will be gathered in a "noise" group. In our study, we are interested in finding the names of the main characters associated to dialogues. Some less frequent characters might not be named in the story, we expect them to be considers as "noise" at this stage, and we will refer to them using interrogation mark symbol "?" in the script (see Fig. 2).

Cluster IDs are used to temporarily identify comic characters in the script until they are replaced by inferred proper names in the next step. For example, character A will be identified as $c0$ and character B as $c1$. We complete the layout-only script with text type information (caption, sound, dialogue) and character clustered IDs as in the following example:

```
# PAGE 1 - 1 PANEL:
## PANEL 1
### SOUND
### CAPTION
### DIALOGUE
c0: ...
c1: ...
```

We are aware that in our study other type of text might be miscellaneously considered as caption but experiments shown that it doesn't alter character identification and character naming, the next main step objective detailed below.

3.2 Character's Name Inference

We propose an automatic method for inferring character names from the generated script. Contrary to [30] which provides names manually or other methods that use previous knowledge e.g. famous comic character found on internet, we automatically infer character's name from the script to allows comics accessibility of lesser-known books or new series as well.

To do so, we leverage the inferring capacity of LLM providing our generated script as context [36]. The script contains spatial-temporal information (panel sequence in reading order and qualified text) and character identifier $c0, c1, etc.$. We propose a chain-of-prompt of four prompts to guide the model throughout its knowledge exploration. LLM is requested to first infer character names using its Named Entity Recognition (NER) capabilities. In this prompt, we also request the model to explain its reasoning to benefit from chain-of-thought approach and minimize eventual hallucinations [36]. Secondly, the LLM is used to associate a proper name to each recurring character ID from the clustering by making inferences from dialogue texts. For instance, if there are two characters in a panel and the first character is mentioning a proper name, it is likely the name of the second character, especially if this is repeated in several panels.

Character names are a first step in high-quality accessibility and we propose to go further by characterizing each character with their relationship, genre, age, and temperament in order to, for instance, select or generate the most appropriated synthetic voice for subsequent speech-to-text processing. We found that many LLM can be used for this task using prompt engineering [23]. Here is the proposed chain-of-prompt providing an extract of our previously generated script as context in place of SCRIPT:

```
USER: This is the script of a comic book: """SCRIPT"""
USER: Please list all character's names.
USER: Please list all corresponding unique identifiers. Itemize.
USER: What are their relationship? Explain  your reasoning
      step-by-step.
```

The chain-of-prompt is guiding the LLM from general context such as "a script of a comic book" from which an overall description will be automatically generated according to the general system prompt. This description forces the network to parse the full script a first time. Then, the second prompt requests basic information about characters (finding proper names and associated IDs from the script) of the story it just discovered in the previous prompt, avoiding looking for "farthest" knowledge. Last, we request their relationship and a last specific information requiring reasoning about character personality. The assistant output can be easily parsed and character identifier e.g. $c0$, $c1$ are associated with proper names e.g. *Curt*, *Cynthia* (see Sect. 4.3). Relationship details can be used to check how well the model understands the story. It can also be included in a character biography section at the beginning of (accessible) books to introduce characters along with story synopsis.

3.3 Panel Description

We propose to include each panel content description in the script to allow the reader to get a better understanding of the story before diving into characters dialogues, such as recommended by the BLV community [19].

Image content description has made great strides in the last two years with the rise of Visual Language Models (VLM). Such models having a great capacity of generalizing, we found that they are able to describe panel images with an important level of detail and give an accurate insight type of lighting, type of view, objects, character, mood, action, etc. Given that these models can also interpret text [39] like OCR, we introduce character name's identifier (see Sect. 3.2) artificially into panels. This facilitates correspondences for the VLM between panel images and script content, as illustrated in Fig. 2.

Character's names are written as clearly as possible to avoid any text recognition error, using a common font with black colour in a white rectangle. We position them carefully, ensuring that important details like facial expressions are not hidden. Our proposal is to centre them on the detected body.

Fig. 2. Panel 5 of page 12 from Boy Loves Girl 41 [10] with identified character's names inserted over corresponding characters.

Here is the proposed prompt sequence providing the script as SCRIPT:

```
USER: Please describe this comic book panel using character's
      names written in white rectangles.
USER: Please rephrase considering character's settings and
      interaction only, naming characters and citing all text
      from the following script extract: """[SCRIPT]"""
```

We formulate two successive user prompts as shown above. First, we ask to describe the scene and then rephrase it by including contextual information from the provided script. This chain-of-prompt approach [36] helps the network to focus progressively on the elements of interest (mitigate missing information and hallucination effect). By indicating "citing all text" in the second prompt, we emphasize the importance of including and referencing all text from the script in the generated description. The limitation "settings and character interaction only" mitigate description about art style that models seems very tempted to describe as well e.g. mid-20th-century American comics, from the 1950s, etc.

We believe that providing a precise panel description, BLV people will also be able to infer invisible actions that are happening between panels (in the gutters) known as "closure" [13]. For example, this is the final description DESC1 for the panel presented Fig. 2 given the below SCRIPT extract:

Script extract SCRIPT given as context :

```
### DIALOGUE
Cynthia: HI, KIDS! WHAT'S UP?
Bill: SPARKLING WIT BRILLIANT CONVERSATION PENETRATING THOUGHT
      WHAT DID YOU EXPECT?
?: HE MEANS IT'S THE USUAL DULL EVENING!
```

Panel description `DESC1` that we insert into the script in the corresponding `PANEL` section:

```
In this comic panel, Curt, Cynthia, and Bill are having a conversation
at a social gathering. Curt, dressed in a brown suit and tie, stands
on the left. Cynthia, wearing a yellow dress, stands in the middle.
Bill, dressed in a gray suit and tie, stands on the right. The background
of the panel features a window with a red curtain and a door. The dialogue
between them goes as follows:
    Cynthia: Hi, kids! Whats up?
    Bill: Sparkling wit... brilliant conversation... penetrating thought...
what did you expect?
    ?: He means its the usual dull evening!
```

4 Experiments

4.1 Dataset

For this exploratory research, we limited our experiments to public domain English comics. A lot of them are available from Digital Comic Museum or Comic Book Plus and have already been annotated in several public datasets. To our knowledge, none of the dataset are proposing text type classification, one dataset (Manga109 [9]) contains character identification annotation and association to text [17] and none of them propose panel description and script. Note that character clustering and name inference require several (ideally all) consecutive pages from consistent episode, albums or series. This ensures homogeneous clustering of as many characters as possible at once and the gathering of associated dialogues to facilitate character name inference. Faced with the lack of appropriated dataset, we chose to manually build and share a toy dataset by considering full albums from which at least some pages were part of eBDtheque dataset [10]. We extended part of the dataset by getting complementary pages of public domain titles like golden age American comics which are also part of several other public dataset, but not annotated according to our experiments. We first focused on "Escape with me" episode of an old title "Boy Loves Girl 41" (1953), from which eBDtheque pages 12 and 15 are both part of. This episode is spread oven page 11 to 17 and composed by 45 panels, 75 character instances and 109 text blocks in total. We exclude advertising content from the last page and do not modify the story analysis. We downloaded all complementary pages from Comic Book Plus.com[4].

We also experimented on a second more recent (2003) publicly available comic book: "Patents" from the World Intellectual Property Organization (WIPO)[5]. It is a free comic book available in English, French, Spanish, Arabic, Chinese and Russian. We chose the English version which is composed of 12 pages, 85 panels, 132 text blocks and 77 character instances.

---

[4] https://comicbookplus.com/?dlid=31803.
[5] https://www.wipo.int/publications/en/details.jsp?id=67.

4.2 Text Type Classification

We evaluated text type classification for each text line generated by an OCR as described in Sect. 3.1. We used Google Vision API to extract text blocks from images using its Optical Character Recognition (OCR) capacity. We combine it with a balloon contour analysis method [33] to detect speech balloons and their tail tips. Each text block detected by the OCR contained by a detected balloon were automatically associated. Text block not contained by a balloon could then only be considered as sound effect if it has important height or slope as described in Sect. 3.1. In episode "Escape with me", all text were detected by the combined text block + balloon contour detection, and we measured the recall (R) and the precision (P) which are 100% and 93% respectively. For "Patents", $P = 100\%$ and $R = 98\%$ respectively. The precision is not at the maximum because some bubbles were detected as two different text blocks due to their complex shape and illustrative text was detected as balloon (e.g. clock numbers and telephone in the last page of the episode). We corrected it manually for the rest of the evaluation to avoid error propagation.

Table 1 gives a summary of correct and confused classification for each evaluated episode.

Table 1. Text type classification confusion matrix comparing reference (ground truth) and predicted text type.

	Escape with me			Patents		
Ref./Pred.	Sound	Dialogue	Caption	Sound	Dialogue	Caption
Sound	**1**	0	0	**5**	1	8
Dialogue	0	**77**	4	1	**78**	0
Caption	0	6	**21**	0	18	**15**

The highest confusion concerns captions that have been classified as dialogue in both experiments. This is due to a known limit of [33]: not designed to detect open and thought balloons. By manually adding a tail to open balloons, we dropped caption false positive down to zero for both titles, which confirms the effectiveness of the proposed method for dialogue and caption text type classification. Regarding sound effects, the confusion with caption is quite high because we include illustrative text into our defined caption category, despite the fact that these are often depicted as onomatopoeia.

4.3 Character Clustering and Name Inference

Clustering. Character clustering is usually preceded by a detection step which supports the consistency in the final textual story reconstruction. To avoid any error propagation, we assume that characters have been detected by any dedicated algorithm from the literature (see Sect. 2.2) and errors are manually curated if needed. Character crops with significant overlap of multiple character have been discarded to avoid confusion in image feature vector computation and clustering. For image feature extraction (embedding), we used CLIP ViT-L/14 model trained with the DataComp-1B[6] and original hyperparameters similarly to [11,39]. We varied the minimum cluster size UMAP parameter from 3 to 10 without any notable impact on the final classification. This might be due to the simplistic style of images, we will extend this evaluation to other styles in the future. We reduced the minimum number of element by cluster of HDBSCAN algorithm to 5 for both titles because of the limited number of pages. Anyway, the algorithm grouped more than 12 instance for each cluster in our experiment, so no impact has been observed. Figure 3 shows cluster result samples of the groups automatically formed by HDBSCAN plus an extra one (misc) not shown in the figure but detailed in Table 2. All clustered images are provided here[7].

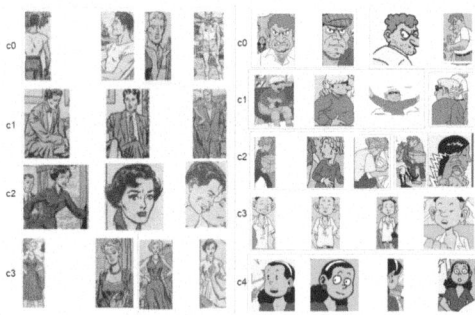

Fig. 3. Example of character clustering output: four and five clusters computed for "Escape with me" (left) and "Patents" (right) respectively.

Table 2. Character clustering confusion matrix "Escape with me" and "Patents".

	Escape				Patents				
Ref./Pred.	c0	c1	c2	c3	c0	c1	c2	c3	c4
c0	**11**	3	0	0	**13**	0	0	0	0
c1	4	**10**	0	0	1	**10**	0	0	0
c2	1	0	**14**	0	0	0	**19**	0	0
c3	0	0	6	**7**	0	0	0	**6**	1
c4	–	–	–	–	0	0	0	0	**13**
Misc	2	4	13	0	2	6	1	2	3

For "Escape with me" episode, we observed the clustering algorithm as no difficulty to separate male and female characters but mixes sometimes two female

---

[6] https://huggingface.co/laion/CLIP-ViT-L-14-DataComp.XL-s13B-b90K.
[7] Character set: https://gitlab.univ-lr.fr/crigau02/publication/2024/MANPU.

characters having only different hair colours or changing clothe a lot e.g. shirtless, in the shadows. The misc category contains a lot of $c2$ character, we think that they are mainly due to a lot of back views, from a distance or with face masked for this main character.

The title "Patents" shows less overall confusion, probably because of better image quality and more colourful characters. However, the first cluster $c0$ contain in reality two different characters that sometimes wear the same hat and are always together in the story (very close with some body part overlaps and sharing same background). We observed the same issue for the last character visible in cluster $c2$ (dark-haired woman with straight hair). She is also very close to the main $c2$ character in the story and her jacket has same colour and zigzag symbols that $c2$ character hairs. We believe that masking the background could improve the clustering. Note that these errors have no impact on the character name inference because it is based on dialogue text only. However, there might be an error propagation during panel description, so we correct them manually before name inference step.

Name Inferences. Character's name inferences are made from the generated script, as introduced in Sect. 3.2. In fact, we discovered that LLM have enough reasoning capacity to find protagonist's names from a sequence of ordered and identified dialogues, as long as they are quoted at least once by one of the characters. Dialogue order and association with comic characters can be computed using automatic methods from the literature (see Sect. 2.2). We used speech balloon and character association method from [34] and we measured its accuracy as an indication, assuming that all visible characters have been correctly detected. They are 77.7% and 88.9% respectively for "Patents" and "Escape with me" respectively. We observed different error types in this method: tail orientation not taken into account (fixed by [16]), balloon with multiple tails, corresponding character not visible (out-of-panel, by phone or in a vehicle, highly overlapping character). We fixed all wrong associations manually before generating the corresponding scripts used for name inferring, in order to avoid any error propagation effects.

Below is an example of LLM output after applying the chain-of-prompts introduced in Sect. 3.2 with the generated script of episode "Escape with me" containing 80 dialogue type texts. All tested LLM outputs are available here [8]:

```
ASSISTANT:
    1. Cynthia Allen (c2) - The protagonist of the story, a young
    woman from a wealthy family who is struggling to break free
    from her parents' expectations.
    2. Bill Patterson (c1) - The gardener who Cynthia meets and
    falls in love with. He is initially portrayed as out of place
    in his job, but later reveals his own dreams and struggles.
```

---

[8] LLM assistant outputs: https://gitlab.univ-lr.fr/crigau02/publication/2024/MANPU.

3. Curt (c0) - Cynthia's boyfriend, who represents the life she is expected to lead by her parents.
4. Cynthia's parents (c4, c3) - They are not named, but are referred to as Cynthia's parents. They want Cynthia to stay in Carleton and lead a comfortable, privileged life.
5. Gloria (c3) - A character who hosts gatherings where Cynthia and Bill first meet. She is portrayed as snobbish and dismissive towards Bill.
6. The unnamed character (?) - There are a few instances where an unnamed character is referenced, likely representing the other guests at Gloria's gatherings.

Primary character's names are correct in this example: Cynthia (c2), Bill (c1) and Curt (c0). Secondary character's names "Gloria" is well predicted as $c3$ but is also wrongly mentioned as not named in the fourth description. The character $c4$ (Cynthia's dad) is effectively not named in this story. The unnamed characters are representing different character throughout the story (sometimes they are just extras). We should find a way to differentiate them even if they are not named, and associate them to characters from other panels if they appear many times.

After character's name inference, we replace IDs by names in the script with a simple find-and-replace function, The full scripts containing character IDs (before inference) and with IDs replaced by names are provided here for both titles[9] named as _script_1_ and _script_2_ respectively.

4.4 Contextual Panel Description

Panel image description are generated following the presented chain-of-prompt approach [36] of visual LLM (VLLM) to extract first general visual elements from the scene, then contextualize it with character's names and text from the image and the script. We randomly selected VLLM from the currently 16 models available in WildVision [20] Arena[10] to show qualitative results among a large panel of VLLM instead of specific ones. We processed panel description using at least two different VLLM with slight variation of the user prompt and selected the most accurate descriptions for inclusion in the final script, named _script_3_ (see here[11]).

Note that for this experiment, we manually fixed errors from previous character clustering step (see Sect. 4.3) to avoid error propagation in image description. We evaluated the panel description by measuring the semantic textual similarity with a human annotated panel description for some pages that we use as ground truth. Text similarity is measured with Sentence Transformers [42] implemented in sbert.net library, based on cosine similarity between text embeddings and

---

[9] Script set: https://gitlab.univ-lr.fr/crigau02/publication/2024/MANPU.
[10] Vision arena: https://huggingface.co/spaces/WildVision/vision-arena.

with the model *mxbai-embed-large-v1* [37]. We evaluated the 14 panel descriptions from the two pages of episode "Escape with me" available in eBDtheque dataset and also the 12 panels from the two first pages of "Patents".

Concerning the pages from the episode "Escape with me", the average similarity score is 60.12% and 69.95% for the second book. The evaluation table containing similarity scores for each panel is available here[11]. We observed that the descriptions depicted character positions, interactions, genres, and attitudes. The different models considered the information from the provided script extract, and we believe this information plays a major role in the similarity measure. We should find a more natural way to identify the unnamed speaking characters, especially for subsequent processing such as text-to-speech. We tried to remove the character label "?" from the image and some models changed the related part of the description by "a voice off-panel says: ...". This is not correct in the case of Fig. 2 because the character is visible in the panel, it is just unnamed.

4.5 Script Generation

The evaluation of the overall quality of the generated script-like description of the story should ideally be done by accessibility experts who are promoting written accessibility[12]. Unfortunately, we have not yet collaborated with such experts to build a public dataset with them. This will be done in a future work. Another way could be to re-generate an artificial comic book images sequence from the script and compare it to the original comics version.

5 Conclusion and Future Work

We show that usual comics content analysis combined with zero-shot transformers prompt engineering are paving the way to accessible comics, even without any fine-tuning on comic-specific datasets. We proposed a straight forward comics script generation based on extracted comics content and enriched with text types, character names and detailed panel description. The generated script can easily be fed into a text-to-speech module to produce an audiobook and also be used for advanced text-based search and indexing purposes.

In the future, we plan to explore additional text categories for illustrative text, etc. aiming to eliminate confusion with onomatopoeia. Additionally, we aim to enhance panel descriptions by incorporating content from previous and next panels, thereby improving coherence and avoiding unnecessary repetition. These enhancements will be validated by accessibility experts. This research as to be extended to other comics genres and LLM understanding capacities should be challenged across languages using multilingual books for instance. Page layout and album summary/synopsis information could also be added by extending the chain-of-prompt, according to blind and low-vision people requirements.

---

[11] Evaluation tables: https://gitlab.univ-lr.fr/crigau02/publication/-/tree/main/2024/MANPU/output/panel_description.

[12] Accessibility experts: https://www.avh.asso.fr/en or https://mangomics-access.fr.

For text-to-speech, complementary information such as balloon contour analysis could be used to include speech tone (e.g. shouted, whispered, thought) into the script and modulate the tone and speed of speech synthesis.

References

1. Augereau, O., Iwata, M., Kise, K.: A survey of comics research in computer science. J. Imaging **4**(87), 87 (2018). https://doi.org/10.3390/jimaging4070087
2. Becht, E., et al.: Dimensionality reduction for visualizing single-cell data using UMAP. Nat. Biotechnol. **37**(1), 38–44 (2019)
3. Campbell-Barner, A.: Reorganizing narratives: increasing accessibility to comic book literature (2021)
4. Carroll, P.J., Young, J.R., Guertin, M.S.: Visual Analysis of Cartoons: A View from the Far Side. In: Rayner, K. (eds.) Eye Movements and Visual Cognition. Springer Series in Neuropsychology. Springer, New York, NY (1992). https://doi.org/10.1007/978-1-4612-2852-3_27
5. Cohn, N.: Navigating comics: An empirical and theoretical approach to strategies of reading comic page layouts. Front. Psychol. **4**, 186 (2013). https://doi.org/10.3389/fpsyg.2013.00186
6. Devi, M.K., Fathima, S., Baskaran, R.: CBCS-Comic book cover synopsis: generating synopsis of a comic book with unsupervised abstractive dialogue. Procedia Comput. Sci. **172**, 701–708 (2020)
7. Dittmar, J.: Comics for the blind and for the seeing. Int. J. Comic Art **16**(1), 458–476 (2014)
8. Fontes, I.V.D.S., dos Santos Miguel, L., Domiciano, C.L.C., Henriques, F.: Aspectos de diseño para la lectura de cómics digitales por personas con discapacidad visual. Cuadernos del Centro de Estudios de Diseño y Comunicación (166) (2022)
9. Fujimoto, A., Ogawa, T., Yamamoto, K., Matsui, Y., Yamasaki, T., Aizawa, K.: Manga109 dataset and creation of metadata. In: Proceedings of the 1st International Workshop on Comics Analysis, Processing and Understanding, pp. 1–5 (2016)
10. Guérin, C., et al.: eBDtheque: a representative database of comics. In: Proceedings of the 12th International Conference on Document Analysis and Recognition (ICDAR) (2013)
11. Guo, H., Wang, B., Bai, J., Liu, J., Yang, J., Li, Z.: M2C: towards automatic multimodal manga complement (2023). arXiv preprint arXiv:2310.17130
12. Huh, M., Lee, Y., Choi, D., Kim, H., Oh, U., Kim, J.: Cocomix: utilizing comments to improve non-visual webtoon accessibility. In: Proceedings of the 2022 CHI Conference on Human Factors in Computing Systems, pp. 1–18 (2022)
13. Iyyer, M., et al.: The amazing mysteries of the gutter: drawing inferences between panels in comic book narratives. In: Proceedings of the IEEE Conference on Computer Vision and Pattern Recognition, pp. 7186–7195 (2017)
14. Laubrock, J., Dunst, A.: Computational approaches to comics analysis. Top. Cogn. Sci. **12**(1), 274–310 (2020)
15. Lee, Y.J., Joh, H., Yoo, S., Oh, U.: AccessComics2: understanding the user experience of an accessible comic book reader for blind people with textual sound effects. ACM Trans. Accessible Comput. **16**(1), 1–25 (2023)
16. Lenadora, D.S., Ranathunge, R.R., Samarawickrama, C.N., De Silva, Y.I., Perera, I., Welivita, A.: Extraction of semantic content and styles in comic books. Int. J. Adv. ICT Emerg. Regions **13**(1), 1–12 (2020)

17. Li, Y., Aizawa, K., Matsui, Y.: Manga109Dialog a large-scale dialogue dataset for comics speaker detection (2023). arXiv preprint arXiv:2306.17469
18. Li, Y., Hinami, R., Aizawa, K., Matsui, Y.: Zero-shot character identification and speaker prediction in comics via iterative multimodal fusion (2024). arXiv preprint arXiv:2404.13993
19. Lord, L.L.G.: Comics: the (not only) visual medium. Ph.D. thesis, Massachusetts Institute of Technology (2016)
20. Lu, Y., Jiang, D., Chen, W., Wang, W., Choi, Y., Lin, B.Y.: WildVision arena: Benchmarking multimodal LLMs in the wild (2024). https://huggingface.co/spaces/WildVision/vision-arena/
21. McInnes, L., Healy, J., Astels, S., et al.: HDBSCAN: hierarchical density based clustering. J. Open Source Softw. **2**(11), 205 (2017)
22. Min, D., et al.: Exploring the impact of table-to-text methods on augmenting LLM-based question answering with domain hybrid data (2024). arXiv preprint arXiv:2402.12869
23. Minaee, S., et al.: Large language models: A survey (2024). arXiv preprint arXiv:2402.06196
24. Nguyen, N.V., Rigaud, C., Burie, J.C.: Comic MTL: optimized multi-task learning for comic book image analysis. Int. J. Doc. Anal. Recogn. (IJDAR) **22**, 265–284 (2019)
25. Nguyen, N.-V., Rigaud, C., Revel, A., Burie, J.-C.: Manga-MMTL: multimodal multitask transfer learning for manga character analysis. In: Lladós, J., Lopresti, D., Uchida, S. (eds.) ICDAR 2021. LNCS, vol. 12822, pp. 410–425. Springer, Cham (2021). https://doi.org/10.1007/978-3-030-86331-9_27
26. Oh, U., Joh, H., Lee, Y.: Image accessibility for screen reader users: A systematic review and a road map. Electronics **10**(8), 953 (2021). https://doi.org/10.3390/electronics10080953
27. Ohnaka, H., Takamichi, S., Imoto, K., Okamoto, Y., Fujii, K., Saruwatari, H.: Visual Onoma-to-wave: environmental sound synthesis from visual onomatopoeias and sound-source images. In: ICASSP 2023-2023 IEEE International Conference on Acoustics, Speech and Signal Processing (ICASSP), pp. 1–5. IEEE (2023)
28. Ponsard, C., Ramdoyal, R., Dziamski, D.: An OCR-enabled digital comic books viewer. In: Miesenberger, K., Karshmer, A., Penaz, P., Zagler, W. (eds.) ICCHP 2012. LNCS, vol. 7382, pp. 471–478. Springer, Heidelberg (2012). https://doi.org/10.1007/978-3-642-31522-0_71
29. Radford, A., et al.: Learning transferable visual models from natural language supervision (2021)
30. Ramaprasad, R.: Comics for everyone: Generating accessible text descriptions for comic strips (2023). https://arxiv.org/abs/2310.00698
31. Rayar, F.: Accessible comics for visually impaired people: challenges and opportunities. In: 2017 14th IAPR International Conference on Document Analysis and Recognition (ICDAR). vol. 3, pp. 9–14. IEEE (2017)
32. Rayar, F., Oriola, B., Jouffrais, C.: ALCOVE: an accessible comic reader for people with low vision. In: 25th ACM International forum for reporting outstanding research and development on Intelligent User Interfaces (ACM IUI 2020), pp. 410–418. ACM : Association for Computing Machinery, Cagliari, IT (2020). https://doi.org/10.1145/3377325.3377510
33. Rigaud, C., Guérin, C., Karatzas, D., Burie, J.C., Ogier, J.M.: Knowledge-driven understanding of images in comic books. Int. J. Doc. Anal. Recogn. (IJDAR) **18**, 199–221 (2015)

34. Rigaud, C., et al.: Speech balloon and speaker association for comics and manga understanding. In: 2015 13th International Conference on Document Analysis and Recognition (ICDAR), pp. 351–355 (2015). https://doi.org/10.1109/ICDAR.2015.7333782
35. Sachdeva, R., Zisserman, A.: The manga whisperer: Automatically generating transcriptions for comics (2024). arXiv preprint arXiv:2401.10224, https://arxiv.org/abs/2401.10224
36. Sahoo, P., Singh, A.K., Saha, S., Jain, V., Mondal, S., Chadha, A.: A systematic survey of prompt engineering in large language models: Techniques and applications (2024). https://arxiv.org/abs/2402.07927
37. Sean, L., Aamir, S., Darius, K., Julius, L.: Open source strikes bread - new fluffy embeddings model (2024). https://www.mixedbread.ai/blog/mxbai-embed-large-v1
38. Sharma, R., Kukreja, V.: Image segmentation, classification and recognition methods for comics: a decade systematic literature review. Eng. Appl. Artif. Intell. **131**, 107715 (2024)
39. Shen, C.T., Yao, V., Liu, Y.: MaRU: A manga retrieval and understanding system connecting vision and language (2023). https://arxiv.org/abs/2311.02083
40. Sousanis, N.: Accessible comics for blind and low-vision readers: An emerging journey. In: 2023 MLA Annual Convention. MLA (2023)
41. Soykan, G., Yuret, D., Sezgin, T.M.: Identity-aware semi-supervised learning for comic character re-identification (2023). https://arxiv.org/abs/2308.09096
42. Thakur, N., Reimers, N., Daxenberger, J., Gurevych, I.: Augmented SBERT: data augmentation method for improving bi-encoders for pairwise sentence scoring tasks. In: Proceedings of the 2021 Conference of the North American Chapter of the Association for Computational Linguistics: Human Language Technologies. pp. 296–310. Association for Computational Linguistics, Online (2021). https://www.aclweb.org/anthology/2021.naacl-main.28
43. Tsubota, K., Ogawa, T., Yamasaki, T., Aizawa, K.: Adaptation of manga face representation for accurate clustering. In: SIGGRAPH Asia 2018 Posters, pp. 1–2 (2018)
44. Vivoli, E., Baeza, J.L., Llobet, E.V., Karatzas, D.: Multimodal transformer for comics text-cloze (2024). arXiv preprint arXiv:2403.03719
45. Wang, Y., Wang, W., Liang, W., Yu, L.F.: Comic-guided speech synthesis. ACM Trans. Graph. (TOG) **38**(6), 1–14 (2019)
46. Yanagisawa, H., Kyogoku, K., Ravi, J., Watanabe, H.: Automatic classification of manga characters using density-based clustering. In: Lau, P.Y., Shobri, M. (eds.) International Workshop on Advanced Imaging Technology (IWAIT) 2020, vol. 11515, pp. 115150F. International Society for Optics and Photonics, SPIE (2020). https://doi.org/10.1117/12.2566845
47. Yanagisawa, H., Yamashita, T., Hiroshi, W.: Manga character clustering with DBSCAN using fine-tuned CNN model. In: Kemao, Q., Hayase, K., Lau, P.Y., Lie, W.N., Lee, Y.L., Srisuk, S., Yu, L. (eds.) International Workshop on Advanced Image Technology (IWAIT) 2019. vol. 11049, pp. 110491M. International Society for Optics and Photonics, SPIE (2019). https://doi.org/10.1117/12.2521116
48. Zhang, Z., Wang, Z., Hu, W.: Unsupervised manga character re-identification via face-body and spatial-temporal associated clustering (2022). https://arxiv.org/abs/2204.04621

Quantitative Evaluation Based on CLIP for Methods Inhibiting Imitation of Painting Styles

Motoi Iwata[✉], Keito Okamoto, and Koichi Kise

Osaka Metropolitan University, 1-1 Gakuencho, Sakai, Osaka, Japan
{imotoi,kise}@omu.ac.jp, sc24688k@st.omu.ac.jp
https://imlab.jp/

Abstract. Image generation AIs with the ability to generate a variety of high-quality images from text and images have been gaining attention, meanwhile, they cause a serious problem where a third party generates AI art that imitates the artist's style by fine-tuning from a specific artist's work. Against this background, a method has been proposed to add small noise perturbations to an artwork to inhibit imitation of the painting style. Currently, two such methods exist: Glaze and Mist, which apply perturbations to artworks so that a false style is learned, and Mist, which makes it difficult to extract features from the work. Glaze and Mist differ from each other in their evaluation manners, and they are not able to evaluate the real problem described above, i.e., the inhibition of imitating the style of a particular artist. Therefore, the purpose of this study is to realize a quantitative evaluation based on the index of artist-likeness. In this paper, we discuss three points: whether the proposed artist-likeness index and the correct prediction rate are suitable for quantitative evaluation of artist-likeness, how much it inhibits the imitation of painting style quantitatively, and whether Glaze or Mist is superior. Our experimental results confirm that the proposed metrics reflect the artist-likeness and quantitatively evaluate, how much Glaze and Mist inhibit the imitation of painting styles quantitatively, and we find that Mist is superior to Glaze according to our approach.

Keywords: Image generation · Adversarial examples · Inhibition of imitation of painting style · Quantitative evaluation

1 Introduction

Since 2022, image generation AI such as Stable Diffusion[1] and MidJourney[2] have been gaining attention for their ability to generate a variety of high-quality images from text and images. Among them, an image generation model called Stable Diffusion is available to the public free of charge, and its source code

---
[1] https://ja.stability.ai/stable-diffusion.
[2] https://www.midjourney.com/.

Quantitative Evaluation Based on CLIP for Methods Inhibiting Imitation

Fig. 1. Incident of training generative AI without artist's permission. These images are from the web article [1].

and trained models can be freely obtained[3] and fine-tuned by individuals to generate images. Fine-tuned image generation models can generate images that closely imitate the style and characters of the original works which are used for fine-tuning. This creates a problem where a third party generates AI art that imitates the artist's style by fine-tuning from a specific artist's work. For example, Fig. 1 shows a case in which a model capable of generating images similar to Hollie Mengert's illustrations was released to the public without her consent [1]. Such AI-based imitation of art style has problems such as commercial use of the artist's model and possible loss of identity. Then some artists have had no choice but to remove their works from the Internet to avoid imitation [7].

Against this background, a method has been proposed to add small noise perturbations to an art work in order to inhibit imitation of the painting style. Currently, two such methods exist: Glaze [7] and Mist [5], which apply perturbations to art works so that a false style is learned, and Mist, which makes it difficult to extract features from the work. Moreover, Nightshade which has the similar purpose to them was released on October 2023 by the same team as Glaze and will be presented on the 45th IEEE Symposium on Security and Privacy on May, 2024 [8].

Glaze and Mist differ from each other in their evaluation manners, and they are not able to evaluate the real problem described above, i.e., the inhibition of imitating the style of a particular artist. Therefore, the purpose of this study is to realize a quantitative evaluation based on the index of artist-likeness. To achieve this quantitative evaluation metric, we will use a model called CLIP, which calculates the similarity between images and texts. In addition, we will perform quantitative evaluation of the generated images in Glaze and Mist to confirm

---

[3] https://github.com/Stability-AI/stablediffusion.

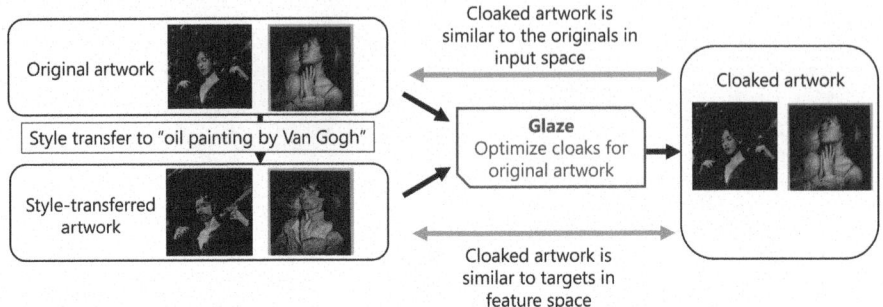

Fig. 2. Overall of Glaze.

the effectiveness of inhibiting style imitation and to compare the performance of Glaze and Mist.

The paper consists of five sections as follows: Sect. 2 describes related work, and Sect. 3 describes the proposed quantitative evaluation of artist-likeness. Section 4 presents the experiments and results, and Sect. 5 concludes with a summary and future works.

2 Related Work

2.1 Glaze

Shan et al. proposed Glaze, the method inhibiting the imitation of painting style [7]. Figure 2 shows how Glaze inhibits imitation. As shown in Fig. 2, the input is an image that a user would like to prevent from being imitated. First, the input image is transformed into an image that is stylistically transformed into another fake painting style, where the image is transformed into a Van Gogh-style image in the figure). Style transformation converts an image to a different style while preserving the objects and persons in the image. Next, a perturbation, which is a weak noise, is calculated and the input image with the perturbation is called a cloaked image. The perturbation is calculated so that the cloaked image is visually similar to the input image, meanwhile it is close to the style-transformed image in latent space. This causes the generation of an image with a different style from the original input image when the cloaked images (hereinafter referred to as Glazed images) are used for fine-tuning.

Shan et al. employed a new CLIP-based index called CLIP-based genre shift as an evaluation metric for quantitative evaluation [7]. The details of the calculation of CLIP-based genre shift are as follows: First, CLIP predicted the top three genres of an input image, where 27 historical and 13 digital image styles are used as genre labels. Here, the input image was assumed as a generated image by the model trained by a particular artist's works. The CLIP-based genre shift score was defined as the percentage of the top three predicted genres that did not include the genre to which the original artist belonged. Therefore, the high

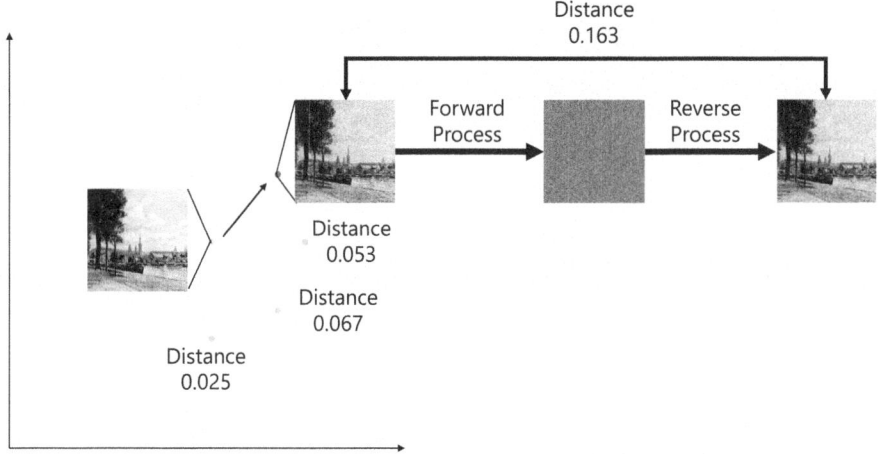

Fig. 3. Process that AdvDM generates perturbations.

CLIP-based genre shift score indicated that most of the imitated images belong to a different genre from the original artist, which meant that the style of the imitated image was successfully inhibited.

2.2 Mist

Liang et al. proposed AdvDM as a method to generate adversarial examples against the diffusion model and published a software program using AdvDM under the name of Mist [5]. Figure 3 shows the process by which AdvDM generates perturbations. The horizontal axis represents the learning process in the diffusion model. AdvDM uses the trained diffusion model to obtain each latent variable with an input image that would like to be avoided from being imitated. Next, multiple perturbations are added to the latent variables in random directions, and each latent variable is then passed through the Diffusion model with forward and reverse processes to predict the corresponding latent variable. Then, the distance between the predicted latent variable and the original latent variable is calculated, and the image with the perturbation that is farthest from the original one is used for the next input image. The above operation is repeated. This operation produces another style image with a higher probability generated by the perturbation by AdvDM, whereas the original image would normally be restored by denoising the latent variable if AdvDM would not be applied. As a result, the Diffusion Model fails to generate images with a similar style to the input images.

The literature that proposed Mist [5] evaluates the quality of the generated image compared to the training source image based on Fréchet Inception Distance (FID) [2] and Precision for objects such as cats and airplanes [4]. However, no such quantitative evaluation has been performed for style, so the effectiveness of Mist in inhibiting style imitation has not been confirmed.

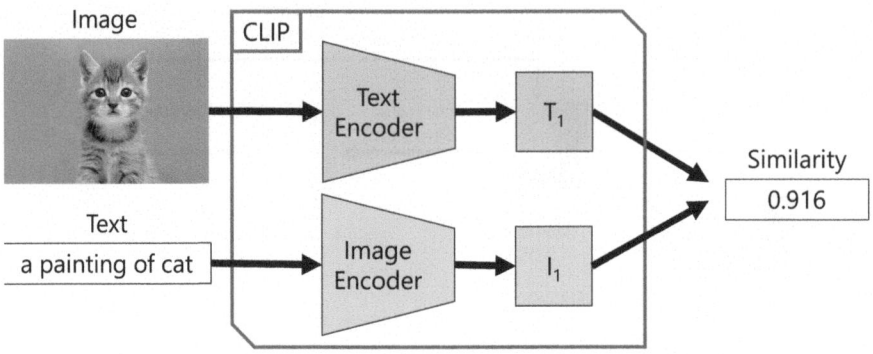

Fig. 4. Example of CLIP in action.

2.3 CLIP

There is a method called CLIP that calculates the similarity between images and text [6]. CLIP is a model developed by OpenAI that uses image-text pairwise contrast learning to learn the relationship between image and text and can perform various image classification tasks with high accuracy without additional learning. Figure 4 shows an example of CLIP in action. Given an image of a cat and the text "a painting of cat" as input, the similarity between the image and text is obtained.

3 Quantitative Evaluation of Artist-Likeness

3.1 Definition of Artist-Likeness Index

In this section, we explain the definition of the artist-likeness index in this experiment. In this experiment, the artist-likeness index is defined as the output of CLIP with the input of a target image and a text including the specific artist name, where we use "a painting by <artist name>" as the text. The artist-likeness index shows us how much an input image is the input artist-like. For example, if the artist-likeness index with artist A is larger than the one with artist B for the same image, it means that the image is artist A-like than artist B-like. Therefore, the artist-likeness index can be used as the evaluation metric for methods inhibiting the imitation of painting styles.

The CLIP-based genre shift score used in the evaluation for Glaze is the similarity of rough image styles, for example, romanticism or impressionism. Therefore, the artist-likeness index defined here is more suitable for evaluating the methods inhibiting the imitation of painting styles.

3.2 Definition of Correct Prediction for Generated Images

This section describes the definition of the correct prediction for generated images, which is used for calculating correct rate described in the next section.

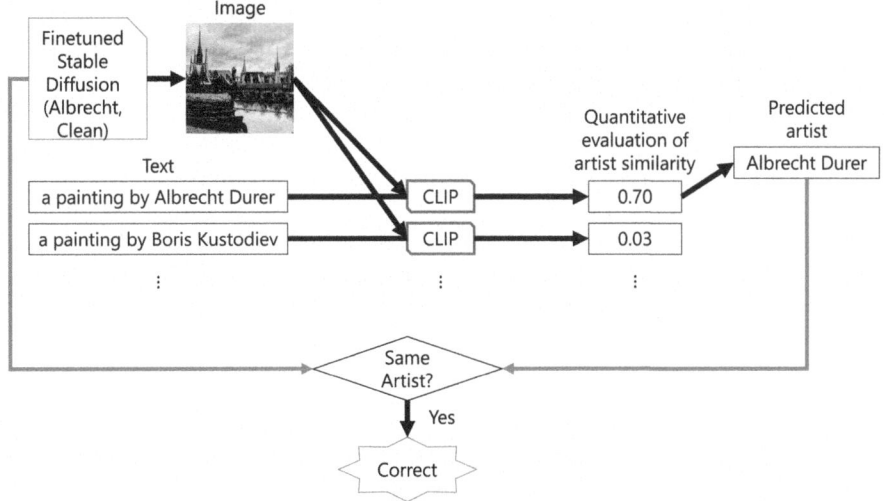

Fig. 5. Process of predicting the artist used for fine-tuning.

Figure 5 shows the process of predicting the artist used for the fine-tuning. First, an image is generated from a fine-tuned model trained by specific artist's images. Next, the artist-likeness index for each artist is obtained, where they are calculated by using "a painting by <artist name>" with each artist's name for one of the inputs to CLIP. Finally, the artist with the highest artist-likeness index is predicted as the artist used for the fine-tuning to generate the image. We define it "correct prediction" when the predicted artist is the same as the artist used for the fine-tuning.

3.3 Definition of Correct Prediction Rate

This section describes the definition of the correct prediction rate, which is used for quantitative evaluation. The correct prediction rate is defined as the percentage of correct predictions described in Sect. 3.2 obtained from multiple images generated by fine-tuned model with a specific artist's images. The correct prediction rate is calculated for each artist. Therefore, by comparing the correct prediction rates of various artists, we can quantitatively analyze how the imitation of painting styles is inhibited. Note that low correct prediction rates mean that methods inhibiting the imitation of painting styles work well.

4 Experiment

4.1 Experimental Condition

In this study, we used the Wikiart Dataset [9], the collection of paintings from WikiArt, the website that compiles publicly available paintings, which contains a total of 52,757 paintings by 195 artists. In addition, 23 artists have more than 500 works in the dataset. In this experiment, we used the images of these 23 artists. First, in order to confirm the validity of the quantitative evaluation of the proposed artist-likeness index, we calculated the correct prediction rate for each artist, with the texts and images of the above 23 artists' names as input to CLIP for all 23 artists' images. We call these correct prediction rates "SOURCE".

Next, in order to quantitatively check the extent of inhibiting the imitation of painting styles, 32 images from 23 artists' images in Wikiart Dataset were randomly selected and used as training data for the fine-tuning of Stable Diffusion. The correct prediction rates were calculated for each artist from the generated images and called "CLEAN". Moreover, we applied Glaze to the above 32 images and then used the Glaze images for fine-tuning. After that, we calculated the correct prediction rates for each artist from the generated images trained by the Glazed images and called them "GLAZE". Also, we employed Mist instead of Glaze in the same manner and called the correct prediction rates "MIST".

For the above fine-tuning, we used Stable Diffusion v1.5 as the base model and LoRA [3], the most mainstream method for imitating painting styles using Stable Diffusion. This method has the advantages of being able to imitate the style from a small number of images (several dozen or so) and the size of the model created by fine tuning is smaller than other methods. This technique was also used to imitate the style of Hollie Mengert's illustrations described in Sect. 1. For the image generation, the parameters of Stable Diffusion were set to default settings, and five images were generated for each prompt using the same prompts that were used for learning to imitate the style of the illustrations. This setting of the prompts ensures that the output image is in the style of the trained image and reduces the number of cases in which the generated image does not resemble the artist's style.

4.2 Experimental Results

Figures 6 and 7 show the correct prediction rates of SOURCE and the correct prediction rates of CLEAN, GLAZE, and MIST, respectively, where the numbers in parentheses in the legend represent the averages of all artists for each correct prediction rate. Appendix A shows a table summarizing the averages of artist-likeness indices used for calculating SOURCE, CLEAN, GLAZE, and MIST.

First, we focus on SOURCE in order to confirm the validity of the quantitative evaluation of the proposed artist-likeness index. Figure 6 shows SOURCE, the correct prediction rates for each artist when the input images to CLIP are the artist's images. As shown in Fig. 6, the average of SOURCE is 0.674 and the lowest is 0.297. They are much higher than the chance rate of 0.043 ($=1/23$). The results confirm that the output of CLIP with the texts used in this experiment reflects the artist's painting style.

Next, we quantitatively check the extent of inhibiting the imitation of painting styles. Figure 7 shows the correct prediction rates of CLEAN, GLAZE, and MIST. As shown in Fig. 7, both GLAZE and MIST are lower than CLEAN for 14 of 23 artists, where it means that both Glaze and Mist inhibit the imitation of

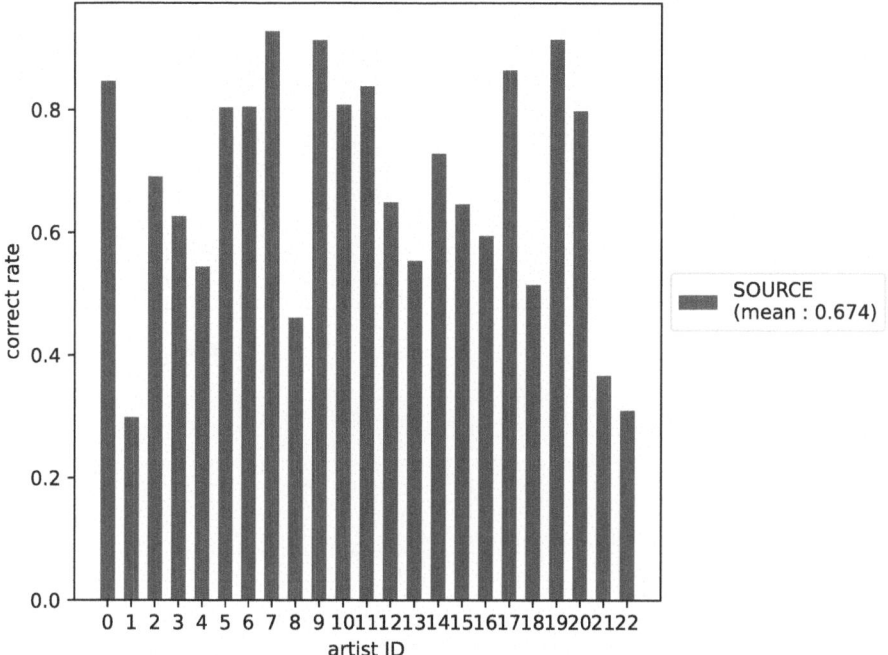

Fig. 6. Correct prediction rates of SOURCE for each artist.

painting styles of 14 artists. Moreover, GLAZE or MIST is lower than CLEAN for 21 of 23 artists. Only for the artist IDs 1 and 5, both GLAZE and MIST are higher than CLEAN. We performed the Wilcoxon signed-rank sum test between CLEAN and GLAZE, and between CLEAN and MIST at the significance level of 0.05 ($p = 0.05$), and found that there was a significant difference between CLEAN and both GLAZE and MIST as follows:

$$p_{\text{glaze}} = 0.0101 < 0.05 \tag{1}$$

$$p_{\text{mist}} = 0.0023 < 0.05 \tag{2}$$

where p_{glaze} and p_{mist} represent p-values of the Wilcoxon signed-rank sum test between CLEAN and GLAZE and between CLEAN and MIST, respectively. This confirms the effectiveness of Glaze and Mist in inhibiting the imitation of the painting style. However, for some artists, the difference between CLEAN and GLAZE and/or MIST is not so significant. This indicates that the style of such artists remains after the process of Glaze or Mist, and it was also confirmed that Glaze and Mist are not always effective for the style of all artists.

Finally, we compare the performances of Glaze and Mist. As shown in Fig. 7, the average of MIST is lower than that of GLAZE. By artist, MIST is lower than GLAZE for 14 of 23 artists. We perform the Wilcoxon signed-rank sum test between GLAZE and MIST at a significance level of 0.05 ($p = 0.05$), and found that there is a significant difference.

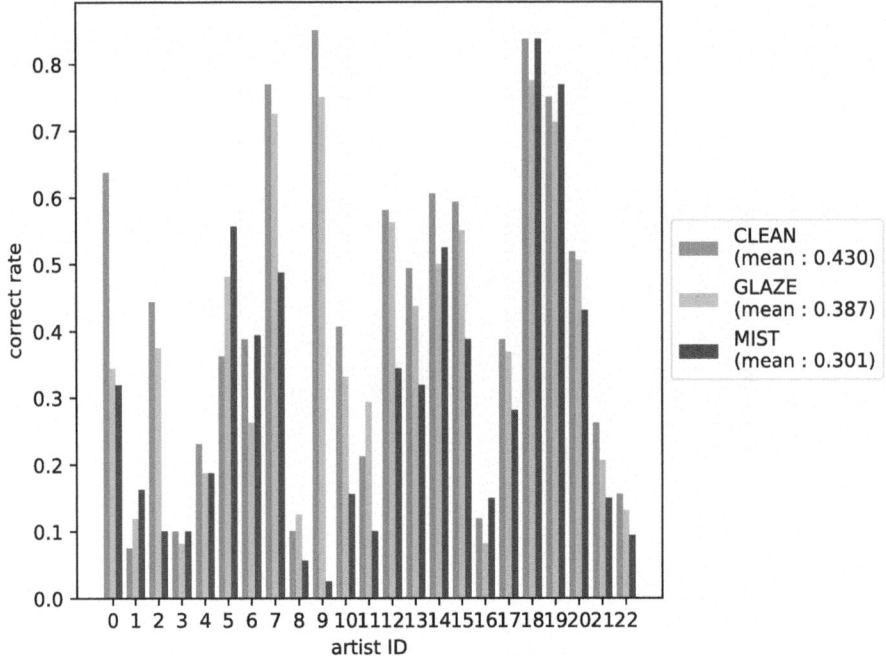

Fig. 7. Correct prediction rates of CLEAN, GLAZE, and MIST for each artist.

$$p_{\text{gm}} = 0.0296 < 0.05 \tag{3}$$

where p_{gm} represents p-value of of the Wilcoxon signed-rank sum test between GLAZE and MIST. We found that Mist was better than Glaze at inhibiting the imitation of the painting style. In addition, MIST is smaller than GLAZE by more than 0.1 for eight artists, while GLAZE is smaller than MIST by more than 0.1 only for one artist ID 6. This means that Mist has higher performance for more than 1/3 of the artists, meanwhile Glaze does for one artist.

5 Conclusion

In this paper, we discussed three points: whether the proposed artist-likeness index and the correct prediction rate are suitable for quantitative evaluation of artist-likeness, how much it inhibits the imitation of painting style quantitatively, and whether Glaze or Mist is superior. Our experimental results confirmed that the proposed metrics reflect the artist-likeness and quantitatively evaluate, how much Glaze and Mist inhibit the imitation of painting styles quantitatively, and we found that Mist was superior to Glaze according to our approach.

Some additional experimental settings are the future works. Firstly, we used only 32 out of 500 images for each artist for fine-tuning Stable Diffusion. Perhaps this is not sufficient for training the artist style. Secondly, we used the pre-trained CLIP model as it is. It would be interesting to fine-tune CLIP on the Wikiart Dataset to improve the performance of the proposed metric. Other future work is needed to visualize and evaluate the element of artist-likeness, i.e., on what basis the artist is predicted to be this type of artist. This would allow us to evaluate the method of inhibiting the imitation of painting style based on which elements were changed by the method.

Acknowledgments. This work was supported in part by JSPS Kakenhi (23KK0188).

A Average Artist-Likeness Indices for Each Artist

Table 1, 2, 3, 4, 5, 6, 7 and 8 show the average artist-likeness index of 160 images for each artist. Each row represents the artist ID used for fine-tuning, while each column represents the artist ID used for the input text to CLIP. We make the average artist-likeness indices bold when the artist ID used for fine-tuning and one for the input text to CLIP are the same.

Table 1. Average artist-likeness indices for SOURCE. The artist ID for each row is from 0 to 11.

	0	1	2	3	4	5	6	7	8	9	10	11
0	**0.754**	0.003	0.002	0.001	0.001	0.008	0.008	0.012	0.005	0.000	0.004	0.028
1	0.031	**0.254**	0.020	0.008	0.007	0.043	0.010	0.005	0.061	0.002	0.014	0.063
2	0.000	0.006	**0.595**	0.029	0.033	0.024	0.095	0.002	0.009	0.002	0.008	0.014
3	0.000	0.004	0.101	**0.515**	0.062	0.011	0.031	0.002	0.014	0.003	0.007	0.045
4	0.000	0.006	0.151	0.060	**0.447**	0.009	0.038	0.001	0.008	0.001	0.004	0.026
5	0.007	0.008	0.016	0.010	0.005	**0.709**	0.011	0.002	0.019	0.005	0.001	0.040
6	0.002	0.002	0.073	0.018	0.042	0.010	**0.697**	0.002	0.038	0.004	0.002	0.036
7	0.048	0.002	0.005	0.001	0.001	0.004	0.011	**0.839**	0.014	0.003	0.012	0.006
8	0.005	0.063	0.019	0.028	0.015	0.050	0.022	0.023	**0.366**	0.014	0.029	0.110
9	0.001	0.004	0.001	0.000	0.002	0.002	0.011	0.031	0.011	**0.878**	0.011	0.004
10	0.021	0.005	0.022	0.013	0.007	0.004	0.032	0.028	0.023	0.005	**0.704**	0.031
11	0.006	0.004	0.007	0.013	0.005	0.025	0.006	0.005	0.043	0.005	0.006	**0.725**
12	0.023	0.007	0.012	0.002	0.002	0.053	0.008	0.006	0.004	0.000	0.000	0.011
13	0.016	0.017	0.017	0.007	0.009	0.018	0.009	0.004	0.004	0.000	0.005	0.035
14	0.025	0.016	0.005	0.004	0.002	0.011	0.005	0.006	0.008	0.002	0.008	0.019
15	0.011	0.011	0.011	0.003	0.003	0.064	0.006	0.002	0.011	0.000	0.000	0.018
16	0.002	0.003	0.051	0.004	0.012	0.025	0.010	0.001	0.008	0.001	0.001	0.052
17	0.001	0.002	0.035	0.006	0.025	0.029	0.018	0.000	0.007	0.001	0.001	0.030
18	0.004	0.025	0.038	0.010	0.021	0.010	0.017	0.001	0.014	0.000	0.011	0.034
19	0.009	0.013	0.002	0.007	0.001	0.028	0.002	0.001	0.003	0.000	0.000	0.015
20	0.090	0.001	0.016	0.004	0.000	0.026	0.023	0.038	0.018	0.010	0.005	0.020
21	0.027	0.008	0.009	0.011	0.020	0.048	0.015	0.013	0.010	0.004	0.004	0.029
22	0.042	0.010	0.096	0.025	0.019	0.090	0.044	0.004	0.019	0.000	0.010	0.028

Table 2. Average artist-likeness indices for SOURCE. The artist ID for each row is from 12 to 22.

	12	13	14	15	16	17	18	19	20	21	22
0	0.002	0.005	0.006	0.024	0.006	0.005	0.037	0.006	0.067	0.008	0.007
1	0.015	0.056	0.019	0.035	0.015	0.025	0.245	0.026	0.013	0.016	0.019
2	0.002	0.006	0.001	0.002	0.017	0.065	0.075	0.000	0.002	0.001	0.010
3	0.001	0.010	0.008	0.002	0.008	0.036	0.132	0.000	0.001	0.001	0.008
4	0.001	0.009	0.002	0.001	0.016	0.059	0.140	0.000	0.001	0.001	0.019
5	0.006	0.013	0.002	0.014	0.016	0.057	0.036	0.003	0.009	0.005	0.005
6	0.000	0.001	0.000	0.003	0.011	0.012	0.028	0.000	0.011	0.001	0.005
7	0.000	0.002	0.007	0.002	0.001	0.004	0.003	0.000	0.030	0.002	0.005
8	0.003	0.028	0.004	0.008	0.013	0.034	0.123	0.001	0.029	0.003	0.009
9	0.000	0.004	0.002	0.000	0.000	0.001	0.018	0.000	0.014	0.003	0.001
10	0.000	0.003	0.002	0.001	0.006	0.007	0.064	0.000	0.010	0.001	0.009
11	0.001	0.022	0.004	0.010	0.011	0.033	0.050	0.001	0.011	0.004	0.002
12	**0.560**	0.037	0.005	0.084	0.015	0.022	0.080	0.012	0.024	0.025	0.008
13	0.021	**0.441**	0.116	0.040	0.030	0.010	0.146	0.002	0.007	0.027	0.020
14	0.006	0.034	**0.680**	0.033	0.005	0.003	0.081	0.007	0.013	0.018	0.010
15	0.045	0.048	0.006	**0.537**	0.059	0.023	0.078	0.012	0.007	0.028	0.015
16	0.010	0.042	0.002	0.029	**0.486**	0.066	0.169	0.000	0.008	0.005	0.012
17	0.003	0.011	0.000	0.003	0.030	**0.767**	0.022	0.002	0.003	0.001	0.004
18	0.020	0.084	0.015	0.058	0.188	0.027	**0.399**	0.000	0.003	0.002	0.019
19	0.003	0.009	0.001	0.013	0.002	0.005	0.013	**0.848**	0.005	0.015	0.007
20	0.001	0.005	0.001	0.017	0.006	0.009	0.008	0.000	**0.693**	0.001	0.008
21	0.067	0.084	0.015	0.108	0.033	0.009	0.131	0.025	0.023	**0.295**	0.012
22	0.009	0.024	0.003	0.069	0.045	0.036	0.120	0.001	0.053	0.007	**0.245**

Table 3. Average artist-likeness indices for CLEAN. The artist ID for each row is from 0 to 11.

	0	1	2	3	4	5	6	7	8	9	10	11
0	**0.482**	0.004	0.002	0.002	0.004	0.019	0.002	0.105	0.002	0.001	0.003	0.016
1	0.013	**0.102**	0.006	0.001	0.008	0.006	0.001	0.004	0.012	0.006	0.014	0.027
2	0.009	0.024	**0.391**	0.007	0.034	0.005	0.017	0.020	0.003	0.001	0.019	0.010
3	0.001	0.027	0.100	**0.105**	0.043	0.018	0.016	0.006	0.008	0.018	0.007	0.024
4	0.001	0.026	0.081	0.015	**0.222**	0.006	0.005	0.004	0.007	0.008	0.002	0.008
5	0.024	0.010	0.022	0.005	0.011	**0.290**	0.003	0.009	0.005	0.028	0.003	0.048
6	0.002	0.018	0.050	0.012	0.048	0.012	**0.318**	0.007	0.011	0.077	0.011	0.026
7	0.073	0.008	0.005	0.001	0.002	0.004	0.001	**0.652**	0.001	0.018	0.012	0.005
8	0.010	0.087	0.009	0.005	0.035	0.021	0.002	0.023	**0.100**	0.036	0.026	0.027
9	0.001	0.003	0.000	0.000	0.003	0.002	0.003	0.007	0.002	**0.795**	0.003	0.003
10	0.007	0.005	0.015	0.004	0.010	0.007	0.004	0.041	0.002	0.004	**0.389**	0.020
11	0.008	0.021	0.010	0.008	0.030	0.029	0.003	0.014	0.043	0.003	0.007	**0.198**
12	0.017	0.013	0.014	0.004	0.003	0.016	0.010	0.010	0.000	0.000	0.001	0.003
13	0.008	0.007	0.002	0.003	0.003	0.006	0.001	0.002	0.000	0.000	0.003	0.004
14	0.016	0.008	0.001	0.003	0.007	0.001	0.003	0.023	0.001	0.007	0.005	0.004
15	0.005	0.012	0.003	0.003	0.002	0.017	0.000	0.001	0.000	0.001	0.000	0.010
16	0.004	0.015	0.028	0.001	0.013	0.010	0.002	0.004	0.004	0.001	0.001	0.008
17	0.007	0.058	0.047	0.014	0.031	0.034	0.001	0.002	0.007	0.000	0.004	0.009
18	0.002	0.017	0.010	0.002	0.009	0.001	0.001	0.000	0.001	0.000	0.002	0.002
19	0.014	0.021	0.002	0.006	0.002	0.012	0.001	0.007	0.001	0.000	0.000	0.015
20	0.116	0.008	0.008	0.002	0.001	0.012	0.008	0.084	0.015	0.011	0.004	0.024
21	0.009	0.007	0.006	0.005	0.045	0.021	0.002	0.017	0.001	0.004	0.000	0.002
22	0.028	0.008	0.031	0.012	0.004	0.031	0.002	0.004	0.003	0.001	0.004	0.018

Table 4. Average artist-likeness indices for CLEAN. The artist ID for each row is from 12 to 22.

	12	13	14	15	16	17	18	19	20	21	22
0	0.005	0.009	0.043	0.077	0.002	0.003	0.073	0.017	0.059	0.055	0.014
1	0.006	0.159	0.038	0.063	0.006	0.004	0.397	0.023	0.005	0.074	0.023
2	0.003	0.018	0.002	0.001	0.003	0.017	0.399	0.000	0.001	0.003	0.013
3	0.007	0.079	0.069	0.005	0.005	0.002	0.400	0.000	0.002	0.022	0.036
4	0.007	0.086	0.038	0.005	0.008	0.012	0.426	0.000	0.004	0.012	0.018
5	0.017	0.136	0.001	0.058	0.008	0.027	0.185	0.008	0.014	0.079	0.011
6	0.000	0.028	0.009	0.007	0.005	0.002	0.299	0.002	0.012	0.031	0.012
7	0.006	0.016	0.024	0.015	0.001	0.001	0.028	0.007	0.009	0.104	0.007
8	0.002	0.137	0.025	0.024	0.011	0.009	0.323	0.002	0.027	0.048	0.008
9	0.000	0.024	0.029	0.000	0.000	0.000	0.031	0.000	0.002	0.091	0.001
10	0.003	0.056	0.014	0.011	0.011	0.002	0.348	0.000	0.002	0.038	0.009
11	0.005	0.133	0.006	0.054	0.011	0.029	0.249	0.006	0.057	0.069	0.005
12	**0.496**	0.070	0.003	0.071	0.003	0.006	0.165	0.006	0.013	0.061	0.017
13	0.008	**0.412**	0.123	0.069	0.004	0.000	0.256	0.002	0.008	0.060	0.020
14	0.015	0.096	**0.532**	0.039	0.002	0.002	0.164	0.002	0.014	0.039	0.018
15	0.034	0.098	0.029	**0.494**	0.017	0.002	0.152	0.022	0.003	0.086	0.010
16	0.016	0.187	0.022	0.045	**0.124**	0.018	0.431	0.004	0.011	0.042	0.009
17	0.017	0.100	0.000	0.016	0.011	**0.310**	0.300	0.001	0.010	0.012	0.008
18	0.007	0.163	0.008	0.052	0.025	0.002	**0.676**	0.000	0.001	0.011	0.010
19	0.007	0.033	0.002	0.071	0.003	0.001	0.065	**0.682**	0.014	0.029	0.013
20	0.008	0.041	0.017	0.067	0.011	0.008	0.048	0.002	**0.423**	0.026	0.054
21	0.032	0.140	0.020	0.145	0.005	0.003	0.231	0.037	0.020	**0.235**	0.013
22	0.012	0.074	0.030	0.087	0.019	0.006	0.354	0.005	0.049	0.058	**0.160**

Table 5. Average artist-likeness indices for GLAZE. The artist ID for each row is from 0 to 11.

	0	1	2	3	4	5	6	7	8	9	10	11
0	**0.311**	0.004	0.002	0.004	0.005	0.017	0.004	0.110	0.002	0.000	0.002	0.016
1	0.013	**0.132**	0.007	0.012	0.007	0.011	0.002	0.006	0.007	0.007	0.008	0.020
2	0.003	0.023	**0.337**	0.010	0.027	0.012	0.029	0.003	0.001	0.001	0.008	0.011
3	0.002	0.016	0.093	**0.083**	0.035	0.023	0.014	0.004	0.007	0.044	0.006	0.022
4	0.003	0.037	0.072	0.012	**0.169**	0.017	0.007	0.002	0.010	0.016	0.001	0.017
5	0.006	0.013	0.018	0.005	0.023	**0.392**	0.003	0.003	0.006	0.014	0.003	0.037
6	0.001	0.011	0.018	0.018	0.033	0.010	**0.255**	0.002	0.013	0.134	0.011	0.061
7	0.047	0.003	0.006	0.006	0.008	0.011	0.008	**0.562**	0.009	0.018	0.014	0.008
8	0.006	0.080	0.008	0.008	0.031	0.029	0.004	0.025	**0.097**	0.012	0.025	0.037
9	0.001	0.003	0.000	0.000	0.011	0.003	0.006	0.005	0.003	**0.603**	0.004	0.005
10	0.010	0.003	0.009	0.006	0.011	0.02	0.014	0.023	0.004	0.008	**0.322**	0.008
11	0.022	0.012	0.006	0.006	0.014	0.022	0.001	0.014	0.022	0.003	0.016	**0.253**
12	0.015	0.008	0.011	0.001	0.003	0.011	0.016	0.005	0.000	0.000	0.000	0.003
13	0.018	0.010	0.004	0.003	0.006	0.005	0.001	0.019	0.001	0.000	0.003	0.003
14	0.012	0.006	0.012	0.001	0.008	0.009	0.002	0.009	0.002	0.001	0.009	0.009
15	0.021	0.007	0.007	0.003	0.002	0.021	0.001	0.011	0.002	0.000	0.000	0.007
16	0.005	0.007	0.019	0.002	0.034	0.030	0.002	0.001	0.003	0.004	0.002	0.011
17	0.017	0.025	0.026	0.013	0.027	0.077	0.004	0.003	0.014	0.000	0.002	0.024
18	0.006	0.031	0.011	0.004	0.012	0.003	0.002	0.000	0.006	0.002	0.008	0.008
19	0.007	0.016	0.002	0.007	0.002	0.037	0.003	0.001	0.003	0.000	0.000	0.018
20	0.094	0.016	0.010	0.007	0.002	0.023	0.008	0.046	0.011	0.005	0.004	0.015
21	0.005	0.005	0.005	0.001	0.045	0.039	0.002	0.011	0.002	0.002	0.001	0.008
22	0.028	0.008	0.034	0.007	0.007	0.042	0.006	0.012	0.004	0.001	0.005	0.010

Table 6. Average artist-likeness indices for GLAZE. The artist ID for each row is from 12 to 22.

	12	13	14	15	16	17	18	19	20	21	22
0	0.009	0.015	0.010	0.119	0.007	0.007	0.220	0.007	0.040	0.074	0.015
1	0.012	0.095	0.017	0.068	0.006	0.003	0.456	0.026	0.004	0.069	0.012
2	0.005	0.039	0.009	0.007	0.010	0.014	0.416	0.001	0.001	0.011	0.022
3	0.008	0.093	0.095	0.005	0.003	0.007	0.351	0.000	0.004	0.041	0.042
4	0.006	0.087	0.027	0.006	0.017	0.014	0.434	0.002	0.005	0.027	0.012
5	0.024	0.110	0.002	0.064	0.008	0.032	0.163	0.012	0.006	0.052	0.007
6	0.000	0.054	0.015	0.005	0.005	0.004	0.286	0.000	0.010	0.049	0.003
7	0.008	0.032	0.038	0.020	0.003	0.001	0.060	0.003	0.034	0.086	0.015
8	0.003	0.133	0.010	0.019	0.014	0.011	0.334	0.008	0.036	0.057	0.013
9	0.003	0.054	0.039	0.001	0.001	0.002	0.090	0.000	0.014	0.150	0.002
10	0.003	0.072	0.032	0.004	0.008	0.003	0.386	0.001	0.007	0.035	0.012
11	0.003	0.165	0.009	0.032	0.010	0.021	0.207	0.023	0.045	0.092	0.004
12	**0.463**	0.067	0.005	0.117	0.005	0.007	0.177	0.005	0.010	0.058	0.013
13	0.009	**0.397**	0.094	0.069	0.004	0.001	0.269	0.002	0.005	0.055	0.022
14	0.020	0.116	**0.426**	0.044	0.004	0.003	0.202	0.000	0.012	0.070	0.022
15	0.046	0.095	0.020	**0.409**	0.016	0.013	0.184	0.035	0.021	0.069	0.012
16	0.021	0.191	0.020	0.056	**0.089**	0.016	0.388	0.001	0.024	0.045	0.027
17	0.013	0.116	0.000	0.014	0.020	**0.309**	0.239	0.003	0.024	0.015	0.015
18	0.011	0.166	0.026	0.067	0.016	0.003	**0.570**	0.000	0.005	0.032	0.011
19	0.011	0.051	0.002	0.064	0.011	0.004	0.063	**0.622**	0.021	0.048	0.008
20	0.011	0.055	0.007	0.097	0.011	0.006	0.054	0.013	**0.388**	0.082	0.035
21	0.012	0.175	0.012	0.162	0.005	0.004	0.227	0.039	0.030	**0.197**	0.011
22	0.019	0.077	0.009	0.111	0.021	0.007	0.292	0.002	0.073	0.067	**0.158**

Table 7. Average artist-likeness indices for MIST. The artist ID for each row is from 0 to 11.

	0	1	2	3	4	5	6	7	8	9	10	11
0	**0.295**	0.005	0.006	0.016	0.034	0.016	0.016	0.069	0.006	0.001	0.003	0.007
1	0.026	**0.150**	0.014	0.006	0.015	0.023	0.005	0.003	0.016	0.001	0.001	0.038
2	0.001	0.018	**0.116**	0.113	0.055	0.031	0.041	0.000	0.009	0.000	0.001	0.002
3	0.002	0.012	0.051	**0.110**	0.036	0.031	0.019	0.001	0.011	0.011	0.009	0.016
4	0.007	0.014	0.021	0.014	**0.189**	0.007	0.009	0.000	0.003	0.003	0.005	0.007
5	0.011	0.003	0.008	0.019	0.020	**0.456**	0.011	0.002	0.008	0.001	0.000	0.018
6	0.003	0.026	0.040	0.035	0.066	0.020	**0.323**	0.001	0.024	0.027	0.002	0.008
7	0.061	0.009	0.008	0.006	0.048	0.022	0.008	**0.396**	0.017	0.020	0.031	0.024
8	0.007	0.102	0.006	0.043	0.064	0.049	0.008	0.008	**0.059**	0.006	0.002	0.013
9	0.007	0.014	0.002	0.042	0.232	0.011	0.003	0.001	0.006	**0.057**	0.000	0.001
10	0.025	0.003	0.007	0.012	0.028	0.026	0.011	0.017	0.007	0.003	**0.181**	0.009
11	0.024	0.007	0.006	0.008	0.034	0.051	0.010	0.003	0.015	0.001	0.001	**0.097**
12	0.014	0.019	0.004	0.002	0.012	0.032	0.006	0.000	0.003	0.000	0.000	0.004
13	0.017	0.009	0.008	0.004	0.027	0.018	0.014	0.003	0.001	0.001	0.003	0.016
14	0.011	0.014	0.002	0.009	0.018	0.011	0.003	0.001	0.001	0.002	0.002	0.003
15	0.023	0.005	0.012	0.010	0.012	0.042	0.008	0.001	0.005	0.000	0.000	0.028
16	0.001	0.007	0.038	0.011	0.023	0.031	0.007	0.000	0.002	0.000	0.000	0.004
17	0.008	0.041	0.024	0.021	0.039	0.041	0.007	0.000	0.011	0.000	0.001	0.007
18	0.002	0.024	0.018	0.012	0.032	0.008	0.008	0.000	0.002	0.000	0.001	0.005
19	0.014	0.020	0.002	0.009	0.005	0.022	0.001	0.000	0.002	0.002	0.000	0.012
20	0.134	0.020	0.021	0.016	0.010	0.033	0.020	0.026	0.015	0.005	0.002	0.021
21	0.019	0.009	0.003	0.006	0.046	0.033	0.004	0.001	0.004	0.001	0.001	0.004
22	0.037	0.022	0.023	0.017	0.013	0.056	0.015	0.006	0.007	0.001	0.002	0.013

Table 8. Average artist-likeness indices for MIST. The artist ID for each row is from 12 to 22.

	12	13	14	15	16	17	18	19	20	21	22
0	0.003	0.007	0.003	0.103	0.012	0.004	0.161	0.035	0.122	0.065	0.008
1	0.017	0.057	0.023	0.049	0.015	0.017	0.430	0.035	0.010	0.033	0.015
2	0.002	0.032	0.007	0.005	0.009	0.016	0.489	0.001	0.014	0.015	0.022
3	0.004	0.086	0.049	0.006	0.011	0.004	0.480	0.000	0.006	0.027	0.020
4	0.007	0.053	0.005	0.022	0.064	0.014	0.465	0.002	0.046	0.024	0.019
5	0.013	0.035	0.001	0.051	0.022	0.019	0.167	0.026	0.057	0.045	0.006
6	0.001	0.019	0.006	0.009	0.005	0.003	0.339	0.000	0.014	0.012	0.018
7	0.004	0.014	0.019	0.021	0.009	0.003	0.095	0.022	0.041	0.111	0.011
8	0.006	0.133	0.034	0.039	0.016	0.013	0.246	0.022	0.062	0.034	0.027
9	0.007	0.039	0.025	0.008	0.004	0.006	0.343	0.000	0.064	0.114	0.014
10	0.001	0.037	0.011	0.009	0.014	0.006	0.538	0.004	0.013	0.032	0.006
11	0.008	0.108	0.003	0.036	0.018	0.048	0.281	0.037	0.134	0.065	0.007
12	**0.303**	0.084	0.011	0.129	0.010	0.005	0.250	0.044	0.028	0.030	0.009
13	0.009	**0.268**	0.032	0.041	0.039	0.003	0.390	0.002	0.028	0.040	0.026
14	0.011	0.083	**0.500**	0.039	0.010	0.003	0.233	0.001	0.013	0.025	0.005
15	0.045	0.095	0.006	**0.304**	0.041	0.009	0.209	0.044	0.028	0.062	0.011
16	0.018	0.144	0.025	0.046	**0.135**	0.063	0.391	0.001	0.021	0.017	0.014
17	0.035	0.104	0.001	0.013	0.024	**0.230**	0.308	0.016	0.052	0.008	0.009
18	0.010	0.076	0.003	0.053	0.024	0.018	**0.668**	0.000	0.002	0.019	0.014
19	0.023	0.025	0.002	0.039	0.003	0.004	0.069	**0.658**	0.033	0.046	0.010
20	0.004	0.031	0.012	0.068	0.021	0.004	0.083	0.039	**0.330**	0.025	0.059
21	0.018	0.114	0.012	0.110	0.011	0.003	0.296	0.103	0.047	**0.143**	0.010
22	0.014	0.056	0.016	0.062	0.045	0.028	0.293	0.020	0.105	0.032	**0.117**

References

1. Baio, A.: Invasive diffusion: How one unwilling illustrator found herself turned into an AI model (2022). https://waxy.org/2022/11/invasive-diffusion-how-one-unwilling-illustrator-found-herself-turned-into-an-ai-model/. Accessed 25 Apr 2024
2. Heusel, M., Ramsauer, H., Unterthiner, T., Nessler, B., Hochreiter, S.: GANs trained by a two time-scale update rule converge to a local nash equilibrium. In: Proceedings of the 31st International Conference on Neural Information Processing Systems, pp. 6629–6640. NIPS'17, Curran Associates Inc., Red Hook, NY, USA (2017)
3. Hu, E.J., et al.: LoRA: low-rank adaptation of large language models. In: International Conference on Learning Representations (2022). https://openreview.net/forum?id=nZeVKeeFYf9
4. Kynkäänniemi, T., Karras, T., Laine, S., Lehtinen, J., Aila, T.: Improved precision and recall metric for assessing generative models (2019). CoRR **abs/1904.06991**
5. Liang, C., et al.: Adversarial example does good: Preventing painting imitation from diffusion models via adversarial examples (2023)
6. Radford, A., et al.: Learning transferable visual models from natural language supervision. In: Meila, M., Zhang, T. (eds.) Proceedings of the 38th International Conference on Machine Learning. Proceedings of Machine Learning Research, vol. 139, pp. 8748–8763. PMLR (2021). https://proceedings.mlr.press/v139/radford21a.html

7. Shan, S., Cryan, J., Wenger, E., Zheng, H., Hanocka, R., Zhao, B.Y.: Glaze: protecting artists from style mimicry by Text-to-Image models. In: 32nd USENIX Security Symposium (USENIX Security 23), pp. 2187–2204. USENIX Association, Anaheim, CA (2023). https://www.usenix.org/conference/usenixsecurity23/presentation/shan
8. Shan, S., Ding, W., Passananti, J., Wu, S., Zheng, H., Zhao, B.Y.: Nightshade: prompt-specific poisoning attacks on text-to-image generative models. In: Proceedings of the 45th IEEE Symposium on Security and Privacy (2024). https://people.cs.uchicago.edu/~ravenben/publications/pdf/nightshade-oakland24.pdf
9. Tan, W.R., Chan, C.S., Aguirre, H., Tanaka, K.: Improved ArtGAN for conditional synthesis of natural image and artwork. IEEE Trans. Image Process. **28**(1), 394–409 (2019). https://doi.org/10.1109/TIP.2018.2866698

Spatially Augmented Speech Bubble to Character Association via Comic Multi-task Learning

Gürkan Soykan[1,2(✉)], Deniz Yuret[1,2], and Tevfik Metin Sezgin[1,2]

[1] Computer Engineering Department, Koç University, Istanbul, Turkey
[2] KUIS AI Center, Istanbul, Turkey
{gsoykan20,dyuret,mtsezgin}@ku.edu.tr
https://ai.ku.edu.tr/

Abstract. Accurately associating speech bubbles with corresponding characters is a challenging yet crucial task in comic book processing. This problem is gaining increased attention as it enhances the accessibility and analyzability of this rapidly growing medium. Current methods often struggle with the complex spatial relationships within comic panels, which lead to inconsistent associations. To address these shortcomings, we developed a robust machine learning framework that leverages novel negative sampling methods, optimized pair-pool processes (the process of selecting speech bubble-character pairs during training) based on intra-panel spatial relationships, and an innovative masking strategy specifically designed for the relation branch of our model. Our approach builds upon and significantly enhances the COMIC MTL framework, improving its efficiency and accuracy in handling the unique challenges of comic book analysis. Finally, we conducted extensive experiments that demonstrate our model achieves state-of-the-art performance in linking characters to their speech bubbles. Moreover, through meticulous optimization of each component-from data preprocessing to neural network architecture-our method shows notable improvements in character face and body detection, as well as speech bubble segmentation.

Keywords: Speech Bubble Association · Speech Bubble to Character Association · Deep Learning for Comics · Comic Book Analysis · Multi-Task Learning

1 Introduction

Comics combine text and images in varying degrees, ranging from purely pictorial to entirely textual, to tell stories or express ideas. This multimodal medium intersects with fields such as education [9], sociocultural studies [4], linguistics, and cognitive sciences [2,11]. Understanding comics can therefore contribute to advancements in these areas. Furthermore, techniques developed for analyzing comics can also be applied to other types of visual languages [5], including cave paintings, mangas, and graphic novels.

In comic books, speech bubbles are frequently used to represent character dialogue. However, matching the right character with their specific speech bubble can be difficult, especially when several characters are on a panel or page. In this study, we investigate the efficacy of employing deep learning models to link the faces and bodies of characters with their corresponding speech bubbles in comics in order to address this issue. This task is crucial in comic processing as it is a prerequisite for character-centric multimodal approaches and is understudied. Without the association of speech bubbles to characters, laying out the structure for comics would be incomplete. To fully comprehend the narrative structure and character interactions in comics, linking speech bubbles with characters is essential, as it enables the understanding of the plot and character development.

Previous studies have explored various methods, such as geometric graph analysis [20], for connecting characters to speech bubbles. However, there has been limited research on the use of deep learning models for this purpose except for the COMIC MTL [17], they propose a deep learning model that can detect and segment components such as faces, bodies, panels, narrative boxes, and speech bubbles. Moreover, the model can associate speech bubbles with the corresponding characters. Therefore, the COMIC MTL framework may provide crucial information about a comic page when integrated with other elements of comic processing, such as speech bubble segmentation with OCR [23] and character face and body detection for character recognition and re-identification tasks [24], Multimodal Multi-Task Transfer Learning [18], persona based comic dialogue generation [1], among others.

We have observed that maximizing the accuracy of character-to-speech bubble associations, as well as improving detection and segmentation outputs, can be achieved by extending the capabilities of existing models. Consequently, our research aims to build upon and significantly augment the COMIC MTL model through a series of strategic enhancements. These enhancements are designed to refine every aspect of the model-from data handling to final output processing. This includes preprocessing and dataset enhancements, advanced sampling methods, backbone model improvements, relation branch augmentation, and enhanced post-processing techniques, along with optimized training. These modifications are evaluated using the extended DCM772 dataset [16], allowing us to validate the efficacy of our improvements in real-world scenarios. The code, instructions and models can be accessed online[1]. Below, we outline the primary contributions of our work:

- We introduce new negative sampling techniques as a data augmentation during training that enhances the model's ability to discriminate between relevant and irrelevant associations between text and characters, thereby improving the accuracy of speech bubble to character linking.
- By refining the pair-pool process, we optimize the selection and evaluation of candidate character-bubble pairs, leading to more precise associations and better utilization of training data.

---

[1] https://github.com/gsoykan/spatially_augmented_comic_mtl.

- We develop a novel masking strategy that effectively filters out irrelevant features in the relation branch, enabling our model to focus on significant elements that contribute to the accuracy of associations.
- We enhance post-processing techniques to refine the outputs of the model, ensuring that the final associations are not only accurate but also coherent within the narrative structure of the comic.

2 Related Work

Rigaud et al. [20] presents a method for associating speech balloons with their respective speakers in comics and manga, enhancing automated content understanding by using image and text analysis techniques. However, it requires prior information about the position of comic page components prior to association. In contrast, Comic MTL, introduced by Nguyen et al. [17], is a deep neural network that can detect and segment various components of comic book images, including panels, speech bubbles, characters, and narrative boxes from pages without needing prior information. The model is trained using a novel loss function that optimizes the network for multiple tasks simultaneously, resulting in improved performance compared to existing methods. This approach was evaluated on benchmark datasets and demonstrated superior performance over other approaches. More recently, a model named "Magi" [21] is proposed to enhance the accessibility of manga for visually impaired individuals by automatically generating transcriptions indicating who said what and when through encoder-decoder architecture by detecting and associating applied to manga panels, characters, and texts.

In the field of object detection, Faster R-CNN [19] and Mask R-CNN [8] are two popular models that have been used for comic book analysis. Faster R-CNN utilizes a region proposal network to generate potential object regions, which are then classified using a convolutional neural network. Mask R-CNN extends Faster R-CNN by adding a segmentation branch to the network, allowing the model to generate pixel-level masks for detected objects. Similarly, Comic MTL is an extension of Mask R-CNN, specifically designed for comic book analysis. More details about Comic MTL can be found in Subsect. 3.1, as it forms the backbone of our proposed approach.

3 Methodology

3.1 Comic MTL Overview

The Comic MTL model [17], an extension of the Mask R-CNN framework [8], is tailored for analyzing comic page images to extract characters, panels, speech balloons, narrative text boxes, and the associations between characters and balloons. The model incorporates enhancements to the Mask R-CNN's loss function and introduces a new relation branch with a Pair-Pool component and binary classifier to detect associations between balloons and characters, offering a sophisticated multi-task learning approach. These modifications enable the model to effectively learn detection and segmentation tasks, and to process binary classification of balloon-character links.

The analysis of relationships between balloons and characters in the Comic MTL model is framed as a binary classification problem, addressed by adding a new branch to the Mask R-CNN model specifically for relation analysis. This binary classifier is supported by the **Pair-Pool component** of the Comic MTL methodology, which processes pairs of bounding boxes together rather than independently. Differing from other branches like detection and mask, which utilize a single shared feature, this new relation branch leverages combined features: (1) the visual features of the two bounding boxes and their union, termed as the *encapsulation box*, and (2) the spatial features. These enhancements enable a more accurate and context-aware analysis of the associations between characters and their dialogue in comics. The model's detailed architecture including the modifications to its core components and the integration of additional features for precise instance segmentation is further elaborated in Appendix A and visualized in Fig. 1.

3.2 Improvements on Comic MTL

These improvements cover various aspects of the methodology, including preprocessing and dataset, training optimization, positive-negative link (association) sampling, backbone model, relation branch, and a post-processing technique called *character consistency*. The enhancements aim to increase the accuracy and efficiency of the Comic MTL model in recognizing and linking balloons and characters in comics.

Preprocessing and Dataset. The Extended DCM 772 dataset [16] utilized in Comic MTL [17] presented some inconsistencies, which prompted a reevaluation of the reported results. To enhance the dataset's quality, several filtering steps were implemented to refine its accuracy and reliability.

First, pages without faces were excluded from the dataset, even if other annotations were present. Pages without characters or speech bubbles were also excluded. Additionally, pages with less than three panels were eliminated, as they often consisted of advertisements or cover pages. Some samples were manually removed if they contained errors in panel annotations.

To enhance the accuracy of the dataset, positive links, where the body or face has a link relationship with the speech bubble, between faces and speech bubbles were evaluated using logical rules, where a face linked to a speech bubble was associated with the corresponding character. Additionally, face and character IDs were matched using intersection ratings, with faces without an ID assigned to the character with the highest intersection rating, provided that the rating was above a certain threshold (0.2).

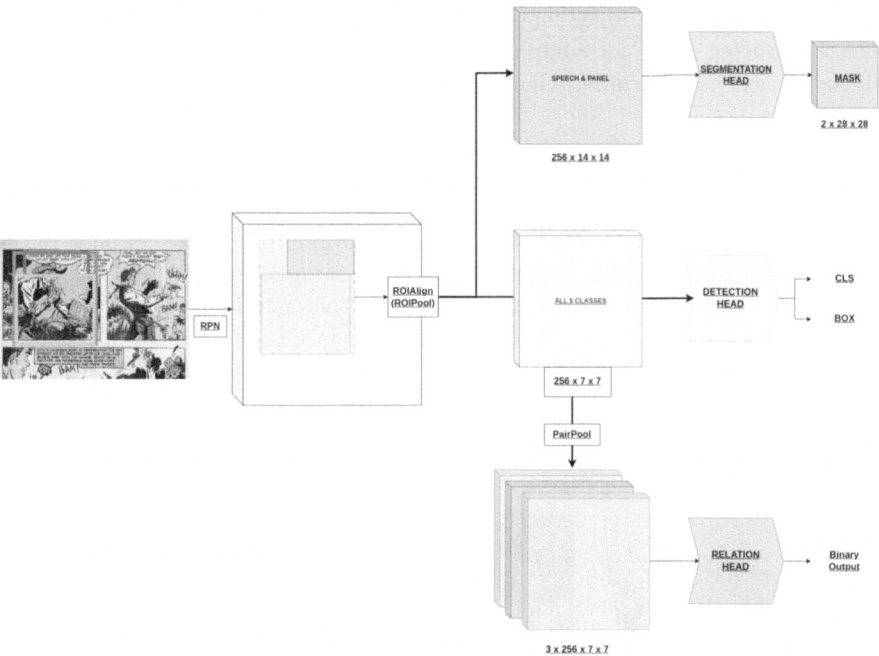

Fig. 1. Overview of Comic MTL framework [17]. The figure shows the architecture of the model, which consists of three main stages: (1) the Region Proposal Network (RPN) stage, (2) the Region of Interest (RoI) Align stage, and (3) Three parallel branches for detection, segmentation, and relation (association) tasks The model takes a comic page as input and generates bounding boxes for characters' faces, bodies, panels, speech bubbles, narrative boxes, segmentations for speech bubbles and panels, and determines the association scores between speech bubbles and characters' faces, bodies, using a relation branch.

After applying these filtering and improvement steps, the final train, validation, and test split sizes were set to 544, 25, and 25 pages, respectively, down from an initial total of 772 pages. The resulting dataset was then used to train and evaluate the proposed model. Some examples of filtered pages can be seen in Appendix B.

Positive and Negative Link (Association) Sampling. A positive link indicates that a face or body is linked to a speech bubble, whereas a negative link indicates the opposite. Our approach aimed to address the imbalance, which may be the location or number of samples, between face and body samples in the existing random sampling methods.

To improve the positive and negative link sampling, we introduced *Sorted PairPool* sampling, which ranks the box proposals according to the intersection rate with ground truth (GT) boxes and uses the best ones. However, we noticed an imbalance between face and body samples in the Sorted Pair Pool Sampling, with most of the top samples coming from bodies. To address this, we introduced *Balanced Sorted PairPool* Sampling and adjusted the parameter a (threshold), which is used to determine whether a bounding box is selected for sampling by checking the Intersection over Union (IoU) with the GT box, to achieve a better balance between face and body samples to get $2N$ samples, N is also a hyperparameter that can be adjusted to indicate number of samples used during the forward pass.

In addition to sorting and balancing the sampling procedure, we also introduced spatial augmentation to generate negative samples. Firstly, **Sliding Window Negative Sampling**, which augments positive samples by converting them to negative samples in a way that by sliding the face or body bounding box up and down, left and right, to create negative samples. For faces, if the sliding window intersects with the body of the face to some degree, that is ignored, see Fig. 2.

Fig. 2. An example of *sliding window negative sampling* method. The positive samples can be seen on the left, and the others are left and right-slide windows to generate negative samples. The red rectangle indicates the speech bubble bounding box proposal, green for the face, and blue for the encapsulation box.

Secondly, **Bubble Mirrored Negative Sampling** is another technique we employed, which involves taking the symmetric face or character bounding box horizontally and vertically relative to the speech bubble center and using it as a negative sample. These negative sampling methods are conceived due to the observation of slid face or character bounding boxes having a high probability of being associated with the speech bubble. We believe this technique helps the model to attend more to the direction of the tail and face or body element.

Finally, as for negative links, two types of negative links are exploited. The first is by pairing a speech bubble with a face or character that does not belong to it but belongs to another speech bubble. Additionally, we used faces or characters that are on the same panel but do not have a link with any speech bubble as negative samples; they are considered as **absent negative links**. These techniques helped to improve the dataset's quality by ensuring a better balance between positive and negative links (see Fig. 3 for bubble mirrored and absent negative sampling).

Fig. 3. On the left, the *bubble mirrored negative sampling* method is shown, where speech bubble and face bounding box proposals are indicated by red and dashed green rectangles, respectively, with green bounding boxes slid in relation to the speech bubble bounding box to generate negative samples. On the right, *Absent negative links* in a single panel. In absent negative links, speech bubbles are associated with a character with no speech bubble associated with it.

The Backbone Model. the Comic MTL [17] framework uses Mask R-CNN [8] network pre-trained on the ImageNet dataset [6] as backbone model. In contrast, we used the following models and weights as the backbone for our experiments. The results showed a significant improvement when using the weights in the third option, which can be seen in Sect. 4, results & discussion.

- *FasterRCNN ResNet50 FPN Weights COCO V1*[2]: This model does not have a mask branch so it is just for testing the effect of detection and relation head.
- *MaskRCNN ResNet50 FPN - Weights COCO V1*[3]: This model is the most similar to that used in Comic MTL. It has the same architecture but is pretrained with COCO.
- *MaskRCNN ResNet50 FPN V2 - Weights COCO V1*[4]: This model relies on vision transformers [12], and produces our best and state-of-the-art results.

Note that the above models are pre-trained on the COCO dataset [13] and are available in the PyTorch Vision library.

Relation Branch. There are two major update areas with the relation branch. One is the encapsulation box masking idea. The other is to update the relation branch network architecture.

Encapsulation Box Masking is a technique used to mask the features of an encapsulation box. Each encapsulation box feature is composed of $256 \times 7 \times 7$

---

[2] FasterRCNN ResNet50 FPN Model Link.
[3] MaskRCNN ResNet50 FPN Model Link.
[4] MaskRCNN ResNet50 FPN V2 Model Link.

feature maps. The 7 × 7 part stands for *Height* and *Width* after the convolution operation; because of that, these maps can be traced back to the image patches of the encapsulation box. It is redundant to use the entire grid of the encapsulation box features, as it contains information that is not related to the face (body) or speech bubble with which the association is being queried. Therefore, this method can be considered a noise-filtering mechanism.

Since the encapsulation box features contain spatial information, a 7 × 7 mask was created in order to be multiplied with the features in an element-wise manner. The mask regions corresponding to the speech bubble and face (body) bounding boxes were set to 1, and the rest of the regions were set to 0, see Fig. 4. Several versions of this method were tried, including connecting the regions between the speech and face (body) bounding boxes using the modified Bresenham line-drawing algorithm [3] and marking connecting regions as 1 as well.

Different **Relation Branch Neural Network Architectures** are implemented and tested. However, most of them failed to give competitive results with the baseline. Some of those models include the implementations and inspirations

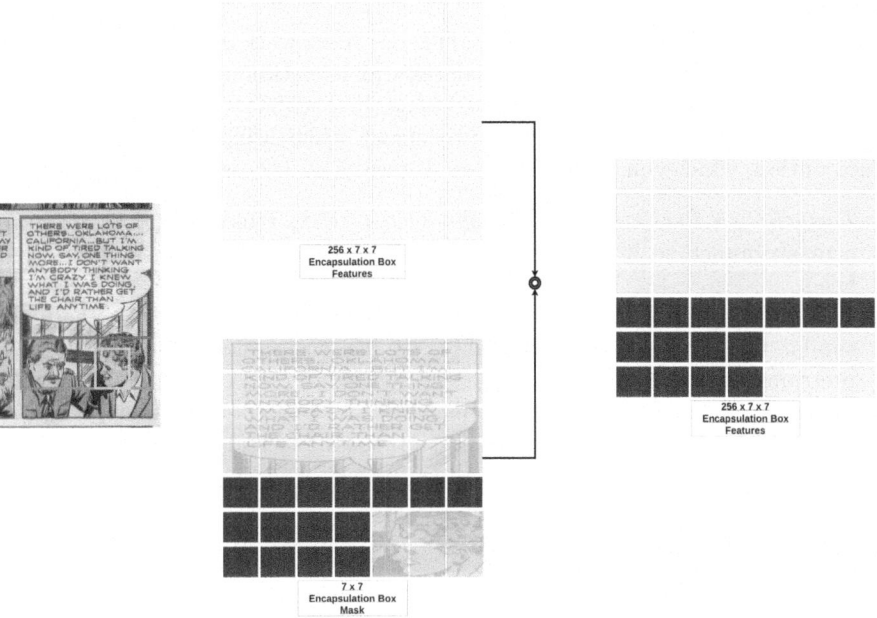

Fig. 4. Example of *Encapsulation Box Masking* applied to ROI-Aligned features of encapsulation box. The 7 × 7 mask is multiplied element-wise with the encapsulation box features, a tensor with dimensions 7 × 7 × 256, with broadcasting, setting only the regions corresponding to the speech bubble and face (body) bounding box to 1 and the rest to 0. This technique helps to filter out the noise and irrelevant information, improving the accuracy of the model's associations.

from proposed architectures from [10,22]. The baseline and the most performant model are named as follows, and they can be seen in Fig. 5:

- **MLP:** This is our baseline model. It is the baseline because it is the most straightforward model, and we assume it was used in Comic MTL. The reason for our assumption is that Comic MTL does not specify relation branch architecture; there is a single figure indicating it is a two-layer MLP.
  After the pair-pool operation, the features are flattened and passed through the two-layered MLP. Before the output layer, spatial features are concatenated, and binary output is obtained.
- **FastRCNNConvFCHead:** We used the box-head structure proposed in the [26] and [12]. The only difference was the output dimension size. We passed the output of this layer through a linear layer and a ReLU before the output layer. Since the final structure of our MTL model is based on [12], we can say that the actual box-head and relation head are composed of common components, but they do not share parameters. For the rest of our experiments, we use this model as the relation branch.

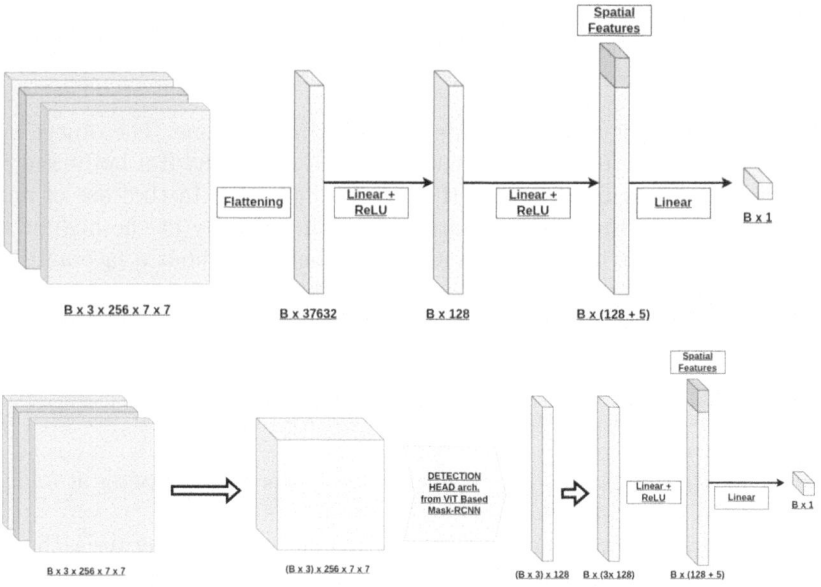

Fig. 5. Overview of baseline and most performant relation branch architectures used in the model. On the top, the MLP head architecture of the relation branch is illustrated. On the bottom, the relation branch utilizes the *FastRCNNConvFCHead* architecture, part of Detectron2's [26] box-head.

Alternative relation branch neural network architectures can be seen in Appendix C along with their performances.

Character Consistency Post-processing and Evaluation. The way we obtain the relation outputs of the model is as follows: We separate the body, face, and speech bubble boxes that the model predicts. Here, we select the predicted boxes with a box score greater than 0.5. Then, we match each speech bubble with all bodies and faces one by one, pass it through the relation branch, and get the relation score, and then it is mapped between 0 and 1 with the sigmoid function. Relations with probabilities between 0 and 1, greater than 0.75, are pooled while others are removed.

Then, for each speech bubble, we divide all its associated boxes into groups according to whether it is a body or a body. We choose those with the highest probability in both groups. Because it is assumed that each speech bubble belongs to only one character, it can only have a single face and/or body.

With this method, it is possible to evaluate the relation performance of the model, and the evaluation was done with this method in Sect. 4. For evaluation, the relevant boxes must match with 0.6 IoU. If there is no match, it is considered that our model has made a wrong prediction for that relation. However, if the boxes are predicted correctly, the relation between them is checked to determine whether there is a GT relation, and according to that, it is true or false.

However, there may be better ways to get the best results from the model, such as **Character Consistency Post-processing** that we proposed. With this method, we wanted to ensure the model gives the closest result to ground truth. This method works as follows:

- The characters given by the model are found one by one. The object mentioned as a character is a combination of face and body. If a body and face intersect at a specific rate, it is the face of that body. In the case of multiple intersections, the face that intersects with the body at the highest rate is assigned to that body. There can also be bodies without a face and faces without a body. Those are still considered character instances.
- Then, the relation score to all speech bubbles for each character is checked.
- If the item has a face and a body, the mean of the character's scores are calculated.
- Then, character components with the highest average score are matched with the speech bubble.
- If the speech bubble used here is not related to another character at a higher level, it can be decided that this is the character's bubble.
- This process continues until no speech bubble meets the criteria that are stated above. The demonstration of the post-processing technique can be seen in Appendix D.

Training Optimization. AdamW [15] was used in place of the stochastic gradient descent (SGD) optimizer to enhance the model's training procedure. The AdamW optimizer is a variation that incorporates weight decay into the optimization process to help avoid overfitting. Weight decay improves generalization performance by regularizing the model. Additionally, it has been demonstrated that the AdamW optimizer converges more quickly than SGD, facilitating accelerated training.

We introduced the *CosineWarmupScheduler* [14] to further optimize the training process. The CosineWarmupScheduler is a learning rate scheduler that gradually increases the learning rate during the warm-up phase and decays it using a cosine annealing schedule. The warm-up phase allows the model to adjust to the learning rate slowly, which leads to faster convergence. The cosine annealing schedule helps the model avoid getting stuck in local minima by gradually reducing the learning rate.

Additionally, we optionally used several augmentations during training, including random cropping, flipping, and rotation. This is intended to increase the diversity of the training data and make the model more resilient to changes in the input data during test time. These improvements to the training optimization process helped to enhance the performance of the Comic MTL model. The model achieved better accuracy and faster convergence during training.

4 Results and Discussion

We have conducted a series of experiments using two different training setups: **slow-training** and **fast-training**. In the slow-training setup, we train the entire dataset consisting of train-val-test splits with 544-25-25 comic pages, respectively, for up to 40 epochs. In contrast, the fast-training setup involves training on a smaller dataset with train-val-test splits of 64-24-24 pages, respectively, for only 10 epochs.

The fast-training setup serves as a helpful approach for serial experimentation, allowing us to quickly evaluate our model's performance with various hyperparameters and configurations. Once we have identified the most promising configurations through fast-training, we then perform slow-training to obtain more reliable results with optimal settings. In both training setups, we first freeze the backbone and only train the heads for the first ten epochs. Afterward, we train the entire model in the following epochs.

4.1 Findings from Fast-Training

In our model optimization, several strategies have been applied to refine the learning process, details of which can be found in Appendix E. The **Pair-pool Sampling** method, particularly the **Sorted** option, effectively uses IoU values to sort and select the most representative samples for training, optimizing the balance between body and face recognition accuracy. For the **Encapsulation Box Mask Results**, the version in which the masked areas of bubble and face or body are connected with line drawing algorithm to mask the connection as well, showed the most promising enhancements by adjusting the masking around relevant areas between the element and the bubble, significantly improving both body and face metrics. The **Sliding Window Negative Samples** technique helped differentiate essential features for association tasks, proving crucial for accurate model predictions. Similarly, **Bubble Mirrored Negative Samples** demonstrated that a balanced use of negative samples and encapsulation masking markedly enhances the model's performance by improving the pair-pool with spatially augmented negative samples. Lastly, the use of **Absent Negative Links** was found to boost body association performance at the expense of face detection, underscoring the importance of careful sample selection and masking strategies in our training methodology.

4.2 Slow-Training

Table 1 provides a comparison of the performance of our approach under different settings during slow-setting training. We explore the impact of negative sampling methods and augmentation on the model's performance. In all of the experiments, encapsulation box masking version 3 is used. As for the pair-pool sampling strategy, sorted-balanced is chosen. As input image size 800 is decided for height, the width is dynamic based on aspect ratio. The number of additional negative samples is fixed at 150 because it resulted in the best body F1 score during the fast-setting experimentation stage.

The results indicate that the choice of the negative sampling method plays a crucial role in the model's performance. Incorporating negative sampling methods improves performance. However, there is a trade-off between the performance of the body and face concerning absent negative links. An increase in absent negative links helps the model differentiate bodies better at a minor face performance cost. However, the gain in body performance is significant.

Additionally, the impact of data augmentation is examined, where the "Aug." column indicates whether augmentation was applied. Although we have tried several different augmentation settings, the results did not improve. Overall, the results emphasize how crucial it is to choose the best mixture of methods to get the best performance out of the model during training.

Table 1. Comparison of performance with different settings during Slow-Training Setting. The *Mixed* option combines absent negative links with negative links and selects negative samples using sorted-balanced Pair-Pooling. The best results for each metric are highlighted in bold.

Negative Sampling Methods				Aug.	BODY			FACE		
Bubble	Sliding	Absent	Additional		Recall	Precision	F1	Recall	Precision	F1
0	0	0	150		62.89	64.35	63.61	59.23	55.39	57.24
5	5	10	150		60.18	64.56	62.29	**65.38**	**64.39**	**64.88**
5	5	10	150	✓	58.82	64.35	61.46	60.00	62.40	61.17
0	0	Mixed	150		61.53	66.99	64.15	63.07	61.19	62.12
0	0	Mixed	150	✓	60.18	62.73	61.43	57.69	58.59	58.14
5	5	Mixed	150		**63.34**	**68.96**	**66.03**	63.84	63.35	63.60
5	5	Mixed	150	✓	60.18	66.16	63.03	57.69	59.52	58.59

4.3 Final Results

Table 2 compares the results obtained from Comic MTL [17] and our approach in various segmentation, detection, and association tasks. The recall, precision, and F1 score performance metrics are reported for each task. For the final results, the best-performing model in terms of body F1 score is chosen, and association scores are reported after character consistency post-processing.

In terms of speech bubble segmentation, our approach is significantly better. However, it should be noted that for speech bubble detection [17] does not provide any results. So, it needs to be clarified how they measured speech bubble segmentation. Namely, whether false positives are included in their evaluation needs to be clarified.

Both Comic MTL and the proposed approach demonstrate high performance regarding panel detection. However, they report higher precision and, thus, higher F1 scores. As for narrative box detection, both Comic MTL perform significantly better. The reason for this discrepancy arguably stems from the fact that the dataset we used was the subset of DCM772 [16], and also, in their work, the image dimensions are not provided. We employed a height of 800 for our approach, while the width was determined according to the aspect ratio.

Additionally, it should be noted that panel and narrative box detections are relatively large components. Additionally, we focused heavily on improving speech bubble association tasks in our experiments. That, in return, might affect the model negatively for panel and narrative box detection tasks. However, for the others, comparatively smaller components, character (body) and face, we have superior scores, although using a smaller dataset. This might be one of the reasons why we report more recall with the association tasks. Our approach outperforms Comic MTL in both tasks, especially showing remarkable progress in body-to-speech bubble association.

The proposed approach consistently achieves competitive or superior results compared to Comic MTL in various segmentation, detection, and association tasks. These findings demonstrate the effectiveness and superiority of our approach to comic analysis.

Table 2. Comparison of the results obtained from the original Comic MTL model [17] and our approach's segmentation, detection, and association tasks. Our spatially augmented approach achieves SOTA results in the association tasks.

	Recall		Precision		F1	
	Comic MTL	Ours	Comic MTL	Ours	Comic MTL	Ours
Speech B. Segmentation	89.50	**97.00**	94.87	**95.80**	92.11	**96.40**
Speech B. Detection	–	92.60	–	95.40	–	94.00
Panel Detection	**98.71**	96.80	**96.84**	92.40	**97.76**	94.60
Char. Detection	67.56	**72.60**	76.21	**82.10**	71.62	**77.00**
Face Detection	72.39	**79.50**	**82.12**	81.00	76.95	**80.30**
Narrative B. Detection	**84.38**	81.40	**91.22**	82.60	**87.66**	82.00
Balloon-Char. Association	45.64	**63.30**	**71.01**	70.00	55.57	**66.50**
Balloon-Face Association	53.47	**61.50**	**72.00**	67.80	61.36	**64.50**

5 Conclusion

In conclusion, we aimed to enhance face and body to speech bubble association tasks by utilizing new negative sampling methods, optimizing the pair-pool process, introducing a novel masking strategy for the relation branch, and developing post-processing methods. Building upon the Comic MTL framework [17], we optimized all components, from the preprocessing of the dataset to the architecture of the neural network in the relation branch, to enhance overall performance.

The significance of this research lies in achieving superior results and advancing the field of comic analysis. The developed methods and strategies can be further explored and extended to address new challenges and tasks within comics. For example, character recognition in comics could enable the tracking of characters across panels in such a way that their speech bubble content is associated with them. This capability, when coupled with OCR models, facilitates the conversion of comics into dialogue-like structures, enhancing accessibility and understanding. Our methods open up new possibilities for character tracking, dialogue extraction, and the semantic understanding of graphic narratives.

Acknowledgments. This project is supported by Koç University & İş Bank AI Center (KUIS AI). We would like to thank KUIS AI for their support.

Disclosure of Interests. The authors have no competing interests to declare that are relevant to the content of this article.

A Detailed Overview of Comic MTL

This section explains the Comic MTL model [17] for analyzing comic page images to extract characters, panels, speech balloons, narrative text boxes, and the associations between characters and balloons. The model builds upon the Mask R-CNN model in instance segmentation. Two modifications are made to the Mask R-CNN model. Firstly, the loss function of the mask branch is modified to simultaneously learn detection and segmentation tasks and consider the origin of annotations because detection and segmentation heads have different inputs. Secondly, an additional branch is added to detect associations between balloons and characters using a Pair-Pool component and binary classifier. The classifier outputs probabilities of a pair being linked or not. The extraction of relevant features for the pairs of balloon-character is required.

The Comic MTL model addresses a detection/segmentation problem with classes that have both bounding box and segmentation mask annotations. Mask R-CNN provides predictions for both bounding boxes and masks, but it requires mask annotations for all classes to be trained effectively. To jointly learn the detection and segmentation tasks from bounding boxes and object masks, the authors enhance Mask R-CNN by only using mask annotations for objects from classes with mask annotations. These classes are speech bubbles and panels.

Instance segmentations of speech bubbles are part of the extended DCM772 dataset. As for panels, it is assumed that the bounding box itself is the segmentation mask. The multi-task loss is defined by a binary cross-entropy loss for mask prediction, and the loss is only applied to regions of interest with mask annotations. The binary cross-entropy loss is a function of the ground-truth mask and the predicted mask for a particular object class. The loss function includes only those classes that have ground-truth segmentation masks.

The relationship analysis between balloons and characters is considered a binary classification problem. Each balloon-character pair is classified as either having a link or not having a link, where a linked pair means that the character speaks the text of the corresponding balloon. It is assumed that a character may link to multiple balloons, but a balloon can only link to one character. The entire image is processed rather than each panel separately, and it is assumed that the model learns the relevant features automatically from training examples. A new branch is added to the Mask R-CNN model to create a binary classifier for relation analysis.

The Pair-Pool component in the Comic MTL methodology processes a pair of bounding boxes instead of taking each box independently. Positive pairs are added to the pool, which are pairs of boxes corresponding to a linked balloon and character in the ground truth, and negative pairs are randomly added to the pool until they reach the same number as positive pairs. The number of pairs is then fixed and used as the input size for the additional branch. The multi-task loss for each sampled Region of Interest (RoI) is defined as the sum of four components: classification loss, bounding box regression loss, segmentation mask loss, and relationship loss. The relationship loss is optimized using binary cross-entropy loss and is calculated based on the predicted and ground-truth relation classes for each pair in the pool.

$$L = L_{cls} + L_{box} + L_{mask} + L_{rel} \qquad (1)$$

Unlike other branches, detection, and mask, that use a single shared feature, the new relation branch uses a combined feature of (1) the visual features of the two bounding boxes and their union, termed the *encapsulation box*, and (2) the spatial features. The model reuses visual features from Mask R-CNN's shared features but encodes spatial features using 5-dimensional vectors that are not affected by translation or scale changes. These features can approximate the relative distance relationship between the two bounding boxes. The visual and spatial features are concatenated at the final layer to form the final features that are fed to the classifier head.

B Filtered Pages

Some examples of filtered pages can be seen in Fig. 6.

Fig. 6. Examples of filtered pages from the Extended DCM 772 [16] dataset. An advertisement page on the left, a cover page in the middle, and an erroneous annotation case on the right.

C Relation Branch Neural Network Architectures

Some examples of other neural network architectures for relation branch:

- **Pair-Triple-Wise Convolutional Network (PTW-ConvNet):** We have a set of feature-maps for speech bubbles, faces, and encapsulation boxes denoted as s_i, f_i, and e_i respectively. We got inspired by [25] to incorporate pair-wise and triplet-wise features of these elements. To compute these features, we multiplied the elements as follows: $sf_i = s_i \cdot f_i$, $se_i = s_i \cdot e_i$, $fe_i = f_i \cdot e_i$, $t_i = s_i \cdot f_i \cdot e_i$ where $\cdot$ denotes element-wise product.
  After fusing the features with one another, we merged them and used them as the input of a Convolutional Neural Network (CNN). The CNN is represented by $d_i = \mathrm{ConvNet}([\mathbf{sf}_i; \mathbf{se}_i; \mathbf{fe}_i; \mathbf{trp}_i])$. The architecture of the model can be seen in Fig. 7.

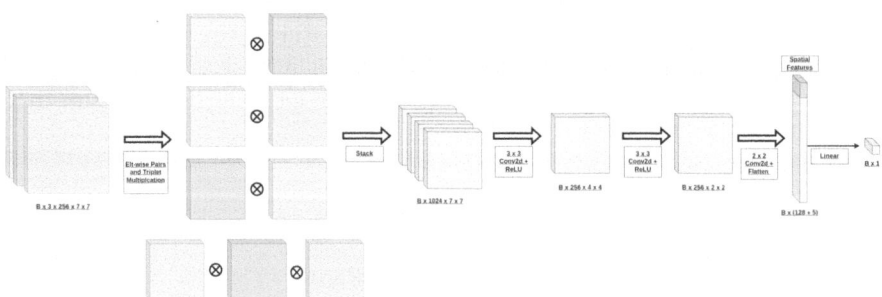

Fig. 7. Relation branch PTW-ConvNet head architecture.

- **MHA on RoI-Align Features:** We base this architecture of the relation head on the core idea of the Vision Transformer (ViT) [7]. However, instead of using a linear projection layer to process image patches, We treat RoI-Aligned features' height and width dimensions as tokens of size 256. This means that the $7 \times 7 \times 3$ tokens are used as input to a transformer encoder model. The outputs of the transformer encoder are concatenated and fed to an MLP to produce the relation head output. Figure 8 demonstrates the model's architecture.

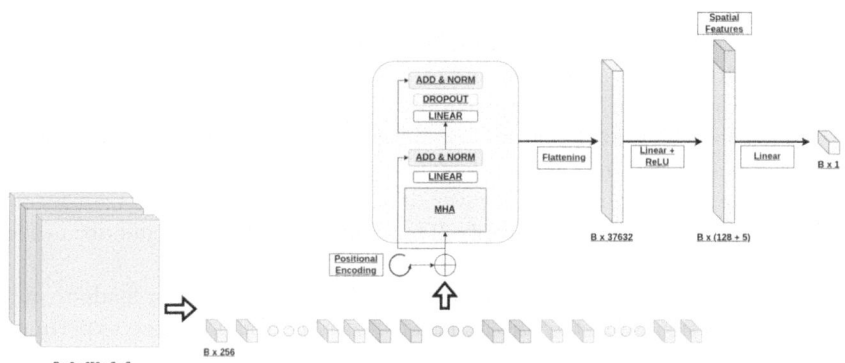

Fig. 8. Relation branch using multi-head attention, treating channel dimension of RoI-Aligned features as tokens.

C.1 Results for Relation Branch Architectures

Table 3 presents the performance comparison results among different neural network architectures with various layer dimensions in the context of relation branch modeling. This comparison aimed to evaluate the effectiveness of these models during fast-setting training. The models considered in the analysis are MLP, PTW-ConvNet, FastRCNNConvFCHead, and MHA on RoI-Align Features; the details of these architectures can be found in paragraph 3.2.

The MLP models serve as the baseline and provide a reference point for comparison. They demonstrate reasonable performance for body association but perform poorly for the face.

The PTW-ConvNet model, inspired by the idea of incorporating pair-wise and triplet-wise features, shows competitive performance in the body and face association tasks. While its results are not the highest in any particular metric, the model consistently performs well across multiple evaluation measures. This shows that using pairs of features for the association task is promising.

The FastRCNNConvFCHead models exhibit different strengths based on their layer dimensions. The model with a layer dimension of 128 achieves high precision in the body association task, indicating its ability to predict positive links accurately. On the other hand, the model with a layer dimension of 512 achieves a notable F1 score in the body association task, demonstrating its overall effectiveness in capturing relations. In the face association task, the FastRCNNConvFCHead model with a layer dimension of 128 achieves the highest F1, implying its proficiency in correctly identifying related elements. Interestingly, when this model's final layer dimension is increased, it loses some of its ability to capture face associations, resulting in lower F1 for face association.

The MHA on RoI-Align Features models, while showing lower performance compared to other architectures, provide insights into the limitations of this approach. These models yield relatively lower recall, precision, and F1 scores in both the body and face association tasks. This suggests that the direct application of the MHA-based approach on RoI-Align Features may not be as effective as the other models in this specific task. On the other hand, with the final layer dimension of 512, it performs competitively with the other models. Using a multi-head attention mechanism and transformer encoders on RoI-Align features is a promising direction. However, it is a known issue to set transformers up and optimize their training stably. Therefore, more experiments are needed, and this should be considered as future work.

Another finding is that increasing the final linear dimension leads to more performant models, almost with all models. This is because these experiments were conducted in the fast-setting. Hence, the training dataset size was 64 comic pages. Because, with slow-setting, this is not the case, possibly leading to overfitting.

The results highlight the importance of selecting the appropriate neural network architecture for relation branch modeling. Each architecture has its strengths and weaknesses, and the choice should be based on the specific requirements and characteristics of the task. Given the complexity of face association tasks and the FastRCNNConvFCHead model's performance, we continued experiments for the slow setting with FastRCNNConvFCHead.

Table 3. Performance comparison of the different relation branch neural network architectures with different layer dimensions during fast-setting training. The best results for each metric are shown in bold.

Model	Final Linear Layer Dim	BODY			FACE		
		Recall	Precision	F1	Recall	Precision	F1
MLP	128	**52.72**	57.42	54.97	31.78	38.31	34.74
MLP	512	**52.72**	62.03	57.00	**43.41**	42.42	42.91
PTW-ConvNet	128	50.45	59.04	54.41	41.08	40.45	40.76
FastRCNNConvFCHead	128	43.18	**65.97**	52.19	39.53	48.57	**43.59**
FastRCNNConvFCHead	512	51.81	64.40	**57.43**	36.43	**48.95**	41.77
MHA on RoI-Align Features	128	45.45	54.05	49.38	18.60	19.83	19.20
MHA on RoI-Align Features	512	50.45	59.67	54.68	41.86	43.54	42.68

D Character Consistency Post-processing

The visualization of our post-processing method can be seen in Fig. 9.

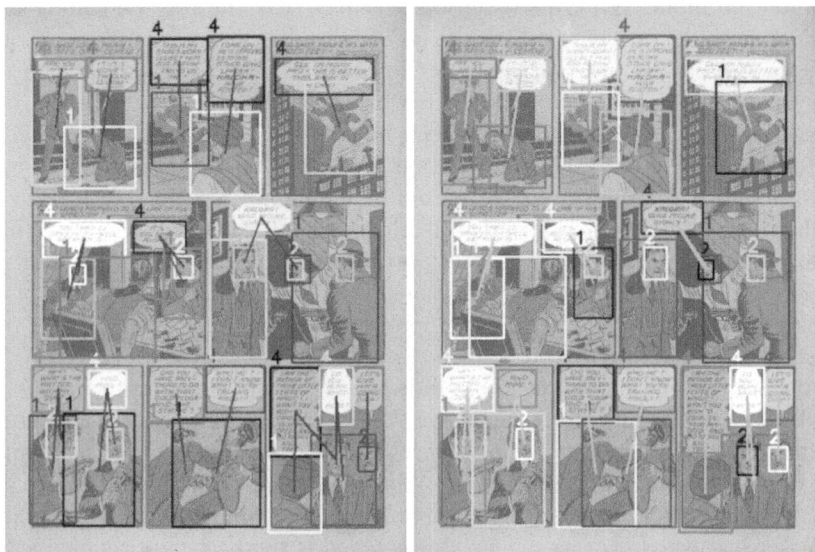

Fig. 9. *Character Consistency Post-processing* applied to a comic page. The before and after images are shown on the left and right. The effect of the post-processing can be seen best in the mid-right and bottom-right panels of the page. Labels 1, 2, and 4 represent body, face, and speech bubble, respectively.

E Fast-Training and Hyperparameter Selection for Our Spatial Augmentations

E.1 Pair-Pool Sampling

We conducted several experiment setups to measure the effects of sampling methods and approaches with fast-setting to understand how they affect overall training performance. The metrics and evaluation strategy are mentioned in paragraph 3.2. Table 4 provides the results of different sampling strategies for face or body and speech bubble association. Experimenting with different Pair-pool strategies was significant in evaluating and selecting training data for the relation branch. Pair-pool curates the training data for the relation branch considering box head output.

The First N strategy is when the boxes to be matched are sorted by box score from the RoI head's outputs. This sample selection is obtained by taking N-positive and N-negative samples with the highest box scores. The First N method is considered to be the baseline for this setup. Following that, **Randomized** is a random selection of N samples among possible positive and negative sample candidates. The randomized method results show that having a higher box score does not necessarily mean training the model better and thus being a good input for the model. **Balanced** method ensures that the face and body have an equal number of elements within positive and negative samples. Namely, there are N//2 face and body samples for N positive or N negative samples. However, class balancing reduces the performance of the model with the fast-setting since solving the face and speech bubble association problem is harder than body and speech bubble association. We argue that this is because faces have less intersection with the speech bubble bounding box; in return, they do not share many features. Additionally, in comparison to bodies, faces can have varying orientations, expressions, and partial occlusions, making it difficult for the model to establish reliable associations. Due to that, the performance of the face is almost always lesser than its body counterpart, not just for pair-pool experiments but in others, too.

The **Sorted** option in the table indicates that the candidate samples are sorted based on their Intersection over Union (IoU) values with their corresponding ground truth (GT) counterparts. The sorting operation is performed, and the top N samples are selected. When combined with the balanced approach, the first N//2 samples are chosen for bodies, and the remaining N//2 samples are selected for faces. This strategy yields the best performance across most metrics, except for face precision.

The superior performance of the Sorted option can be attributed to the fact that when the selected samples are closer to their target boxes, the relation branch can be effectively trained, leading to improved results. By prioritizing the samples with higher IoU values, the model focuses on associations that are more likely to be correct, resulting in better overall performance. You can also refer to Appendix F for how we adjusted the number of positive and negative samples in our experiments.

Table 4. Performance comparison of Pair-Pool sampling strategies with fast-setting.

Method	BODY			FACE		
	Recall	Precision	F1	Recall	Precision	F1
First N	50.00	52.88	51.40	40.31	34.21	37.01
Randomized	50.90	54.10	52.45	41.08	**43.44**	42.23
Randomized & Balanced	51.36	53.81	52.55	37.20	36.09	36.64
Sorted	**51.36**	**54.32**	**52.80**	**48.06**	40.00	**43.66**
Sorted Balanced	50.45	53.11	51.17	46.51	36.14	40.67
Sorted & Weighted Loss	46.36	48.11	47.22	36.43	31.54	33.81

E.2 Encapsulation Box Mask Results

First, we evaluated the performance without any masking (Without Mask). The results showed moderate body recall, precision, and F1 score performance. However, the face scores are still higher than the masked versions. We then explored different versions of encapsulation masking. **Version 1** marked only the relevant regions, which correspond to face or body and speech bubbles, as one while masking the rest. This approach improved the body metrics but slightly degraded face performance. In **version 2**, we introduced additional masking by drawing lines from box centers. Although it decreased body and face scores, it provided valuable insights that led to version 3. In **version 3**, we relaxed the masking by marking surrounding areas within a 1-unit distance from the lines drawn in version 2. This further improved body and face scores. Finally, **version 4** modified version 1 by assigning masked areas a value of 0.25 instead of 0. This adjustment yielded improved face performance in return, resulted in a slight decrease in body scores. This makes sense of how much space a possible face occupies in the 7 × 7 grid compared to a body.

The different masking strategies had mixed effects on the model's performance, with some strategies improving specific metrics while affecting others. Careful selection of the appropriate masking strategy is crucial, considering the task requirements and trade-offs. Though these results justify why encapsulation mask is crucial, mask versions' performance does not directly translate to slow-setting. In the slow-setting, we found that mask version 3 performs best. The results are summarized in Table 5.

Table 5. Performance comparison of various encapsulation masking strategies with fast-setting.

Method	BODY			FACE		
	Recall	Precision	F1	Recall	Precision	F1
Without Mask	51.36	54.32	52.80	**48.06**	40.00	**43.66**
1	**54.54**	**57.41**	**55.94**	42.63	38.19	40.29
2	49.54	53.69	51.53	40.31	37.14	38.66
3	53.63	51.52	52.56	45.73	35.32	39.86
4	53.18	56.52	54.80	44.18	**40.14**	42.06

E.3 Sliding Window Negative Samples

We varied the count of sliding window negative samples and examined their impact on the performance metrics for both body and face association with speech bubbles. We increased the count of sliding window negative samples to 5 without applying encapsulation box masking. This adjustment led to an improvement in body recall and F1 score. However, face recall, precision, and F1 score experienced a significant drop. Then, we introduced encapsulation box masking to count five sliding window negative samples to address the decline in face metrics. This combination yielded much better results compared to the previous method. Specifically, extreme drops in the face scores were mitigated. This also shows the importance of encapsulation box masking and the sensitivity of face and speech bubble association problems. Additionally, body recall and F1 score were slightly increased.

We further increased the count of sliding window negative samples to 10 without masking and observed a decline in performance across all metrics. However, when encapsulation box masking was applied along with ten sliding window negative samples, the performance improved for both body and face association. Continuing the experimentation, we increased the count of sliding window negative samples to 15 without masking, increasing scores compared to counts of 5 and 10. As the number of sliding window samples increases, they lose their status as outliers. Nevertheless, when encapsulation box masking was applied with 15 sliding window negative samples, the performance improved compared to the previous method. Body and face scores both exhibited improvements except body precision.

Including sliding window negative samples during training surely does not boost the model's performance in the fast-setting. It may seriously harm the model's performance. However, that can be blocked by encapsulation box masking. Thus, it can help the model to differentiate which parts of the feature provide the association information since it was our observation that a model without any guidance could output a non-face area as an association. These findings highlight the importance of carefully selecting the count of sliding window negative samples and incorporating encapsulation box masking to strike a balance between body and face association. The results of the experiments comparing the number of sliding window negative samples during training, along with the inclusion of encapsulation box masking, are presented in Table 6.

Table 6. Performance comparison of the number of sliding window negative samples during training with fast-setting. Results with encapsulation box masking are also included in the table to show the combined effect of both techniques and as mitigation for drops in results of face and speech bubble association.

Count	Encapsulation Box Masking	BODY			FACE		
		Recall	Precision	F1	Recall	Precision	F1
0		51.36	54.32	52.80	**48.06**	40.00	**43.66**
5		53.18	54.93	54.04	34.88	24.19	28.57
5	✓	**53.63**	54.63	**54.12**	46.51	37.26	41.37
10		43.18	51.91	47.14	31.78	33.06	32.41
10	✓	52.27	52.99	52.63	47.28	30.80	37.30
15		42.72	**57.66**	49.08	38.76	35.46	37.03
15	✓	50.45	54.68	52.48	43.41	**41.79**	42.58

E.4 Bubble Mirrored Negative Samples

A similar trend with sliding window negative samples can be seen in the results. Without the encapsulation mask, there is no improvement in any of the F1 scores. However, when encapsulation box masking is introduced, the F1 score surpasses all other cases, including the case of having encapsulation box masking without bubble-mirrored negative samples. That shows a 55.94% F1 score in its maximum performance. Notably, the same positive performance increase cannot be seen when the bubble-mirrored sample count is increased beyond 5. We think that is because providing five samples is sufficient to teach the model bubble symmetry. More than that would limit the samples with actual faces or bodies.

The findings suggest that encapsulation box masking and the quantity of bubble-mirrored negative samples significantly impact how well the body and face association model performs. While encapsulation box masking and a moderate amount of bubble-mirrored negative samples enhance the model's performance, a higher amount of bubble-mirrored negative samples could make it more difficult to distinguish between bodies and faces. These results suggest improvements to the training procedure and preventing performance declines in the speech bubble association task. The performance comparison of the quantity of bubble-mirrored negative samples during training with the fast-setting is shown in Table 7.

Table 7. Performance comparison of the number of *bubble mirrored negative samples* during training with fast-setting. Results with encapsulation box masking are also included in the table to show the combined effect of both techniques and as mitigation for drops in results of face and speech bubble association.

Count	Encapsulation Box Masking	BODY			FACE		
		Recall	Precision	F1	Recall	Precision	F1
0		**51.36**	54.32	52.80	48.06	40.00	43.66
5		46.36	50.49	48.34	34.88	29.22	31.80
5	✓	53.63	**59.89**	**56.59**	**49.61**	39.50	**43.98**
10		50.00	57.29	53.39	34.10	**47.82**	39.81
10	✓	50.45	53.11	51.74	44.18	39.04	41.45
15		45.45	53.19	49.02	38.76	29.41	33.44
15	✓	45.90	54.59	49.87	37.98	40.49	39.20

E.5 Absent Negative Links

Table 8 compares the base model's performance in body and face association tasks. Different experimental settings were explored, including varying counts of *absent negative links* and the application of encapsulation box masking. The results demonstrate that incorporating *absent negative links* and utilizing encapsulation box masking techniques impact the model's performance positively in terms of body and speech bubble association.

Table 8. Performance comparison of the number of *absent negative links* during training with fast-setting. Results with encapsulation box masking are also included in the table to show the combined effect of both techniques and as mitigation for drops in results of face and speech bubble association.

Count	Encapsulation Box Masking	BODY			FACE		
		Recall	Precision	F1	Recall	Precision	F1
0		51.36	54.32	52.80	**48.06**	40.00	**43.66**
5		47.27	56.83	51.61	38.76	37.59	38.16
5	✓	48.63	57.52	52.70	40.31	**45.21**	42.62
10		44.54	58.33	50.51	31.00	31.49	31.25
10	✓	50.90	**65.11**	**57.14**	31.78	41.83	36.12
15		48.18	48.84	48.51	40.31	35.86	37.95
15	✓	**52.27**	53.99	53.11	36.43	38.84	37.60

In the body association task, including 5 *absent negative links* along with encapsulation box masking yields improved precision and F1 score. Additionally, in the face association task, encapsulation box masking has a positive effect on precision. These findings suggest that the careful inclusion of negative samples and the use of appropriate masking techniques enhance the model's performance. The phenomenon of declined face and speech bubble association performance continued with this data augmentation technique. This again shows the difficulty of solving the face and speech bubble association problem.

This technique is important in effectively addressing the face or body and speech bubble association problem. The outcomes highlight the significance of thoughtful data preparation and the implementation of suitable methodologies to achieve improved performance in these association tasks. Using absent negative links improves the overall body and speech bubble association performance but leads to a decline in the face. This can also be seen in Table 1. In that, when absent negative links are used more than ten and mixed with other negative samples to be selected for which pair-pool sampling strategy is applied, it has the same effect.

F Impact of Adding More Negative Samples

We evaluated the performance of our model on the association of bodies and faces with speech bubbles by adding more negative samples. The base case was when N samples constituted positive and negative; N was 76. The results in Fig. 10 showed that the model achieved reasonable precision, recall, and F1 scores for body associations. However, as we discussed, it faced more significant challenges in accurately associating faces with speech bubbles.

The rationale behind measuring the effect of adding more negative samples was related to how the model is used for inference. During inference, the model has to identify the relation of a speech bubble with an element correctly. However, in most cases, there are more wrong options than correct ones since there can only be one face and one correct body. When additional negative samples were introduced during training, the model improved body precision and recall. It peaked when the additional samples were 150, almost equivalent to a 1:3 ratio between positives and negatives. However, this improvement was not consistent for face associations.

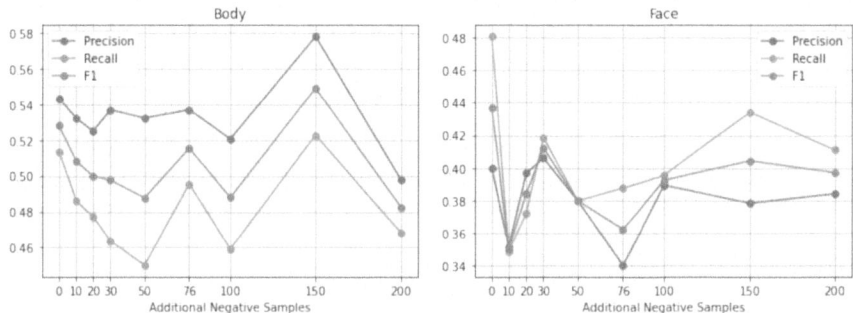

Fig. 10. The impact of increasing the number of negative samples on precision, recall, and F1 scores for body and face association with speech bubbles, using the fast-training setup in the relation network. The base number of negative samples is 76.

References

1. Agrawal, H., Mishra, A., Gupta, M., et al.: Multimodal persona based generation of comic dialogs. In: Proceedings of the 61st Annual Meeting of the Association for Computational Linguistics (Volume 1: Long Papers), pp. 14150–14164 (2023)
2. Augereau, O., Iwata, M., Kise, K.: An overview of comics research in computer science. In: 2017 14th IAPR International Conference on Document Analysis and Recognition (ICDAR). vol. 3, pp. 54–59. IEEE (2017)
3. Bresenham, J.E.: Algorithm for computer control of a digital plotter. IBM Syst. J. 4(1), 25–30 (1965)
4. Brienza, C.: Producing comics culture: a sociological approach to the study of comics. J. Graph. Novels Comics 1(2), 105–119 (2010)
5. Cohn, N.: The Visual Language of Comics: Introduction to the Structure and Cognition of Sequential Images. A&C Black (2013)
6. Deng, J., Dong, W., Socher, R., Li, L.J., Li, K., Fei-Fei, L.: ImageNet: a large-scale hierarchical image database. In: 2009 IEEE Conference on Computer Vision and Pattern Recognition, pp. 248–255. IEEE (2009)
7. Dosovitskiy, A., et al.: An image is worth 16 × 16 words: Transformers for image recognition at scale (2020). arXiv preprint arXiv:2010.11929
8. He, K., Gkioxari, G., Dollár, P., Girshick, R.: Mask R-CNN. In: Proceedings of the IEEE International Conference on Computer Vision, pp. 2961–2969 (2017)

9. Herbst, P., Chazan, D., Chen, C.L., Chieu, V.M., Weiss, M.: Using comics-based representations of teaching, and technology, to bring practice to teacher education courses. ZDM **43**(1), 91–103 (2011)
10. Hu, H., Gu, J., Zhang, Z., Dai, J., Wei, Y.: Relation networks for object detection. In: Proceedings of the IEEE Conference on Computer Vision and Pattern Recognition, pp. 3588–3597 (2018)
11. Laubrock, J., Dunst, A.: Computational approaches to comics analysis. Top. Cogn. Sci. **12**(1), 274–310 (2020)
12. Li, Y., Xie, S., Chen, X., Dollár, P., He, K., Girshick, R.B.: Benchmarking detection transfer learning with vision transformers (2021). CoRR abs/2111.11429, https://arxiv.org/abs/2111.11429
13. Lin, T.-Y., et al.: Microsoft COCO: common objects in context. In: Fleet, D., Pajdla, T., Schiele, B., Tuytelaars, T. (eds.) ECCV 2014. LNCS, vol. 8693, pp. 740–755. Springer, Cham (2014). https://doi.org/10.1007/978-3-319-10602-1_48
14. Loshchilov, I., Hutter, F.: SGDR: Stochastic gradient descent with warm restarts (2016). arXiv preprint arXiv:1608.03983
15. Loshchilov, I., Hutter, F.: Decoupled weight decay regularization (2017). arXiv preprint arXiv:1711.05101
16. Nguyen, N.V., Rigaud, C., Burie, J.C.: Digital comics image indexing based on deep learning. J. Imaging **4**(7), 89 (2018). https://doi.org/10.3390/jimaging4070089, http://www.mdpi.com/2313-433X/4/7/89
17. Nguyen, N.V., Rigaud, C., Burie, J.C.: Comic MTL: optimized multi-task learning for comic book image analysis. Int. J. Doc. Anal. Recogn. (IJDAR) **22**, 265–284 (2019)
18. Nguyen, N.-V., Rigaud, C., Revel, A., Burie, J.-C.: Manga-MMTL: multimodal multitask transfer learning for manga character analysis. In: Lladós, J., Lopresti, D., Uchida, S. (eds.) ICDAR 2021. LNCS, vol. 12822, pp. 410–425. Springer, Cham (2021). https://doi.org/10.1007/978-3-030-86331-9_27
19. Ren, S., He, K., Girshick, R., Sun, J.: Faster R-CNN: towards real-time object detection with region proposal networks. In: Advances in Neural Information Processing Systems, vol. 28 (2015)
20. Rigaud, C., Le Thanh, N., Burie, J.C., Ogier, J.M., Iwata, M., Imazu, E., Kise, K.: Speech balloon and speaker association for comics and manga understanding. In: 2015 13th International Conference on Document Analysis and Recognition (ICDAR), pp. 351–355. IEEE (2015)
21. Sachdeva, R., Zisserman, A.: The manga whisperer: Automatically generating transcriptions for comics (2024). CoRR abs/2401.10224. https://doi.org/10.48550/ARXIV.2401.10224
22. Santoro, A., et al.: A simple neural network module for relational reasoning. In: Advances in Neural Information Processing Systems, vol. 30 (2017)
23. Soykan, G., Yuret, D., Sezgin, T.M.: A comprehensive gold standard and benchmark for comics text detection and recognition (2022)
24. Soykan, G., Yuret, D., Sezgin, T.M.: Identity-aware semi-supervised learning for comic character re-identification (2023)
25. Tang, K., Zhang, H., Wu, B., Luo, W., Liu, W.: Learning to compose dynamic tree structures for visual contexts. In: Proceedings of the IEEE/CVF Conference on Computer Vision and Pattern recognition, pp. 6619–6628 (2019)
26. Wu, Y., Kirillov, A., Massa, F., Lo, W.Y., Girshick, R.: Detectron2 (2019). https://github.com/facebookresearch/detectron2

ComicBERT: A Transformer Model and Pre-training Strategy for Contextual Understanding in Comics

Gürkan Soykan[1,2](✉) ⓘ, Deniz Yuret[1,2] ⓘ, and Tevfik Metin Sezgin[1,2] ⓘ

[1] Computer Engineering Department, Koç University, Istanbul, Turkey
{gsoykan20,dyuret,mtsezgin}@ku.edu.tr
[2] KUIS AI Center, Istanbul, Turkey
https://ai.ku.edu.tr/

Abstract. Despite the growing interest in digital comic processing, foundational models tailored for this medium still need to be explored. Existing methods employ multimodal sequential models with cloze-style tasks, but they fall short of achieving human-like understanding. Addressing this gap, we introduce a novel transformer-based architecture, *Comicsformer*, and a comprehensive framework, *ComicBERT*, designed to process and understand the complex interplay of visual and textual elements in comics. Our approach utilizes a self-supervised objective, *Masked Comic Modeling*, inspired by BERT's [6] masked language modeling objective, to train the foundation model. To fine-tune and validate our models, we adopt existing cloze-style tasks and propose new tasks - such as scene-cloze, which better capture the narrative and contextual intricacies unique to comics. Preliminary experiments indicate that these tasks enhance the model's predictive accuracy and may provide new tools for comic creators, aiding in character dialogue generation and panel sequencing. Ultimately, *ComicBERT* aims to serve as a universal comic processor.

Keywords: Digital Comics Processing · Transformer Architectures · Self-Supervised Learning · Cloze-Style Tasks · Neural Comic Understanding

1 Introduction

Comics are a multimodal form that blends text and imagery to narrate stories or express ideas, spanning a range from purely visual representations to text-dominated formats. This medium intersects various disciplines including education [9], sociocultural studies [3], and the cognitive sciences [2,12]. Enhancing our understanding of comics could significantly contribute to advancements in these areas. Moreover, the techniques developed for analyzing and interpreting comics could also be extended to other types of *Visual Language* [5], such as cave paintings, mangas, and graphic novels.

However, a foundation model for comic processing has not yet been proposed. The most similar study to this is the method suggested in [10]. A multimodal sequential model is proposed with cloze-style tasks, but it is far from human-level performance due to both the dataset and method used. Considering these points, we are trying to establish a method that can serve as a foundation model and a self-supervised objective to pre-train it. The base model can be called a *Comicsformer*. Comicsformer is a transformer encoder structure that can process visual information of panels, speech bubble texts, narrative text, and ultimately character visual and identity information for sequences of comic panels. The self-supervision task that can train Comicsformer to become a foundation model is called *Masked Comic Modeling*, inspired by the masked language modeling task in BERT [6] and contrastive learning [11]. Thus, overall, the general framework can be referred to as *ComicBERT*. To the best of our knowledge, we are the first to propose these tasks and models for comics in the literature.

We use the cloze-style tasks, visual-cloze, text-cloze, and character coherence, introduced in [10] to fine-tune the model and measure its success. In addition to these, we propose two novel tasks that will be useful in comic book understanding. One of them is scene-cloze; unlike the cloze-style tasks just mentioned, it tries to predict all of the elements in the next scene, not just some of them. The other is contextual character to speech attribution; this task allows us to understand whether a given speech bubble context belongs to the existing character when the sequential panel context is given. We believe these new tasks, contextual character-to-speech attribution and scene-cloze, can become tools that will be useful to comic illustrators; they can be utilized to decide what a particular character has to say or what the next panel should be given context information from previous n panels.

The Comicsformer structure and ComicBERT can be trained later with different tasks and information. They can be used to predict or process visual grammar information [5]. Alternatively, the ComicBERT framework can be considered a universal comic processor and used as a foundation model, which is its primary motivation. The code, instructions and models can be accessed online[1]. Our main contributions can be summarized as follows:

- *Comicsformer*, a multimodal transformer-encoder architecture, is tailored for analyzing comic panels and their elements.
- *Masked Comic Modeling (MCM)*, a novel contrastive self-supervised pre-training strategy, enhances the learning process without labeled data.
- *ComicBERT*, derived from *Comicsformer* and *MCM*, serves as a foundation model for comprehending comics.
- By using the ComicBERT framework we achieved state-of-the-art results on cloze-style tasks, especially getting closer to human-level comprehension in text-cloze and visual-cloze tasks.

---

[1] https://github.com/gsoykan/comicbert

2 Related Work

2.1 Multimodal Comic Processing

Iyyer et al. [10] propose a multimodal neural network architecture to address downstream tasks specific to comics, such as text cloze, visual cloze, and character coherence, which they also introduce. Additionally, they provide the COMICS dataset, which contains the largest collection of comic books and serves as the dataset for our study. *Manga-MMTL* [15] combines a comic character's visual features of faces and their corresponding text to extract multimodal character features. The learned character features are then evaluated by information retrieval methods to prove their effectiveness. However, in order to train their framework, one of the tasks is to classify comic albums, which may be unavailable for some datasets. Also, it is focused on the faces of the characters and dialogues only. Thus, it cannot provide context for the panel or the sequence of panels. *MPDialog* [1] is a persona-based dialogue generation framework for comic strips. They use a multimodal transformer decoder with panel image features, a detailed persona description text, and speech bubble content associated with identities. So, instead of using character embeddings as an input model, they are mapped to textual descriptions, which are hard to annotate. Yet, this can be a good candidate for the downstream task for comics. However, the evaluation is particularly hard and does not necessarily result in context embeddings out of the model to be used for other tasks. Whereas the cloze-style tasks proposed in [10] and in this study force the model to learn contextual information of panel sequences and can be combined to extract generalized context features for comics. Recently, "Magi" [17] has been proposed to enhance the accessibility of manga for visually impaired individuals. It does so by automatically generating transcriptions that identify who said what and when, utilizing an encoder-decoder architecture that detects and associates manga panels, characters, and texts. Our most significant difference from them is that we offer a universal comic processor framework with a pre-training task. Indeed, ideas from ComicBERT and Magi can be combined to create improved models.

2.2 Contrastive Learning with Encoder Models

TaCL (Token-aware Contrastive Learning) [21] employs two instances of the same pre-trained BERT variant: one instance is frozen and designated as the teacher model, while the other is trained using masked language modeling and next-sentence prediction tasks in conjunction with their proposed TaCL loss. TaCL incorporates contrastive learning between the representations of the student and teacher models; however, the input tokens of the teacher model are left unmasked. The authors suggest this approach leads to a more discriminative distribution of token representations. Sunder et al. [22] applies tokenwise contrastive learning to align the representations of a BERT model with a speech encoder augmented with cross-attention mechanisms and non-contextual word embeddings. the BERT model serves as a teacher to transfer token-level embeddings to speech representations.

3 Methodology

Our research relies on three preceding studies: first, quality OCR from comic pages [19]; second, component detection from comic pages and associating speech bubbles with characters via the augmented ComicMTL framework [14], third, generating character embeddings that include identity information from the faces and bodies of characters, using an identity-aware comic character Re-Identification backbone [20]. This integrated effort facilitated the creation of the required dataset, which involved processing comic book pages sourced from the COMICS dataset [10].

3.1 Dataset

The overall data preparation process involves several steps. First, the MTL model [14] is applied to detect speech bubbles that score above a 0.5 threshold, character faces and bodies, narrative boxes, and panels. This model is also capable of segmenting instances of speech bubbles and associating them with characters. Following detection, panels are ordered by their z-order, and components within these panels are similarly associated and ordered. Speech bubble context extraction is carried out using an end-to-end OCR model trained with the Comics Text+ datasets [19], which utilizes speech bubble segmentations. At the conclusion of this process, panels are systematically organized by their associated components. The order of speech and the speakers associated with each speech bubble are clearly identified within the panels, while detected faces and bodies are grouped as characters. Further details of these steps can be found in Appendix A. As a result of the data preparation method mentioned above, we present the sizes of the resulting dataset's components to provide an understanding of its scope. The dataset comprises 388,079 narrative texts, 1,525,553 speech bubble texts, 2,201,437 character instances, and 1,145,654 panels. These elements are sourced from approximately 4,000 (3959 to be exact) comic magazines. For a brief textual analysis including word frequency distributions see Appendix B.

3.2 Comicsformer Architecture

Comicsformer is a multimodal transformer encoder architecture, based on the transformer model introduced by Vaswani et al. [24], that processes sequential comic book panels' units. The hyperparameters used in the experiments are as follows: input dimension is 384, headcount is 4, hidden layer dimension is 1536, and 6 transformer sub-encoder layers are used. The dropout rate is 0.1, and the activation function is ReLU. Additionally, layer normalization is performed with an epsilon value of $1e-5$, and it is applied after attention and feedforward operations.

On average, Comicsformer operates on a relatively small scale compared to other models in the literature [26]. The main factor enabling this is the construction of the input embeddings. Unlike standard NLP architectures that use word or subword token embeddings, comic processing cannot rely on a fixed vocabulary. Therefore, our input embeddings are derived from other pre-trained models that operate on comic components, extracting embeddings for each "comic unit." The primary goal of Comicsformer's output is to obtain contextualized comic unit embeddings. The success or failure of our experiments is directly tied to the effectiveness of these contextualized comic unit embeddings. Both the self-supervised pre-training method ComicBERT and the cloze-style tasks depend on this. In summary, what distinguishes Comicsformer as a special type of transformer encoder is the nature of the inputs, the pre-training strategy, and the application of the outputs in downstream tasks.

Encoding Modalities and Inputs of the Comicsformer. The comic units used in Comicsformer include panels, characters (face and body detection pairs), speech bubble texts, and narrative texts. As these are different data types, we treat them as separate modalities. The input consists of sequential comic book panels, and the units, as described, are utilized for each panel. We begin by selecting a pre-trained model for each modality. During experiments, these selected models are kept frozen. Then, the output from each pre-trained model is projected to the Comicsformer input dimension using a linear layer. These projection layers are trainable.

The pre-trained models used for each modality are as follows:

- **Panels**: EfficientNet [23] processes the panel images. The baseline model B0 is chosen for its low computational cost. Transformations applied during the panel image processing include resizing to the longest side at 256, padding to make the image square, then center cropping to 224×224, followed by normalization.
- **Characters**: Soykan et al.soykan2023identityaware describes a model that enables the *Character modality* in their character re-identification (ReID) framework. Character instances comprise the character's face and body images, which may not necessarily appear together due to occlusions or the character facing away. The Character ReID model uses an identity-aware self-supervised ResNet-50 backbone [7], fine-tuned with an identity network. For character processing, we concatenate the network's final output, a 256-dimensional identity embedding, with the 2048-dimensional backbone embedding before the identity head. This concatenated vector is then passed through a projection layer. This process provides Comicsformer with both the character's self-supervised backbone features and the identity information. The transformations for body images include resizing by the longest size to 128, padding to make the image square, and normalization. Face images, already square, are resized to 96 and normalized with a mean and standard deviation of 0.5.
- **Texts of Speech Bubbles and Narrative Boxes**: The approach for language processing involves using sentence or paragraph embeddings instead of separate embeddings for each token, which would increase the context window

size significantly. Therefore, BERT [6] and its derivatives, specifically Sentence Transformers [16], were considered. The models used in experiments include 'bert-base-uncased', 'sentence-transformers/all-MiniLM-L6-v2', and 'sentence-transformers/all-distilroberta-v1' from Hugging Face's transformers library [25]. The DistilRoBERTa [18] variant showed the best performance without significant computational overhead, thanks to the distillation process. The self-supervised contrastive learning objective used in Sentence Transformers aligns well with our objectives, making it successful. The text of speech bubbles and narrative boxes is processed using the same pre-trained language model to obtain text embeddings.

Further, after obtaining each modality's so-called *token embeddings* from a panel, the token embeddings for a panel consist of embeddings for the panel itself, three character-speech pairs, and the narrative. These collectively form the *panel unit tokens*. It is important to note that if the MTL model does not establish a character association for a speech bubble, the character token is represented using a padding embedding. Similarly, padding tokens are used in place of speech tokens for characters without speech. Moreover, if a character has multiple speech bubbles and corresponding text, the speech is paired based on speech order (e.g., the embedding for the character, followed by the embedding for the first speech, then the embedding for the character again, and the second speech, and so on, meaning character embedding can be reused in case of multiple speeches). Comicsformer employs a transformer architecture, which allows the use of any number of sequential panels within the context window. We adopt the approach of the referenced work [10], which uses three context panels.

To separate each panel's unit tokens, [SEP] (Separator) token embeddings are inserted, similar to how BERT [6] handles the "Next Sentence Prediction" task. In this scenario, a total of 26 token embeddings are used for each sequence. Before feeding the token embeddings to Comicsformer as input, we sum them with positional embeddings [24] and modality embeddings. While positional embeddings are constant values and not trained during training, modality embeddings are specific to each modality (panel, character, speech, narrative) and are trainable embeddings initialized with random values sampled from a uniform distribution within the range of -0.1 and 0.1. After these steps, layer normalization is applied to complete the input formation. The whole process is depicted in Fig. 1.

3.3 Masked Comic Modeling and ComicBERT

Masked Comic Modeling (MCM) is a key aspect of our methodology in the Comicsformer framework. Let S be the sequence of units in a comic page comprising panels, characters, speech, and narrative texts. Each unit u within the sequence S is represented by a vector $\mathbf{u}$ in an input embedding space. The MCM task aims to contextualize these embeddings.

Let T be a set of indices representing tokens within S. For each token index t in T, we define a masking operation M_t, which masks the corresponding token's embedding in $\mathbf{u}_t$. The objective is to recover or retrieve the masked embeddings

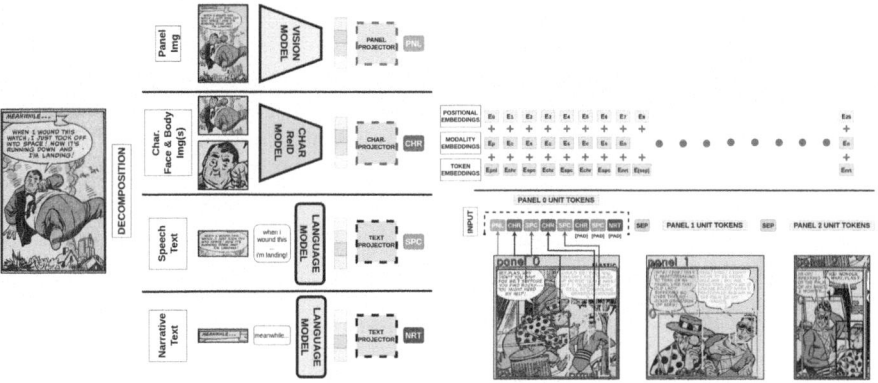

Fig. 1. On the left the decomposition of a panel as Comicsformer input is shown. The sequential panels are the primary inputs to the Comicsformer, where each panel is used and transformed into a fundamental subunits. On the right the formation of the Comicsformer input is illustrated. For each panel, modality projections are used as token embeddings. In the experiments, three sequential panels were used, and special separator ([SEP]) tokens were employed to separate the panels. The sequential structure of the comics was reflected in the model using positional embeddings. Trainable special modality embeddings were used on tokens coming from modalities to facilitate the discrimination abilities of the model.

using contextual information. The masking rate in our experiments is 0.2 of the sequence S.

Given the Comicsformer model, the corresponding output token embedding, o_t, of a masked token u_t is passed through a feed-forward network (FFN) to obtain a transformed embedding h_t. This transformation can be expressed as:

$$o_{1:n} = \text{Comicsformer}(u_1, u_2, \ldots, u_n)$$

$$h_t = \text{FFN}(o_t)$$

The FFN utilized in the model comprises two linear layers with identical input and output dimensions, aligned with Comicsformer's output dimension. The FFN incorporates a GELU activation function [8] and layer normalization in between linear layers, contributing to the augmentation of the model's internal representation transformation capabilities and acting as a projection head.

MCM aims to maximize the similarity between h_t and the corresponding unit's encoding while minimizing the similarity with other masked unit encodings within the sequence and batch. Let M be all masked input embeddings. Then, this can be formulated as an optimization problem for a single masked embedding:

$$\text{maximize} \sum_{t \in M} \left(\text{similarity}(h_t, u_t) - \lambda \sum_{u' \in M \setminus \{u_t\}} \text{similarity}(h_t, u') \right)$$

where similarity($\cdot$) measures the similarity between two embeddings and λ is a balancing parameter.

As the problem is formulated, it transforms into a contrastive learning problem. Therefore, contrastive learning can be applied to train Comicsformer using NT-Xent loss [4] or similar contrastive losses. In our experiments, the NT-Xent loss was employed. Thus, the NT-Xent loss for the MCM problem can be conceptualized as follows for ith masked unit in the batch among a total of N masked units where cosine similarity is used as the similarity function during our experiments: :

$$\mathcal{L}_{\text{MCM},i} = -\log\left(\frac{\exp(\text{sim}(h_i, u_i)/\tau)}{\sum_{k=1}^{2N} \mathbb{1}[k \neq i] \exp(\text{sim}(h_i, u_k)/\tau)}\right)$$

When MCM is employed with the Comicsformer architecture, the entire framework is termed ComicBERT. The contextualized embeddings o_t obtained through MCM enhance the understanding of the relationships between different units on the comic page. The resulting pre-trained model, ComicBERT, can be employed for various downstream tasks in comics understanding and processing. The framework can be visualized as shown in Fig. 2, and the pseudocode for MCM algorithm is detailed in Appendix F.

3.4 Downstream Tasks Overview

Cloze-style tasks for comics were first proposed in [10] to measure the ability of deep learning models to capture context and meaning from comics. These tasks utilize features from context panels $p_{i-3}, p_{i-2}, p_{i-1}$ to predict the features of panel p_i from a single modality. This setup can be extended to incorporate an arbitrary number of context panels, a potential area for future exploration.

The features of the final panel, p_i, are predicted using a cloze-style framework. In all experiments, three candidates are presented as options, and models leverage context information to assign a higher score to the correct candidate. The tasks are described as follows:

- **Text Cloze:** Predicting the text of a single speech bubble in the final panel p_i. This task also considers the final panel's visual information with models, including the panel modality. However, the textbox areas are masked to focus on artwork features while considering the context for scoring.
- **Visual Cloze:** Predicting the artwork of the final panel p_i. Unlike *text cloze*, text features of the final panel are not provided.
- **Character Coherence:** Reordering two dialogue text boxes in the final panel.

Task Variations: Each task has two difficulty levels, except for character coherence. In the *easy* case, incorrect candidates are chosen from different comic books, while in the *hard* case, candidates are selected from the same comic book.

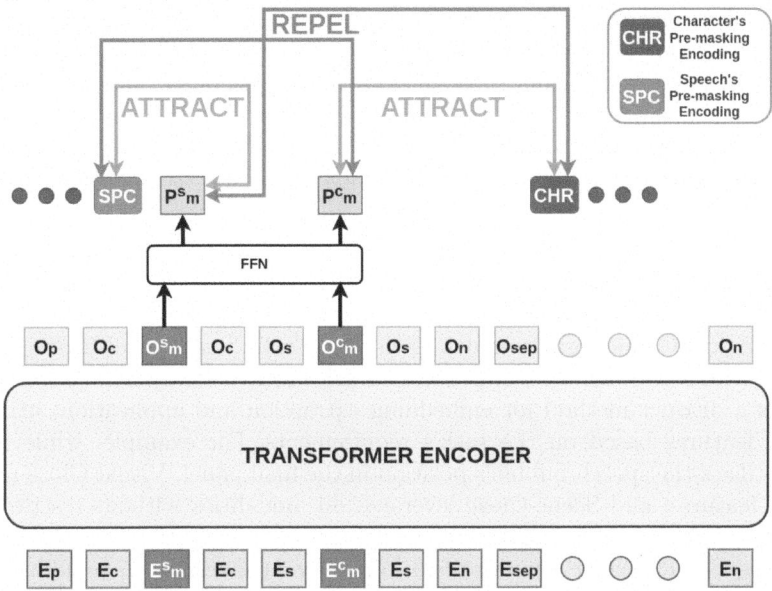

Fig. 2. Overview of the *Masked Comic Modeling (MCM)* task used in *Comicsformer* for comic representation learning. This self-supervision approach, inspired by BERT [6], aims to obtain contextualized embeddings for panels, characters, speech, and narrative texts. Contrastive learning methods are employed to enhance these embeddings by maximizing similarity with the correct modality while minimizing it with others. The resulting model, termed ComicBERT, effectively encodes diverse comic elements.

Additional Tasks. In addition to the aforementioned tasks, we propose two new tasks in our study: *Scene Cloze* and *Contextual Character to Speech Attribution*.

- **Scene-Cloze:** Unlike other cloze-style tasks, this task does not predict a single feature of the final panel, p_i, such as image features or speech bubble text. Instead, it utilizes all the information present in the final panel. The context of p_i is extracted and compared with the context of previous panels to measure their similarity. The task involves predicting the next scene based on the context from the previous n panels.
- **Contextual Character to Speech Attribution:** In this task, by fusing the features of any character from the final panel with context embeddings, it aims to measure the similarity between context-enhanced character embedding and any given text embedding, ideally belonging to a speech bubble. It attempts to maximize the similarity of the speech bubble content associated with the character. This task differs from the Character Coherence task, as it does not involve character speech reordering within a panel, but rather, it scores the likelihood of a character saying any speech bubble content given the context from the previous panels.

Context Extraction. When this input is provided to the Transformer encoder, we obtain output features for each position. We then perform mean pooling on these features, specifically those of non-padded positions, to obtain a single context precursor vector. This precursor is passed through the context projector layer to obtain the final context features. The context projector consists of a linear layer followed by an L2 normalization operation, see Appendix G for iilustration.

Scoring and Option Embeddings. In our research, we detail the processes involved in obtaining and scoring option embeddings across various cloze-style tasks using the Comicsformer model. Each task, Text-Cloze, Visual-Cloze, Scene-Cloze, Character Coherence, and Contextual Character to Speech Attribution, employs a distinct method for embedding extraction and application, utilizing specific features based on the task's requirements. For example, while Text-Cloze utilizes the speech bubble's position in the final panel, Visual-Cloze masks textual features, and Scene-Cloze leverages all modalities without restrictions. Character Coherence and Contextual Character to Speech Attribution similarly adapt the Comicsformer inputs to focus on relevant modalities, including character and speech features. These embedding processes involve specific masking strategies to align with the unique requirements of each task, supported by various model types like Text-only, Char-only, Panel-only, and combinations thereof. The complete details of these methodologies are elaborated in Appendix C.

3.5 Pre-training and Fine-Tuning Details

Self-supervised ComicBERT Pre-training. The training process for the ComicBERT framework involves two settings. In the fast setting, the model is trained for 50 epochs on the first 20,000 comic pages, while in the slow setting, training extends to 500 epochs on the first 100,000 comic pages until the 3000th series of Comics [10], with early stopping triggered at 100 epochs. Gradient clipping is applied with a value of 0.5 to stabilize training. The model is configured to use panel, character, speech bubble, and narrative information, and a masking rate of 0.2 is used along with the NT-Xent loss function. None of the token types are exempt from the masking strategy. A learning rate of 0.0001 is employed with a weight decay of 0.01 to control overfitting. The AdamW [13] optimizer is used with a cosine schedule with warmup (0.1 of max. epochs as warmup). The Comicsformer properties include an input dimension of 384, 4 heads in multi-head attention, a feedforward dimension of 1536, 6 encoder layers in the encoder, and a dropout of 0.1. The ComicBERT framework processes context from three panels, and testing is performed on a separate set of 10,000 pages; these pages come from series above 3500 from the Comics dataset, which has close to 4000 series. A batch size of 128 is used, and early stopping is implemented with a patience of 20 epochs, monitoring minimum validation loss. Some comments about the pretraining can be found in Appendix H.

As mentioned earlier, the generated sequences from the pages undergo filtration before training. This filtration process removes sequences with fewer than two speech bubbles and two characters. This is a modality regularization technique. Additionally, it eliminates irrelevant sequences originating from advertisement pages, as the Comics dataset might include such pages. The first 3000 series are allocated for training, series 3000–3500 for validation, and from series 3500 to the end for testing. In the fast setting, the number of sequences used in training is as follows: train dataset size: 57,250, val dataset size: 5,818, test dataset size: 30,525. On the other hand, for the slow setting, the train dataset size is 292,288, the validation dataset size is 15,594, and the test dataset size is 30,525.

Task Specific Fine-Tuning of Comicsformer. The Comicsformer properties are the same as those used in the pre-training of ComicBERT. Most of the hyperparameters are the same. Those that differ are as follows: The maximum number of epochs is 20. The projection layer output dimension is 256 if ComicBERT weights were used in training and they were not frozen. The Masked Comic Modeling objective was active if transfer learning from ComicBERT was applied. Early stopping was active with a patience of 5 epochs based on the maximum validation accuracy value during training.

For every task-specific training, page-wise training, validation, and test split counts were 20,000, 2,000, and 10,000. Sequence filtering strategies were employed. For every task, like ComicBERT, sequences with fewer than two speech bubbles and two characters in their context panels are removed. Additionally, every task has an additional filter for the final panel based on the number of characters associated with a speech bubble. Text-cloze requires 1; character coherence requires 2; scene-cloze and character-to-speech attribution require at least 1; whereas, for visual-cloze, no such filter exists. Furthermore, for text-cloze and character coherence tasks, speech bubble areas for the final panel are masked.

4 Results and Discussion

4.1 Comparison of ComicBERT and The COMICS Architecture

Table 1 presents the results of this comparison. The model, referred to as "Ours," is the ComicBERT model pretrained with the fast setting, then fine-tuned with transfer learning for task-specific adaptation. The results can be interpreted from three perspectives: No-Context (NC), see Appendix E, excluding NC for text-cloze along with visual-cloze, and character coherence results.

- **NC Results:** These can be observed specifically for the text-cloze task. Their results are better than those of ComicBERT in both easy and hard settings. From this, it can be inferred that our model generally depends more on context. Learning does not occur thoroughly without context information. This could be attributed to parameter size, as their model is LSTM-based and thus has significantly fewer parameters.

- **Text-Cloze and Visual-Cloze:** In this case, ComicBERT performs notably better, especially with panel information. However, the improvement in the text-only model is minimal. In the hard setting, while the improvement is not as significant as in the easy setting, it is still noticeable. Nevertheless, the results remain distant from the human baseline. One of the reasons we wanted to incorporate character information is due to this upper bound. It can be said that approaching the human baseline using only panel and text information is challenging. Nonetheless, ComicBERT is still able to surpass previous results.
- **Character Coherence:** This task can be arguably considered controversial. When the task is proposed, it's framed around characters, yet Iyyer et al. [10] does not use any character information. They attempt to perform text ordering only, which is dependent on the z-order. Thus, this task may merely reflect z-order accuracy. While our results are close to theirs, they are lower, and regardless, the results do not improve much further with character information or longer pre-training.

Table 1. Comparison of the results with ComicBERT pre-trained Comicsformer architecture and Iyyer et al.'s [10] results (comics) based on accuracy on cloze-style tasks. Human baseline is also shared for only hard tasks since the easy task is trivial for humans. *NC-Image-text* models use only the final panel's image without any information from preceding context panels. No context model results strike the usefulness of the context for the cloze style tasks and narrative understanding.

MODEL	Text Cloze				Visual Cloze				Char. C.	
Task Difficulty	EASY		HARD		EASY		HARD			
	comics	ours	comics	ours	comics	ours	comics	ours	comics	ours
Text-only	63.4	**66.4**	52.9	**57.8**	55.9	55.9	**48.4**	47.0	**68.2**	65.1
Image-only	51.7	**80.9**	49.4	**56.6**	85.7	**88.7**	63.2	**64.8**	**70.9**	64.7
Image-text	68.6	**84.0**	61.0	**62.7**	81.3	**89.1**	59.1	**65.8**	**69.3**	66.1
NC-Img-text	**63.1**	52.3	**59.6**	48.5	–	–	–	–	**65.2**	65.0
Human	–		84		–		88		87	

4.2 Multi-modal Results of ComicBERT and Comicsformer

In this section, we present the results for the text-cloze, visual-cloze, scene-cloze, and character coherence tasks, as shown in Tables 2 and 3. The results encompass the outcomes of two approaches: Comicsformer, which was trained directly with task-specific training without pre-training, and ComicBERT, which was finetuned from a model pre-trained with Masked Comic Modeling. Regarding the *contextual character and speech attribution task*, we did not obtain significant results, thus leaving it as a potential avenue for future work.

Table 2. Performance Comparison between Comicsformer and ComicBERT across Text-Cloze, Visual-Cloze, and Scene-Cloze tasks. ComicBERT is pre-trained using the Masked Comic Modeling task before and during its training for each task.

Model	Text Cloze			
	easy		hard	
	Comicsformer	ComicBERT	Comicsformer	ComicBERT
Text-only	34.1	**66.4**	33.7	**57.8**
Char-only	55.6	**88.3**	33.6	**60.7**
Panel-only	63.0	**80.9**	33.5	**56.6**
Panel-Text	59.2	**83.9**	32.9	**62.7**
Panel-Char	68.1	**91.1**	33.6	**65.0**
Char-Text	55.2	**89.0**	33.4	**66.6**
Panel-Char-Text	69.8	**92.3**	33.2	**68.2**

Model	Visual Cloze			
	easy		hard	
	Comicsformer	ComicBERT	Comicsformer	ComicBERT
Text-only	52.9	**55.9**	33.5	**46.9**
Char-only	84.8	**91.9**	54.9	**71.6**
Panel-only	83.5	**88.7**	33.7	**64.8**
Panel-Text	82.6	**89.1**	33.4	**65.8**
Panel-Char	87.0	**93.7**	56.2	**75.9**
Char-Text	83.9	**91.9**	58.1	**74.3**
Panel-Char-Text	85.6	**93.9**	51.2	**74.9**

Model	Scene Cloze			
	easy		hard	
	Comicsformer	ComicBERT	Comicsformer	ComicBERT
Text-only	66.7	**72.4**	33.0	**62.4**
Char-only	85.1	**93.3**	51.8	**73.9**
Panel-only	82.4	**88.9**	33.3	**64.8**
Panel-Text	83.3	**91.1**	33.0	**70.1**
Panel-Char	85.2	**94.9**	51.7	**76.2**
Char-Text	84.6	**94.2**	52.0	**77.6**
Panel-Char-Text	86.9	**95.6**	51.3	**79.3**

During the evaluation, apart from character coherence, the focus should be on the cloze-style tasks. Regarding character coherence, the situation mentioned in the previous Subsect. 4.1 still holds, even to the extent that it can be considered as a validation of its improper setup. Despite pre-training, the results show only a very slight improvement.

Table 3. Performance Comparison of Character Coherence Task between Comicsformer and ComicBERT. ComicBERT is pre-trained using the Masked Comic Modeling task before and during its training for the character coherence task.

Model	Char. Coheren.	
	Comicsformer	ComicBERT
Text-only	65.1	**66.7**
Char-only	64.4	**66.4**
Panel-only	64.8	**66.7**
Panel-Text	66.1	**67.1**
Panel-Char	**66.0**	65.9
Char-Text	64.9	**66.1**
Panel-Char-Text	66.1	**66.4**

Conversely, looking at the cloze-style tasks yields promising outcomes for the proposed approach, using character information, Comicsformer architecture as a base model, and Masked Comic Modeling pre-training. In the previous section, it was noted that the ComicBERT approach exhibited improvement compared to previous results. With the addition of character information, it can be confidently stated that the results in the *Panel-Char-Text* model have significantly improved over the *Panel-Text* model. Across all cloze-style tasks, the *Panel-Char-Text* model surpasses 90% accuracy, showcasing state-of-the-art performance. However, it should be noted that the most crucial improvement is observed in the hard setting, which requires the most contextual information due to the hard setting setup. Remarkable improvement is observed with the ComicBERT framework. The model achieves an accuracy of 68.2 for text-cloze and 74.9 for visual-cloze. Examining these findings, it can be said that our results are getting closer to the human baseline. Moreover, it demonstrates the importance of character information in comprehending comics. As the scene-cloze task is novel, the results here will serve as a baseline for future work. In the hard setting, we achieved nearly 80% accuracy in this task. This implies that when the model is provided with context, it can successfully identify the possible content of the 4th panel within the same comic book with an accuracy of nearly 80%. This displays the potential for using this framework in creating comics and highlights how deep learning models can successfully process complex media like comics.

Considering the contributions of different modalities, when examined individually for the three cloze-style tasks, text consistently yields the lowest accuracy, while character consistently exhibits the highest performance. Additionally, when comparing pairs of modalities, it is evident that *Panel-Text* consistently achieves lower results than the other combinations. This reaffirms the critical importance of integrating character information for effective comic understanding. Our best results are presented in Table 4.

Table 4. The best results of ComicBERT on cloze-style and Character Coherence tasks are achieved using information from panels, characters, and text. The model was pre-trained for 100 epochs on nearly 300,000 sequences. These results currently represent the SOTA for text-cloze and visual-cloze tasks and also provide a baseline for the scene-cloze task.

Task	Task Difficulty	
	easy	hard
Text-Cloze	93.7	69.5
Visual-Cloze	95.7	77.1
Scene-Cloze	96.7	80.6
Char. Coheren.		67.1

5 Conclusion and Future Work

The evaluation of the proposed framework, ComicBERT, revealed its promising potential in extracting contextual information from comic pages, specifically from the sequences of panels. We achieved SOTA results in various cloze-style tasks by incorporating the Comicsformer architecture and pre-training it using Masked Comic Modeling. The integration of character information proved crucial, consistently improving the model's performance and pushing it closer to human-level understanding. Despite these achievements, certain aspects require further attention and exploration. For instance, the contextual character-to-speech attribution task presented challenges and yielded limited results.

Future work may focus on enhancing contextual understanding through refined task designs, exploring multimodal interactions within comics for deeper narrative insights, developing interactive comic creation tools, extending our models to other visual media, and adapting our framework to various comic styles and cultural contexts. In conclusion, the development of the ComicBERT framework and its success in cloze-style tasks, along with the insights gained from multimodal analysis open new avenues for the intersection of technology and art regarding comics.

Acknowledgments. This project is supported by Koç University & İş Bank AI Center (KUIS AI). We would like to thank KUIS AI for their support.

Disclosure of Interests. The authors have no competing interests to declare that are relevant to the content of this article.

A Data Preparation Details

A.1 Handling Characters and Associating with Speech Bubbles

First, we performed pairing for bodies and faces. The algorithm used for pairing works as follows: we calculate the intersection rate of faces within bodies. If it is higher than 20%, they are added to the candidate list. From the candidates, we select the one with the highest intersection rate. However, in some cases, there can be multiple faces within a body bounding box. In such situations, the y-coordinate of the faces is used to choose the one higher up in the body bounding box. If there is no intersection, faces or bodies alone can represent the entire character instance.

After extracting character instances, finding their associations with the existing speech bubbles is necessary. The Multi-Task Learning (MTL) model provides relation scores between all speech bubbles and all faces and bodies.

We filtered out the ones with relation scores below 0.2. Then, we successfully attempted to assign relations to character instances based on these scores.

Since we had already determined which panel each component belonged to, we leveraged this information. Therefore, we filtered the relations for speech bubbles, faces, and bodies based on whether they belong to the same panel. As a result, the possibility of a speech bubble being assigned to a character outside of its panel was eliminated.

Next, we checked the relation scores for the components that make up each character (i.e., face and body). We selected the highest score associated with the character instance's chosen speech bubble.

While associating each speech bubble with a character, initially, each character should be assigned the highest-scoring speech bubble that corresponds to them. Then, any remaining speech bubbles may belong to the same character as their second or third speech bubble.

At this point, we employed the same algorithm again. However, if a speech bubble has already been assigned to a character, we added a penalty of up to 0.35 in subsequent calls for that character's possible next assignments. The purpose of this penalty is to prevent the model from making incorrect associations, as the likelihood of one of two characters owning both speech bubbles in the same panel is lower than the possibility of them both sharing a single speech bubble.

A.2 Z-Order Calculation

To simulate the Z-order, we performed the following steps. First, we obtained the union box of the bounding boxes and then partitioned this union box into an $n \times n$ grid. The center points of the bounding boxes are then snapped to the grid points. Next, each bounding box is considered as a point with integer values for the x and y coordinates on the grid. Then these points are sorted first based on the column values and then based on the row values[2] to calculate the Z-order. The value of n is 4, which applies to both panels and speech bubbles.

A.3 Component Association with Panels

Every detected component is assigned to a panel using the following logic: their intersection rate with each panel is calculated, along with their distance based on

---

[2] https://numpy.org/doc/stable/reference/generated/numpy.lexsort.html

the center coordinates. Then, the panel with the maximum intersection rate is chosen for assignment. In cases where a bounding box has the same intersection rate with multiple panels, the one with the minimum center distance is selected. If the intersection rate is below 0.2, then that bounding box is left unassigned and is not used for our experiments.

A.4 OCR and Text Extraction

To extract text from comic books, the speech bubble segmentations from the Multi-Task Learning (MTL) model are utilized. Instead of directly using the text from *Comics Text+*, which is obtained from speech bubble detections, the decision was made to use segmentations. This choice results in improved text extraction because the image used with Optical Character Recognition (OCR) models represents the pure content of the speech bubble without any background or overlapping speech bubbles, ideally. As a result, the extracted text is of even higher quality compared to *Comics Text+*, which already outperforms the original COMICS dataset's texts.

The segmentation mask is extracted for every speech bubble instance if it surpasses the 0.6 score threshold. Subsequently, morphological transformations are applied to the binary mask using the cv2 library[3] to smooth out the edges and fill small gaps. Some examples can be seen in Fig. 4.

B Analysis Based on Word Frequencies

For a brief textual analysis, we would like to share the word frequency distributions. The most frequent words in both speech and narrative text can be observed in Figs. 6 and 5.

C Details of Scoring and Option Embeddings

We explain the process of obtaining option embeddings separately for each task, following the explanation of the scoring process. Considering the method of context embedding extraction, the scoring process for an option is as follows:

$$\text{option\_logit} = \mathbf{O} \cdot \mathbf{c}$$

Since we perform L2 normalization on both the option and the context vectors, the option logit equals the cosine of the angle between those two vectors:

$$\text{option\_logit} = \cos(\theta)$$

Given that the cross-entropy loss is used as the loss function and there are three options, which is consistent across all cloze-style tasks, the final scores are calculated as follows:

$$\text{scores} = \text{softmax}([\text{option\_logit}_0, \text{option\_logit}_1, \text{option\_logit}_2])$$

---

[3] https://docs.opencv.org/4.x/d9/d61/tutorial_py_morphological_ops.html.

Fig. 3. A sample comic page is used to demonstrate the finalized inference process of the MTL model for the data preparation step in the Comicsformer model. The panels are z-ordered, and speech bubbles within a panel are displayed with their order indicated at the bottom left. Characters' components are colored with the same color, and speech bubble associations are shown with lines originating from their centers.

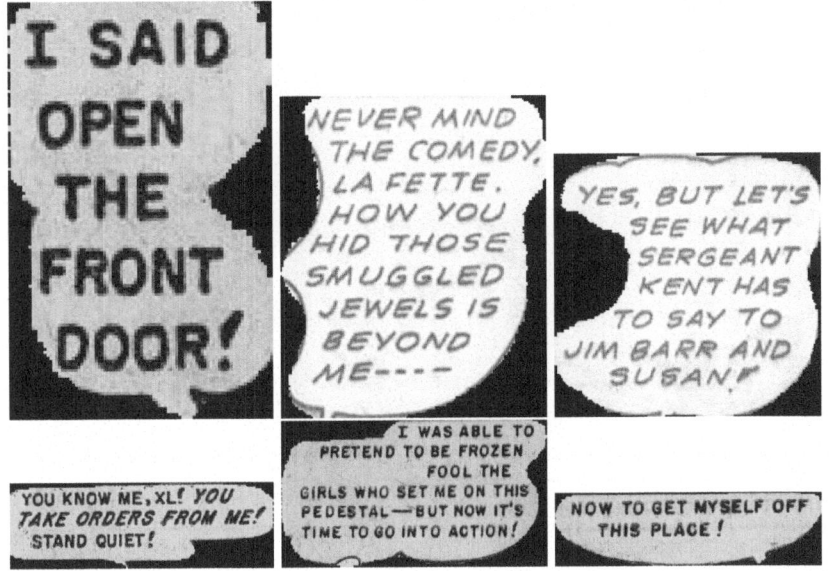

Fig. 4. The *Comic Text+ OCR models* use examples from speech bubble segmentations as shown above. Utilizing speech bubble segmentations instead of detections is particularly useful for irregularly shaped speech bubbles.

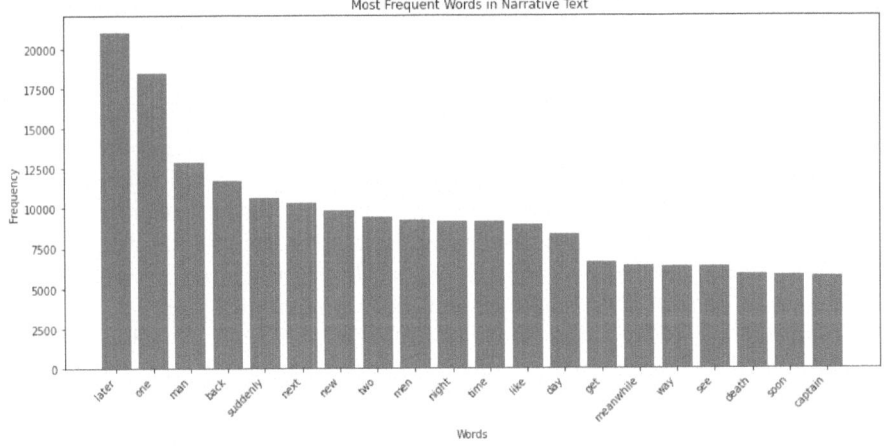

Fig. 5. Frequency distribution of the top 20 common words, apart from stopwords and punctuation, in narrative boxes.

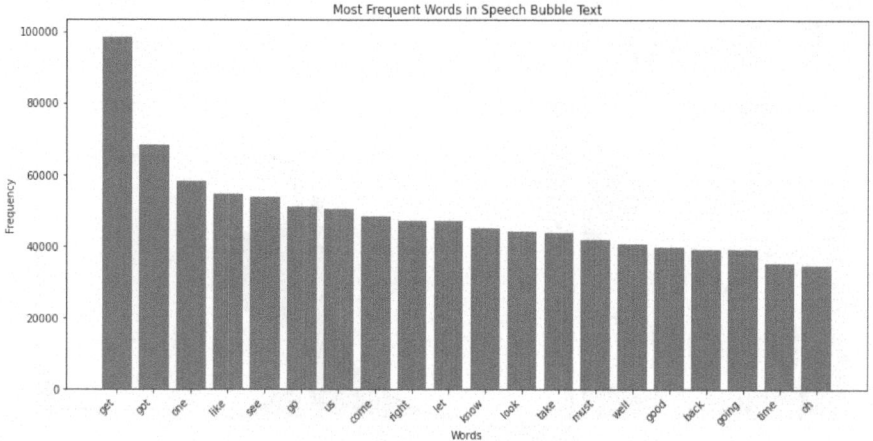

Fig. 6. Frequency distribution of the top 20 common words, apart from stopwords and punctuation, in speech bubbles.

The process for extracting option embeddings for each task is as follows:

- **Text-Cloze:** The final panel is encoded as Comicsformer input and fed to the Comicsformer. The output feature at the position of the speech bubble, which is the third position (assuming the final panels should only have a single speech bubble text for the text cloze), is obtained and processed with the output projector, which has the same architecture as the context projector, i.e., a linear layer followed by L2 normalization. Depending on the model type, encoded panel features are masked; for instance, if it is the panel-only (image-only) model, then character input features are masked in the input.
- **Visual-Cloze:** Similar to Text-Cloze, but with textual (speech bubble, narrative) input features for the final panel masked. Since the task requires excluding text information from the final panel apart from the visual features, when character modality is involved, it is not masked in the input. However, only the output feature of the panel token is used from the outputs.
- **Scene-Cloze:** There are no restrictions for Scene-Cloze, thus all modalities of the model type are used. Similar to context extraction, the mean pooling operation is applied to get the final output features before the output projector.
- **Character Coherence:** Similar to Scene-Cloze, all available modality information is used in the input of the Comicsformer. However, only the character and speech output features are used during the mean pooling operation. The options construction differs from the other tasks since they use information from other panels. The correct options use panel information as is, whereas for the wrong option, the speech input token feature positions are swapped between the first and second.
- **Contextual Character to Speech Attribution:** This task differs from the others as it does not use the Comicsformer architecture for options. Instead,

context features are fused with the character projections coming from pretrained models. Their similarity with speech projections is measured after both projections are processed through a linear layer and L2 normalization.

Proper masking is applied for each model type. Based on the modality available during the context and option feature extraction, masking was applied accordingly. The model types are as follows:

- **Text-only**: Uses only text modality, speech, and narrative.
- **Char-only**: Uses only character modality.
- **Image-only (Panel-only)**: In the original task definition, it was called Image-only when the model uses only panel visual features. However, we used character information based on the characters' images. Hence, the modal is changed to Panel-only.
- **Image-Text (Panel-Text)**: Uses panel visual information and text modality.
- **Panel-Char**: Uses panel visual information and character visual features.
- **Char-Text**: Uses character visual features with text features.
- **Panel-Char-Text**: Uses all available information from the three modalities.

D ComicBERT Pretraining Results

Table 5 present the pretraining results.

Table 5. Pretraining Results of the Masked Comic Modeling (MCM) Task for ComicBERT. Two distinct settings were employed during the experiments. The upper setting involved fewer sequences and shorter training duration, optimizing for time efficiency (fast setting). The fast setting was utilized for most experiments, except for the results in Table 4. For each setting, the batch size was 128.

Seq. Count	# Epoch	NT-Xent Loss	Top-1 (%)	Top-5 (%)	Mean Pos.
57,250	50	3.3	26.3	48.6	23.7
292,288	100	2.8	37.1	58.4	18.9

E No-Context (NC) Models

The No-Context (NC) model is a model type proposed by Iyyer et al. [10] to measure the impact of context. The NC model has only been employed in the Panel-Text (Image-Text) model. It can be regarded as an ablation study specific to context. Instead of context, the visual features of the final panel p_i are used in place of context for this model.

F Masked Comic Modeling Algorithm

The main learning algorithm for *Masked Comic Modeling(MCM)* is detailed in Algorithm 1.

Algorithm 1. Main learning algorithm of *Masked Comic Modeling* (MCM)

Require: Batch size B, constant τ, modality encoder e, masking function m, Comicsformer model c, feed-forward network f, NT-Xent loss function nt.
1: **for** sampled minibatch $\{S_k\}_{k=1}^{B}$ **do**
2:     $u \leftarrow e(S)$ ▷ Encode input sequences
3:     $u_m, i \leftarrow m(u)$ ▷ Apply masking to encoded units and get masked indices
4:     $o \leftarrow c(u_m)$ ▷ Generate Comicsformer representations
5:     $h \leftarrow f(o)$ ▷ Obtain FFN features
6:     $u_{um} \leftarrow u[i]$ ▷ unmasked panel unit encodings
7:     $h_m \leftarrow h[i]$ ▷ masked representations
8:     $u_{um} \leftarrow \text{FLATTEN}(u_{um})$
9:     $h_m \leftarrow \text{FLATTEN}(h_m)$
10:    $\mathcal{L} \leftarrow nt(u_{um}, h_m)$ ▷ Compute NT-Xent loss
11:    Update networks e, c, f to minimize $\mathcal{L}$ ▷ Backpropagate and update
12: **end for**
13: **return** Comicsformer $c(\cdot)$, Modality encoder $e(\cdot)$, and discard feed-forward network $f(\cdot)$

G Obtaining Sequential Context Embedding from Comicsformer

Figure 7 illustrates the process of context extraction from Comicsformer.

H ComicBERT Pre-Training Details

During the ComicBERT pre-training phase, we employed two distinct settings, namely slow and fast, based on the duration of training. The fast setting was trained for 50 epochs and utilized a smaller dataset. This training was completed within approximately 24 h on an NVIDIA V100 GPU. In contrast, the slow setting took nearly a week on the same GPU for training. A substantial difference becomes evident when observing the pre-training outcomes for both settings (see Appendix D for pretraining results). However, this contrast is not equally reflected in the results of the cloze-style tasks. The results are presented in Table 4, which were achieved using ComicBERT trained with the slow setting. Nevertheless, its contribution over the fast setting remains marginal. Sharing the results of these two experimental settings could guide future research involving similar experiments, saving time and serving as a reference.

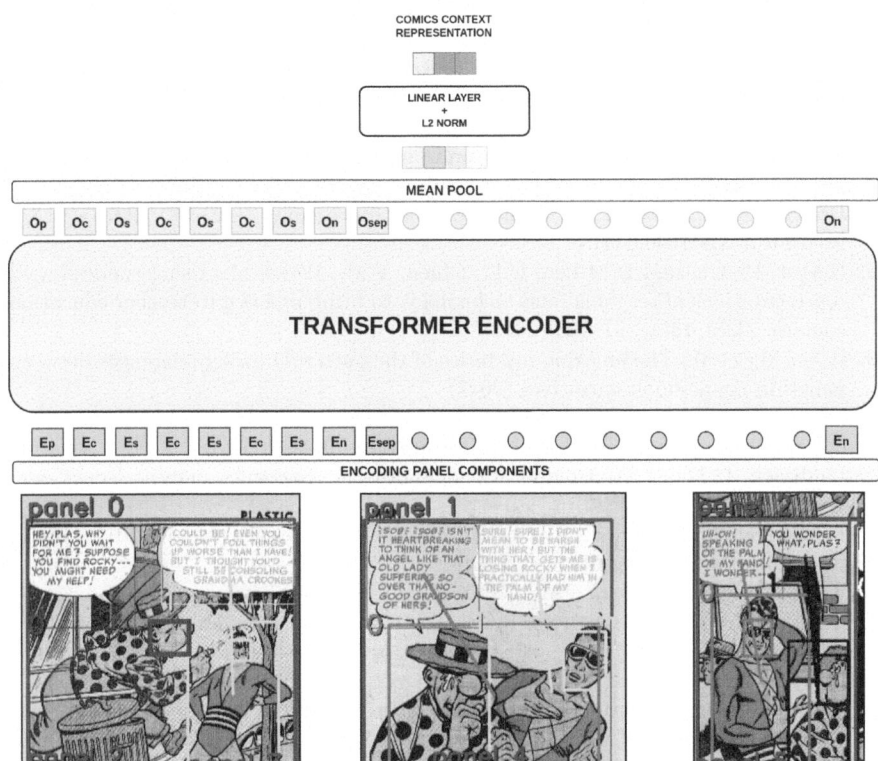

Fig. 7. Obtaining the context embedding with Comicsformer from comic sequences. The transformer encoder outputs at the core of Comicsformer are fused using mean pooling. During mean pooling, the outputs corresponding to padding tokens are not considered. The resulting mean-pooled output is passed through a linear layer (in the experiments, the input-output dimensions are 384×256). Finally, L2 normalization is applied. L2 normalization lets us directly obtain the cosine similarity since downstream tasks use dot product.

References

1. Agrawal, H., Mishra, A., Gupta, M., et al.: Multimodal persona based generation of comic dialogs. In: Proceedings of the 61st Annual Meeting of the Association for Computational Linguistics (Volume 1: Long Papers), pp. 14150–14164 (2023)
2. Augereau, O., Iwata, M., Kise, K.: An overview of comics research in computer science. In: 2017 14th IAPR International Conference on Document Analysis and Recognition (ICDAR). vol. 3, pp. 54–59. IEEE (2017)
3. Brienza, C.: Producing comics culture: a sociological approach to the study of comics. J. Graph. Novels Comics **1**(2), 105–119 (2010)
4. Chen, T., Kornblith, S., Norouzi, M., Hinton, G.: A simple framework for contrastive learning of visual representations (2020). arXiv preprint arXiv:2002.05709

5. Cohn, N.: The Visual Language of Comics: Introduction to the Structure and Cognition of Sequential Images. A&C Black (2013)
6. Devlin, J., Chang, M.W., Lee, K., Toutanova, K.: BERT: Pre-training of deep bidirectional transformers for language understanding (2019)
7. He, K., Zhang, X., Ren, S., Sun, J.: Deep residual learning for image recognition. In: Proceedings of the IEEE Conference on Computer Vision and Pattern Recognition, pp. 770–778 (2016)
8. Hendrycks, D., Gimpel, K.: Gaussian error linear units (GELUs) (2016). arXiv preprint arXiv:1606.08415
9. Herbst, P., Chazan, D., Chen, C.L., Chieu, V.M., Weiss, M.: Using comics-based representations of teaching, and technology, to bring practice to teacher education courses. ZDM **43**(1), 91–103 (2011)
10. Iyyer, M., et al.: The amazing mysteries of the gutter: Drawing inferences between panels in comic book narratives (2017)
11. Jaiswal, A., Babu, A.R., Zadeh, M.Z., Banerjee, D., Makedon, F.: A survey on contrastive self-supervised learning. Technologies **9**(1), 2 (2020)
12. Laubrock, J., Dunst, A.: Computational approaches to comics analysis. Top. Cogn. Sci. **12**(1), 274–310 (2020)
13. Loshchilov, I., Hutter, F.: Decoupled weight decay regularization (2017). arXiv preprint arXiv:1711.05101
14. Nguyen, N.V., Rigaud, C., Burie, J.C.: Comic MTL: optimized multi-task learning for comic book image analysis. Int. J. Doc. Anal. Recogn. (IJDAR) **22**, 265–284 (2019)
15. Nguyen, N.-V., Rigaud, C., Revel, A., Burie, J.-C.: Manga-MMTL: multimodal multitask transfer learning for manga character analysis. In: Lladós, J., Lopresti, D., Uchida, S. (eds.) ICDAR 2021. LNCS, vol. 12822, pp. 410–425. Springer, Cham (2021). https://doi.org/10.1007/978-3-030-86331-9_27
16. Reimers, N., Gurevych, I.: Sentence-BERT: sentence embeddings using siamese BERT-networks. In: Proceedings of the 2019 Conference on Empirical Methods in Natural Language Processing. Association for Computational Linguistics (2019). https://arxiv.org/abs/1908.10084
17. Sachdeva, R., Zisserman, A.: The manga whisperer: Automatically generating transcriptions for comics. CoRR **abs/2401.10524** (2024). https://doi.org/10.48550/ARXIV.2401.10524
18. Sanh, V., Debut, L., Chaumond, J., Wolf, T.: DistilBERT, a distilled version of BERT: smaller, faster, cheaper and lighter (2019). ArXiv **abs/1910.01108**
19. Soykan, G., Yuret, D., Sezgin, T.M.: A comprehensive gold standard and benchmark for comics text detection and recognition (2022)
20. Soykan, G., Yuret, D., Sezgin, T.M.: Identity-aware semi-supervised learning for comic character re-identification (2023)
21. Su, Y., et al.: TaCL: Improving BERT pre-training with token-aware contrastive learning (2021). arXiv preprint arXiv:2111.04198
22. Sunder, V., Fosler-Lussier, E., Thomas, S., Kuo, H.K.J., Kingsbury, B.: Tokenwise contrastive pretraining for finer speech-to-BERT alignment in end-to-end speech-to-intent systems (2022). arXiv preprint arXiv:2204.05188
23. Tan, M., Le, Q.: EfficientNet: Rethinking model scaling for convolutional neural networks. In: International Conference on Machine Learning, pp. 6105–6114. PMLR (2019)
24. Vaswani, A., et al.: Attention is all you need. In: Advances in Neural Information Processing Systems, vol. 30 (2017)

25. Wolf, T., et al.: Transformers: state-of-the-art natural language processing. In: Proceedings of the 2020 Conference on Empirical Methods in Natural Language Processing: System Demonstrations, pp. 38–45 (2020)
26. Xu, P., Zhu, X., Clifton, D.A.: Multimodal learning with transformers: a survey. IEEE Trans. Pattern Anal. Mach. Intell. **45**(10), 12113–12132 (2023). https://doi.org/10.1109/TPAMI.2023.3275156

Author Index

A

Akundi, Prathyusha II-175
Allen, Jonathan Parkes II-87
Anger, Jérémy I-40
Anquetil, Eric I-3
Armenakis, Yiannis I-103
Atamni, Nour II-119
Aubry, Mathieu II-3

B

Berg-Kirkpatrick, Taylor II-87
Bertini, Marco I-154
Biescas, Nil I-27
Biondi, Niccolò I-154
Biswas, Sanket I-27
Bizais-Lillig, Marie II-37
Borkar, Jaydeep II-57
Bottaioli, Natalia I-40
Burie, Jean-Christophe I-198

C

Campaioli, Irene I-154
Camps, Jean-Baptiste II-140
Chao, Yu II-184
Chavallard, Pauline II-163
Chen, Danlu II-87
Chen, Yung-Hsin I-12

D

De Gregorio, Giuseppe II-71, II-102
Decours-Perez, Aliénor II-22
Dupin, Boris II-37

E

El-Sana, Jihad II-119

F

Facciolo, Gabriele I-40
Ferretti, Lavinia II-102
Fitsilis, Fotios I-103

G

Gardella, Marina I-40
Gatos, Basilis I-103
Georgoulea, Maria-Eleni I-103

I

Imbert, Florent I-3
Iwata, Motoi I-216

K

Kaddas, Panagiotis I-103
Kamilaki, Maria I-89
Karatzas, Dimosthenis I-154
Katsouros, Vassilis I-138
Kermorvant, Christopher I-40
Kiousi, Eleni I-103
Kise, Koichi I-216
Klut, Stefan I-73
Konstantinidou, Maria II-102
Koornstra, Tim I-73
Kouletou, Eleanna I-138
Kubade, Ashish II-175
Kyrkos, Charalambis I-103

L

Liu, Changsong II-184
Lladós, Josep I-27

M

Maas, Martijn I-73
Madi, Boraq II-119
Marthot-Santaniello, Isabelle II-71, II-102
Mikros, George I-103
Mohd, Bilal Arif Syed II-175
Morel, Jean-Michel I-40
Mouchère, Harold II-71
Mowlavi, Seginus I-40
Murel, Jacob I-125, II-87

N
Nardoni, Mariateresa I-154

O
Okamoto, Keito I-216
Oparnica, Milena I-59

P
Palaiologos, Konstantinos I-103
Papavassiliou, Vassilis I-138
Pavlopoulos, John II-102
Pena, Rodrigo C. G. II-71, II-102
Peng, Liangrui II-184
Perrin, Simon II-71
Peters, Luke I-73
Petit, Samuel I-198
Pilligua, Maria I-27
Preciozzi, Javier I-40

R
Rabaev, Irina II-119
Real, Thibaud II-163
Rigaud, Christophe I-198
Rozenberg, Olivier I-103

S
Sezgin, Tevfik Metin I-168, I-231, I-257
Shahid, Taimoor II-87
Siglidis, Ioannis II-3
Smith, David A. I-125, II-57, II-87
Soullard, Yann I-3

Soykan, Gürkan I-168, I-231, I-257
Ströbel, Phillip B. I-12
Stutzmann, Dominique II-3

T
Tadros, Antoine I-40
Tarride, Solène I-40
Tasouli, Christina I-103
Tavenard, Romain I-3
Thiée, Lukas-Walter II-199
Tsitrinovich, Vasily II-119

V
Valveny, Ernest I-27
van Koert, Rutger I-73
Vasyutinsky-Shapira, Daria II-119
Vazquez-Corral, Javier I-27
Vidal-Gorène, Chahan II-22, II-37, II-140
Vivoli, Emanuele I-154
Vlachou-Efstathiou, Malamatenia II-3
von Gioi, Rafael Grompone I-40

W
Wang, Yanwei II-184

Y
Yuret, Deniz I-168, I-231, I-257

Z
Zhang, Xiang II-87

SPRINGER NATURE

GPSR Compliance

The European Union's (EU) General Product Safety Regulation (GPSR) is a set of rules that requires consumer products to be safe and our obligations to ensure this.

If you have any concerns about our products, you can contact us on ProductSafety@springernature.com

In case Publisher is established outside the EU, the EU authorized representative is:

Springer Nature Customer Service Center GmbH
Europaplatz 3
69115 Heidelberg, Germany

The manufacturer's authorised representative in the EU is Springer Nature Customer Service Centre GmbH, Europaplatz 3, 69115 Heidelberg, Germany. If you have any concerns regarding our products, please contact ProductSafety@springernature.com

Printed and bound by CPI Group (UK) Ltd, Croydon, CR0 4YY

26/03/2026

02078984-0003